BOSCH

Automotive electrics and electronics

UPDATED EDITION

INTERNATIONAL ®

Imprint

Published by:
© Robert Bosch GmbH, 1999
Postfach 30 02 20
D-70442 Stuttgart
Automotive Equipment Business Sector,
Product-Marketing software products (KH/PDI).

Editor-in-Chief:
Dipl.-Ing. (FH) Horst Bauer.

Editors:
Dipl.-Ing. Karl-Heinz Dietsche,
Dipl.-Ing. (BA) Jürgen Crepin,
Dipl.-Holzw. Folkhart Dinkler.

Layout:
Dipl.-Ing. (FH) Ulrich Adler,
Berthold Gauder, Leinfelden-Echterdingen.

Translation:
Peter Girling.

Technical graphics:
Bauer & Partner, Stuttgart.

Printed in Germany. Imprimé en Allemagne.
3rd edition, September 1999.
SAE Society of Automotive Engineers
400 Commonwealth Drive
Warrendale, PA 15096-0001 U.S.A.

(3.0 N)

ISBN 0-7680-0508-6

Authors

Vehicle electrical system and circuit diagrams
Dipl.-Ing. W. Gansert, Dr. Ing. T. Bertram.

EMC and interference suppression
Dr.-Ing. W. Pfaff.

Starter batteries
Dr.-Ing. G. Richter.

Traction batteries
Dr.-Ing. B. Sporckmann, Dipl.-Ing. E. Zander,
RWE Energie AG, Essen.

Alternators
Dr.-Ing. K.-G. Bürger.

Starting systems
Dr.-Ing. K. Bolenz.

Lighting technology
Dr.-Ing. M. Hamm, Dipl.-Ing. T. Spingler,
Dipl.-Ing. D. Boebel, Dipl.-Ing. B. Wörner,
Dipl.-Ing. H.-J. Lipart.

Washing and cleaning systems
Dr.-Ing. J.-G. Dietrich.

Theft-deterrence systems
Dipl.-Ing. (FH) H. Hennrich.

Comfort and convenience systems
Dipl.-Ing. F. Jonas, Dipl.-Ing. R. Kurzmann,
Dr.-Ing. G. Hartz.
Dipl.-Ing. G. Schweizer, Behr GmbH & Co.

Information systems
Dipl.-Ing. P. Rudolf, Dr. rer. nat. D. Elke,
Ing. (grad.) D. Meyer.

Occupant-safety systems
Dipl.-Ing. B. Mattes.

Driving-safety systems and drivetrain
Dr.-Ing. G. Schmidt, Dipl.-Ing. (FH) D. Graumann.

Unless otherwise stated, the above are all employees of Robert Bosch GmbH, Stuttgart.

Foreword

All the manuals from the Bosch "Technical Instruction" publication range dealing with "Automotive Electrics/Automotive Electronics" were combined to form this reference book. In order to do justice to the great increase in the number of subjects which must be dealt with, the control systems for spark-ignition engines were removed and published separately in their own reference book "Gasoline-engine management".

The extent of the equipment installed in the vehicle has increased immensely in the past years as a result of the rapid developments which have taken place in electrical and electronic components and systems.

The alternator in its role as the energy generator, the battery as the energy accumulator, the starter, and other loads such as the management system for the spark-ignition engine must not only function reliably, but they must also be perfectly matched to each other. Modern headlamp systems such as the "Litronic" with integrated headlight levelling control and cleaning system, guarantee long-distance illumination and perfect light distribution. Ingenious theft-deterrence systems protect vehicles against break-in or unauthorized use. Comfort and convenience systems keep driver fatigue down to a low level, and provide stress-free and comfortable surroundings. Information and navigation systems keep the driver informed on his/her whereabouts or guide the driver to his/her destination by means of automatic route calculation. Parking-aid systems make parking and maneuvering an easy matter. Dynamic driving-safety systems (ABS, TCS, ESP) help prevent accidents in the first place and, should an accident occur, occupant-safety systems (e.g. airbags) mitigate the accident results and reduce the danger of injury.

The wide variety of subjects covered by this reference book means that the reader who is interested in automotive engineering technology is provided with a wide range of easily understood descriptions of the vehicle's most important electrical and electronic systems and components.

The editorial staff

Contents

Vehicle electrical systems

The history of vehicle electrical systems

This is an integral part of the history of Bosch itself. Within the span of a hundred years, it demonstrates the development of the electrical and electronic equipment in the automobile. A development in which Bosch always played a leading role, starting with its high-voltage magneto in 1902, and proceeding up to the present day with the introduction of the Electronic Stability Program (ESP) in 1995.

Magneto-ignition system
In 1902, the first electrical system was installed in a vehicle in the form of the magneto ignition system. Although this consisted of the magneto itself, an ignition distributor, ignition coils, and spark plugs (Fig. 1), there was no question of regarding these few components together with their cables and wires as an on-board electrical system.

Complete automotive systems
It only took another 11 years, and Bosch had the first complete automotive electrical system ready for installation (Fig. 2). This was comprised of the magneto-ignition system with spark plugs, starter, DC generator, headlamps, battery, and regulator switch. This marked the start of progress towards a genuine on-board system.

The birth of the vehicle electrical system
The already very extensive scope of the electrical installation in the vehicle can be seen from the Bosch wiring diagram from 1958 (Fig. 3). The devices are not represented by symbols but by schematized line drawings. The arrangement of the electrical lines has started to resemble a wiring diagram. With the introduction of the electronically controlled D-Jetronic gasoline-injection system in 1967, electronics entered the automotive world.

Today's vehicle electrical system
In the past few years, the increasing complexity and scope of the vehicle's equipment led to a rapid rise in the number of electrical loads. Formerly, the starter, the ignition, and the lighting system were the main points of interest. In the course of the years though, the vehicle's basic equipment was extended continually by the addition of electronic ignition and fuel-injection systems, together with comfort and convenience systems with their wide range of drive motors, and safety and security systems.

Today, the electrical and electronic equipment in the automobile has become so extensive that it would be totally impossible to show this using such a diagram as that in Fig. 3. The diagrammatic presentation showing equipment and wiring in their relative positions to each other was abandoned in favor of a method which showed the on-board electrical system in the form of schematic diagrams. The idea was to present a schematic overview of the installation featuring a high level of clarity, notwithstanding the large number of individual systems and components concerned (refer to the Chapter "Schematic diagram of a gasoline-engine passenger car").

On conventional on-board networks, the interplay of the different subsystems is defined by the allocation of individual lines/wires to individual signals.

Today, the immense increases in the amount of data exchanged between the electronic components can no longer be handled with such cabling techniques, the wiring harness becomes too complicated, and too many pins are needed on the ECU's. This applies particularly on upper-class vehicles. These problems have been solved by using CAN (Controller Aided Network) a special serial bus system developed by Bosch for automotive applications.

Examples of developments in vehicle electrical systems

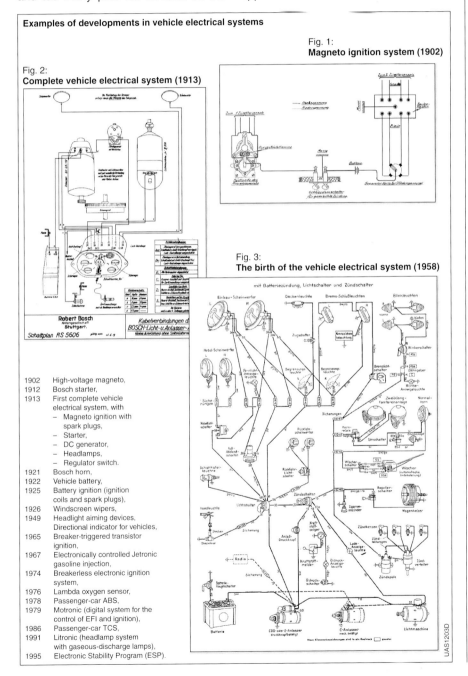

Fig. 1:
Magneto ignition system (1902)

Fig. 2:
Complete vehicle electrical system (1913)

Fig. 3:
The birth of the vehicle electrical system (1958)

1902	High-voltage magneto,
1912	Bosch starter,
1913	First complete vehicle electrical system, with
	– Magneto ignition with spark plugs,
	– Starter,
	– DC generator,
	– Headlamps,
	– Regulator switch.
1921	Bosch horn,
1922	Vehicle battery,
1925	Battery ignition (ignition coils and spark plugs),
1926	Windscreen wipers,
1949	Headlight aiming devices, Directional indicator for vehicles,
1965	Breaker-triggered transistor ignition,
1967	Electronically controlled Jetronic gasoline injection,
1974	Breakerless electronic ignition system,
1976	Lambda oxygen sensor,
1978	Passenger-car ABS,
1979	Motronic (digital system for the control of EFI and ignition),
1986	Passenger-car TCS,
1991	Litronic (headlamp system with gaseous-discharge lamps),
1995	Electronic Stability Program (ESP).

UAS1203D

Data transmission between the systems

System overview

Increasingly widespread application of electronic control systems for automotive functions such as
- Electronic engine management (Motronic),
- Electronic transmission control,
- Electronic vehicle immobilizers,
- Antilock braking system (ABS),
- Traction control system (TCS),
- Electronic Stability Program (ESP), and
- On-board computers

has made it vital to interconnect the individual control circuits by means of networks. Data transfer between the various control systems reduces the number of sensors while also promoting exploitation of the performance potential in the individual systems.

The interfaces can be divided into two categories:
- Conventional interfaces, with binary signals (switch inputs), pulse-duty factors (pulse-width modulated signals), and
- Serial data transmission, e.g., Controller Area Network (CAN).

Conventional interfaces

In conventional automotive data-communications systems each signal is assigned to a single line. Binary signals can only be transmitted as one of two conditions: "1" or "0" (binary code). An example would be the a/c compressor, which can be "on" or "off."
Pulse-duty factors can be employed to relay more detailed data, such as throttle-valve aperture.
Increasing data traffic between various on-board electronic components means that conventional interfaces are no longer capable of providing satisfactory performance. The complexity of current

wiring harnesses is already difficult to manage, and the requirements for data communications between ECUs are on the rise (Figure 1).

Serial data transmission (CAN)

These problems can be solved with a CAN, that is, a bus system (bus bar) specially designed for automotive applications.
Provided that the ECU's are equipped with a serial CAN interface, CAN can be used to relay the signals from the sources listed above.

There are three basic applications for CAN in motor vehicles:
- To link ECU's,
- Body-related and convenience electronics (multiplex), and
- Mobile communications.

The following is limited to a description of communications between ECUs.

ECU networking

This strategy links electronic systems such as Motronic, electronic transmission-shift control, etc. Typical transmission rates lie between approximately 125 kBit/s and 1 MBit/s, and must be high enough to maintain the required real-time response. One of the advantages that distinguishes serial data transfer from

Fig. 1

Conventional data transmission

GS Transmission-shift control,
EGAS Electronic Throttle Control (ETC),
ABS Antilock Braking System,
TCS Traction Control System,
MSR Engine drag-torque control.

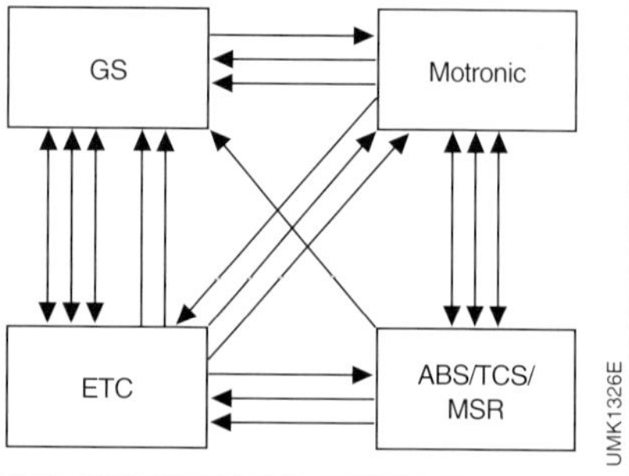

conventional interfaces (pulse-duty factors, switching and analog signals, etc.) is the high speeds achieved without placing major burdens on the central processing units (CPU's).

Bus configuration
CAN works on the "multiple master" principle. This concept combines several ECU's with equal priority ratings in a linear bus structure (Figure 2).
The advantage of this structure is that failure of one subscriber will not affect access for the others. The probability of total failure is thus substantially lower than with other logical configurations (such as loop or star structures).
With loop or star architecture, failure in one of the subscribers or the central ECU will provoke total system failure.

Content-keyed addressing
The CAN bus system addresses data according to content. Each message is assigned a permanent eleven-bit identifier tag indicating the contents of the message (e.g., engine speed). Each station processes only the data for which identifiers are stored in its acceptance list (acceptance check). This means that CAN does not need station addresses to transmit data, and the interfaces do not need to administer system configuration.

Bus arbitration
Each station can begin transmitting its highest priority message as soon as the bus is unoccupied.
If several stations initiate transmission simultaneously, the resulting bus-access conflict is resolved using a "wired-and" arbitration arrangement. This concept grants first access to the message with the highest priority rating, with no loss of either time or data bits.
When a station loses the arbitration, it automatically reverts to standby status and repeats the transmission attempt as soon as the bus indicates that it is free.

Message format
A data frame of less than 130 bits in length is created for transmissions to the bus. This ensures that the queue time until the next – possibly extremely urgent – data transmission is held to a minimum. The data frames consist of seven consecutive fields.

Standardization
The International Organization for Standardisation (ISO) has recognized a CAN standard for use in automotive applications with data rates of over 125 kBit/s, and along with two other protocols for data rates of up to 125 kBit/s.

Fig. 2

Linear bus structure

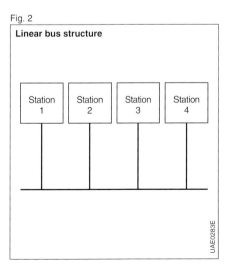

UAE0283E

Cartronic

The order concept for system networking

The development of the electronic systems in the vehicle is determined by the ever-increasing demands for more and more performance regarding safety, security, fuel economy, and comfort and convenience. Increasingly severe legislation, and the demand for the incorporation of information and entertainment technology also play an important role. Individual systems in the vehicle, which to a great extent previously operated independently of each other, can be networked to form a vehicle-wide alliance, and the standardisation of alliance components can make a valuable contribution to achieving the above objectives.

Technological state-of-the-art
The traction control system (TCS) is a good example of a system alliance which is already in operation in today's vehicles. Communication of the TCS control unit (ECU) with the engine ECU enables drive torque to be closed-loop controlled.

Concept
Cartronic is an order concept for all the vehicle's closed and open-loop control systems. The concept incorporates modular, expandable architectures for "Function", "Safety & Security", and "Electronics", all of which are based on specified formal configuration and modelling conventions.

Architectures
Here, architecture is understood to mean both the structuring system and its transformation into a concrete structure.
Function architecture comprises all the closed-loop and open-loop control assignments present in the vehicle. The system alliance's assignments are allocated to logical components, and the component interfaces and their interplay are defined.

The safety and security architecture extends the functional architecture by the addition of elements which guarantee the reliable operation of the system alliance. Finally, a system is defined for the electronics, stipulating how the system alliance is to be implemented using demand-aligned hardware topologies (Fig. 1).

Architecture conventions
Function architecture conventions serve system-alliance organization. This is independent of special hardware topology and is the sole result of logical and functional viewpoints.
The conventions define components, the permitted interaction between them via communication relationships, and the modelling samples for similar, frequently repeated assignments.

Function analysis
From the technical viewpoint, the first step in ascertaining the advantages of a system alliance is to carry out a functional analysis of the previously self-sufficient individual systems. The observation of the functions at such an abstract level is independent of implementation using a special hardware topology, and therefore leads to one and the same function architecture for various hardware topologies. This permits the variety of software and hardware types to be limited, and the utilisation of the electronic units for the basic functions of a large number of vehicle types.

Structural elements
On the one side, the architecture elements are systems, components, and communications relationships, and on the other, structuring and modelling conventions.

Systems, components
Within the framework of stucturing, a system is defined as the formation of a whole by combining components which are in interaction with each other via some form of communication.

The term component is not simply applied to some form of physical unit in the sense of a part, but rather is understood to stand for a functional unit.

As far as Cartronic is concerned, one differentiates between three different types of component:

– Components with predominantly coordinating assignments,
– Components with mainly operative assignments,
– Components which exclusively generate information and make it available for use.

Structuring conventions

These structuring conventions define the permitted communications relationships within the architecture of the vehicle as a whole. A difference is made between structuring conventions which regulate the communications relationships at the same level, and those which do so at higher or lower levels, taking into account the given outline conditions. Furthermore, the structuring conventions regulate the passing-on of communications from one system to another.

Modelling conventions

The modelling conventions contain samples which combine the components and communicatons relationships needed for solving special, frequently occurring assignments. These samples can then be applied at various points within the vehicle's structure (example: Electronic Energy Management).

Architecture characteristics

A structure that has been developed in accordance with the modelling and structuring conventions features the following characteristics:

– Declared, uniform structuring conventions and modelling patterns,
– Hierarchical order flow,
– High level of autonomy of the individual components,
– Controls, sensors, and estimators are regarded as being information providers of the same value, and
– Each component is presented to the other components as visibly (or as invisibly) as required.

Fig. 1
Cartronic: Hardware topology (example)

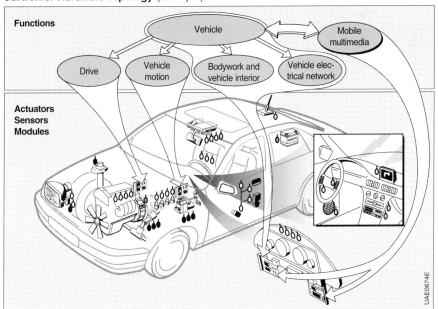

UAE0674E

Vehicle electrical-system circuits

Conventional vehicle electrical systems

In the vehicle, in addition to other variables, the cabling method used between the alternator, battery, and loads influences the voltage level and therefore the battery's state of charge. If all loads are connected to the battery side of the system, the total current I_G flows in the charging line. I_G is comprised of battery current I_B and load current I_V.

The high voltage drop due to the total current flowing in the charging line results in a lower charge voltage. If, on the other hand, the loads are all connected to the alternator side of the system, the voltage drop is lower and the charge voltage higher. Here though, loads which are sensitive to voltage peaks or high levels of ripple (electronics) are at a disadvantage. These facts are behind the OEM's and the service workshops being recommended to connect voltage-insensitive loads with high power inputs to the alternator side of the system, and voltage-sensitive loads with low power inputs to the battery side.

Adequate conductor cross-sections, and connecting points and junctions whose contact resistances do not increase even after long periods of operation, are imperative if voltage drops are to be kept to a minimum.

Future vehicle electrical systems

In series-production vehicles with 12 V electrical systems, the battery is a compromise between in some cases mutually antagonistic requirements. It must be large enough to provide the power needed for cranking the engine, and it must be able to supply the on-board electrical system.

During cranking, very high currents (300...500 A) are drawn from the battery. Such high currents result in a voltage drop which can be dangerous for certain loads (e.g. units equipped with microprocessors).

Fig. 1

Future vehicle electrical system

1 Lighting system (vehicle electrical system), **2** Starter, **3** Engine management (vehicle electrical system),
4 Starting battery, **5** Further loads on the vehicle electrical system (for instance power roof),
6 System-supply battery, **7** Alternator, **8** Charging and disconnect module.

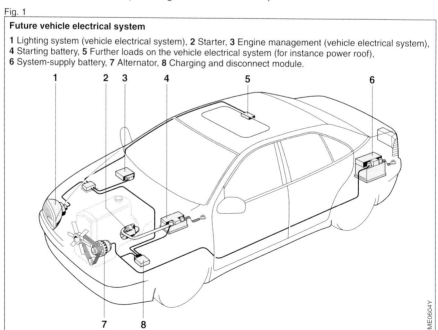

UME0604Y

On the other hand, while the vehicle is actually being driven, only low currents flow. Here, the battery's nominal capacity is decisive. It is practically impossible to combine both these requirements (on the one hand for cranking, and on the other for actual driving) in a single battery.

On future vehicle electrical system designs, such as "On-board system 2000", equipped with two batteries (one as the starter battery and the other as the system-supply battery for the other components), the battery functions of "Provision of high powers for cranking the engine" and "Provision of power for the on-board electrical system" are separate from each other. This means that on the one side the voltage drop in the on-board system is avoided when the engine is cranked, and on the other that the vehicle can be sure to start in the cold even though the supply-system battery has discharged to 30%.

Starter battery

The starter battery only needs to deliver very high currents during engine cranking, and this only lasts for a limited period. Since it is very compact, the battery can be mounted near to the starter and connected to it with a short cable so that voltage drop can be kept to a minimum. Battery nominal capacity can also be reduced, and in order that it is always well charged, the battery has a nominal voltage of 10 V. The voltage difference to 12 V results in the starter battery having charging priority.

Charge and disconnect module

During the cranking (starting) process, and with the engine stationary, the charge and disconnect module separates starter battery and starter from the remainder of the vehicle electrical system. In doing so, it not only prevents voltage drop when cranking takes place, but also prevents starter-battery discharge when loads are switched on with the engine stopped.

System-supply battery

The system-supply battery is exclusively for the on-board electrical system (that is, without the starter). Although it provides relatively low currents (e.g. approx. 20 A for the engine-management system), it is extremely deep-cycle resistant In other words, provided that its nominal capacity is high enough and that permissible discharge depth is not exceeded, this battery is able to provide large amounts of energy and then recharge again to the same level. The battery's dimensions for the most part depend upon the capacity reserves needed for the switched-on loads (e.g. parking lamps, hazard-warning flasher), the loads which are always on irrespective of whether the vehicle is in use or not, and the permissible degree of battery discharge.

System variants

Variants of this system are available for vehicles with very high load consumptions. Such variants can be in the following forms:

− The starter battery (instead of the system-supply battery) provides the power for the engine-management system, or the system is switchable,
− The starter battery is also of the 12 V type, although here the circuitry involved in ensuring charging priority for the starter battery goes hand in hand with higher costs,
− Nominal voltage higher than 12 V,
− Multi-voltage networks which, in parallel to the +12 V system-supply circuit, also have a −12 V circuit (or −24 V), so that in addition, 24 V (or 36 V) is available on the outer conductors of both circuits,
− Installation of 2 alternators.

The objective of the particular on-board electrical system concept (for instance, avoiding voltage drop when starting, minimisation of weight, or an extremely high level of starting reliability) is decisive when selecting the system variant to be used.

Calculation of conductor sizes

Quantities and units

Quantity		Unit
A	Conductor cross section	mm²
I	Current	A
l	Conductor length	m
P	Power required by load	W
R	Resistance (load)	Ω
S	Current density in conductor	A/mm²
U_N	Nominal voltage	V
U_{vl}	Permissible voltage drop in insulated conductor	V
U_{vg}	Permissible voltage drop in entire circuit	V
ϱ	Resistivity	Ω·mm²/m

Calculation

In determining the conductor cross section, allowance must be made for voltage drop and the effect of elevated temperatures.

Berechnungsschritte
1. Determine the current I of the load:
$I = P / U_N = U_N / R$,
2. Calculate the conductor cross section A using the U_{vl} values given in Table 2 (ϱ = 0.0185 Ω·mm²/m for copper):
$A = I \cdot \varrho \cdot l / U_{vl}$
3. Round-off the value for A to the next larger conductor cross section in accordance with Table 1.
Individual conductors which have cross sections less than 1 mm² are not recommended because their mechanical strength is inadequate.
4. Calculate the actual voltage drop U_{vl}:
$U_{vl} = I \cdot \varrho \cdot l / A$ and
5. Check the current density S in order to avoid excessive conductor temperatures (in brief operation, $S < 30$ A/mm²; see Table 1 for values for continuous operation).
$S = I/A$.

Table 1

Electrical copper conductors for motor vehicles
Single-core, untinned, PVC-insulated. Permissible working temperature: 70 °C.[2]

Nominal conductor cross-section mm²	Approximate number of individual wires[1]	Maximum resistance per meter [1] at +20 °C mΩ/m	Maximum conductor diameter[1] mm	Nominal thickness of insulation[1] mm	Maximum cable outer diameter[1] mm	Permissible continuous current (standard value)[2] at ambient temperature	
						at +30 °C A	at +50 °C A
1	32	18.5	1.5	0.6	2.7	19	13.5
1.5	30	12.7	1.8	0.6	3.0	24	17.0
2.5	50	7.60	2.2	0.7	3.6	32	22.7
4	56	4.71	2.8	0.8	4.4	42	29.8
6	84	3.14	3.4	0.8	5.0	54	38.3
10	80	1.82	4.5	1.0	6.5	73	51.8
16	126	1.16	6.3	1.0	8.3	98	69.6
25	196	0.743	7.8	1.3	10.4	129	91.6
35	276	0.527	9.0	1.3	11.6	158	112
50	396	0.368	10.5	1.5	13.5	198	140
70	360	0.259	12.5	1.5	15.5	245	174
95	475	0.196	14.8	1.6	18.0	292	207
120	608	0.153	16.5	1.6	19.7	344	244

[1] As per DIN ISO 6722, part 3.   [2] As per DIN VDE 0298, part 4.

The values given for U_{vl} in Table 2 are used to calculate the dimensions of the positive conductor. The voltage drop in the ground return is not taken into account. In the case of insulated ground cables, the total length in both directions should normally be used.

The U_{vg} values given in the table are test values and cannot be used for conductor calculations because they also include the contact resistance of switches, fuses, etc.

Table 2
Recommended max. voltage drop

Type of conductor	Recommended max. voltage drop in positive conductor U_{vl}		Permissible voltage drop in entire circuit U_{vg}		Notes
Nominal voltage U_N	12 V	24 V	12 V	24 V	
Lighting conductors					
from terminal 30 of light switch	0.1 V	0.1 V	0.6 V	0.6 V	Current at
to lamps < 15 W					nominal voltage
to trailer socket					and nominal
from trailer socket					power
to lamps					
from terminal 30 of light switch	0.5 V	0.5 V	0.9 V	0.9 V	
to lamps > 15 W					
to trailer socket					
from terminal 30 of light switch	0.3 V	0.3 V	0.6 V	0.6 V	
to headlamps					
Charging cable					
from terminal B+ of alternator	0.4 V	0.8 V	–	–	
to battery					
Main starter cable	0.5 V	1.0 V	–	–	Starter short-circuit current at +20 °C (notes 1 and 2)
Starter control lead					Maximum
From starter switch to terminal 50					control current
of starter					(notes 3 and 4)
Solenoid switch with single winding	1.4 V	2.0 V	1.7 V	2.5 V	
Solenoid switch with pull-in and	1.5 V	2.2 V	1.9 V	2.8 V	
hold-in windings					
Other control leads					Current at
from switch to relay, horn, etc.	0.5 V	1.0 V	1.5 V	2.0 V	nominal voltage

Notes
1. If necessary, in special cases, when a very long main starter cable is concerned, the U_{vl} value can be exceeded at reduced minimum starting temperatures.
2. In case of a main starter cable with insulated return, the voltage loss in the return line shall not exceed that in the outgoing line. In each case, max. 4 % of the nominal voltage is permissible, that is a total of max. 8 %.
3. The U_{vl} values apply for solenoid-switch temperatures of 50 ... 80 °C.
4. If necessary, take into account the line up to the starter switch.

Plug-in connections

Assignments and requirements

It is the job of electrical plug-in connections to provide a reliable connection between the different system components and thus contribute to ensuring the system's efficient functioning no matter what the operating conditions. They are designed to withstand the very varied range of loads and stresses which they are subjected to during the vehicle's service life. Examples for such loading/stressing are:
– Vibration acceleration,
– Temperature fluctuations,
– Very high and very low temperatures,
– Effects of dampness and humidity,
– Aggressive liquids and toxic gases,
– Microscopic motion at the contact points, leading to fretting corrosion.

This loading can lead to an <u>increase</u> in contact resistance which in the worst case can result in an open circuit. Insulation resistances can also <u>decrease</u> to the point at which neighboring conductors can short-circuit to each other.

Electrical plug-in connectors must have the following characteristics:
– Low contact resistances at the current-carrying parts,
– High insulation resistance between current-carrying parts which are at different voltages,
– High level of resistance to leakage to prevent entry of water, humidity, and saline fog.

Depending upon their area of application, in addition to their physical characteristics plug-in connectors must comply with further stipulations such as:
– Uncomplicated, faultless handling in automotive-assembly applications, reliable reverse-polarity protection,
– Reliable, and perceivable locking and unlocking processes,
– Ruggedness coupled with automatic-machine-capability during wiring-harness manufacture and transportation.

Design and types

Bosch has a number of standard ranges available for the different areas of application in which its plug-in connectors are used. Specially selected contacts are used in these standard ranges in line with the different operating conditions which will be encountered by the connections during actual service. In the following, two examples are given to demonstrate their characteristics.

Bosch BMK micro-contacts

These tin or gold-plated contacts for 0.6 mm contact pins were developed specifically for a 2.5 mm grid dimension, and feature high levels of thermostability (up to 155 °C) and vibration strength. Since it permits extremely space-saving designs, the Bosch micro-contact is suitable for multi-pole plug-in connections.

The contact itself is comprised of two parts, one of which carries the current, while the other (outside steel spring) is responsible for generating the contact-making force.

Due to the outside steel spring (Fig. 1), the contact-making force is maintained not only during extremes of temperature, but also throughout the whole of the vehicle's useful life. The actuating forces are unavoidably higher due to this outside steel spring, but they are reduced by a special insert aid which also ensures the correct axial alignment of the plug-in

Fig. 1

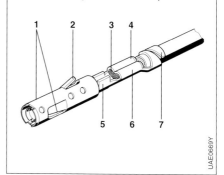

Micro-contact

1 Outside steel spring, **2** Locating spring (primary spring), **3** Single conductor (wire), **4** Entrance radius, **5** Contact body, **6** Conductor crimp, **6** Insulation crimp.

connection, so that damage to contacts or pins due to tilting or inserting at an angle is ruled out.

The complete plug-in connection is sealed at the connector strip of the respective ECU by means of a continuous radial gasket in the plug housing which, with its three seal lips, ensures reliable electrical connection at the ECU seal collar.

A seal plate, through which the contacts complete with crimped-on conductor are inserted, is used to protect the contact point against dampness entering along the cable (Fig. 2). The seal plate is in the form of a silica-gel plate instead of the conventional, commercially-available single-conductor sealing, and has the further advantage of permitting far smaller constructions and variations in the pin configuration (differences in the numbers of poles used). The seal plate is in close contact with the cable insulation and therefore provides a highly efficient seal.

When the plug is assembled, the contacts and lines are pushed through the already mounted seal plate so that the contact slides into its final position in the contact carrier where, by means of the locating spring, it locks tightly of its own accord. As soon as all contacts are in their final positions, a slide pin is used to establish the so-called secondary locking. This is an additional locking function and increases the resistance against unintentional pulling out of the cable and contacts.

Bosch BSK sensor/actuator contacts

These contacts (Fig. 3) are used for the 2...7-pole compact plug-in connections employed in the engine compartment for connecting the components to the ECU. The 5 mm grid dimension ensures the mechanical ruggedness which is imperative in such applications. Thanks to the wave-shaped design of its interior, the BSK reliably prevents the transmission of vibration from the cable area to the contact zone. This ensures that relative motion at the contact surface is avoided so that no fretting corrosion can take place.

This compact plug-in connection features single-conductor sealing which prevents dampness entering the contact zone. The contact-making force ensures that the three seal lips in the plug housing form a reliable seal against splashwater and other forms of dampness.

The self-latching snap connections with supplementary unlocking function are the guarantee for uncomplicated handling during vehicle assembly and customer-service work. To unlock, pressure must be applied to a point on the surface identified by a corrugation.

These plug-in connections are used typically for diesel-engine components (for instance, rail-pressure sensor, injectors), or for gasoline engines (for instance, injectors and knock sensor).

Fig. 2

Multi-pole plug-in connection with micro-contacts (section)

1 Pressure plate, 2 Seal plate, 3 Radial seal, 4 Slide pin (secondary lock), 5 Contact carrier, 6 Contact.

Fig. 3

Sensor contact

1 Outside steel spring, 2 Single conductor (wire), 3 Conductor crimp, 4 Insulation crimp, 5 Wave-shaped interior design.

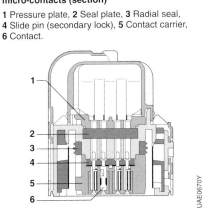

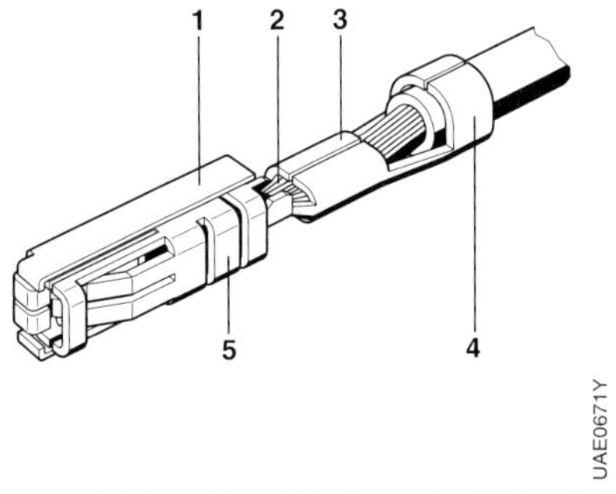

Electrical symbols
and circuit diagrams

While the electrical systems in current vehicles are still responsible for furnishing lighting, today's modern systems also embrace an extended array of electric and electronic devices for open and closed-loop engine-management systems as well as numerous accessories for comfort and convenience.

Clear and concise circuit diagrams using standardized symbols are needed to provide an overview of these complex vehicle circuits. Circuit and terminal diagrams are a help during trouble-shooting. They also facilitate field installation of accessories, and furnish support for troublefree installations and modifications on the vehicle's electrical equipment.

They are an excerpt from the range of standardized symbols used in vehicle wiring diagrams. With few exceptions, these symbols correspond to the definitions of the International Electrotechnical Commission (IEC).

In some DIN standards the symbols have been modified to correspond more closely to the IEC recommendations, e.g., the symbols for inductance and electrical machinery. While both rectangular and semicircular representations are valid options, the semicircular representations are recommended, as they are more easily understood internationally and more accurately reflect the options provided by modern drawing and duplication technology.

Electrical symbols

The electrical symbols on the following pages have been taken from the section in this manual entitled "Circuit diagram for a passenger vehicle with spark-ignition engine."

Requirements

Symbols are the smallest components of a circuit diagram, and are the simplest way to represent electrical devices and their component parts. They illustrate how a device operates, and are used

Figure 1

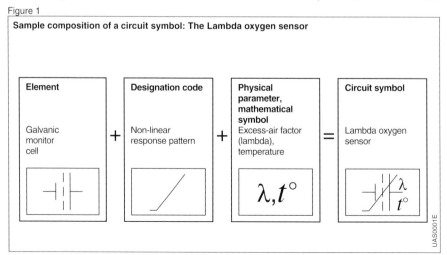

Sample composition of a circuit symbol: The Lambda oxygen sensor

Element		Designation code		Physical parameter, mathematical symbol		Circuit symbol
Galvanic monitor cell	+	Non-linear response pattern	+	Excess-air factor (lambda), temperature $\lambda, t°$	=	Lambda oxygen sensor

UAS0001E

together with circuit diagrams to illustrate how technical sequences proceed. Symbols do not indicate the shapes and dimensions of the devices they represent, nor do they show the locations of their terminal connections. This abstract representation format represents the only practicable option for illustrating how the devices are connected using a circuit diagram.

Every symbol should satisfy the following criteria: it must be easy to remember and identify, easy to understand, easy to draw and should clearly indicate the type of device which it represents.

Symbols comprise symbol elements and designations (Figure 1).

Designation codes can take the form of letters, numbers, mathematical and special-purpose symbols, abbreviations of units, characteristic curves, etc.

If a circuit diagram showing the internal circuitry of a device becomes too complex, or if the function of the device can be illustrated without showing all of the details, the circuit diagram for this specific device can be replaced by a single symbol (without internal circuitry, refer to Figures 1 and 2b).

Simplified representations are usually used for integrated circuits, with their typically high levels of compactness (synonymous with high levels of functional integration within an individual component).

The symbols defined in DIN 40 900, Parts 12 and 13, are prescribed for binary and digital circuits; and show both circuit and function. Symbols from DIN 40 900, Part 13 are also used for analog circuitry in computer and control technology.

Figure 2

Circuit diagram of an alternator with voltage regulator

a With internal circuitry, b Circuit symbols.

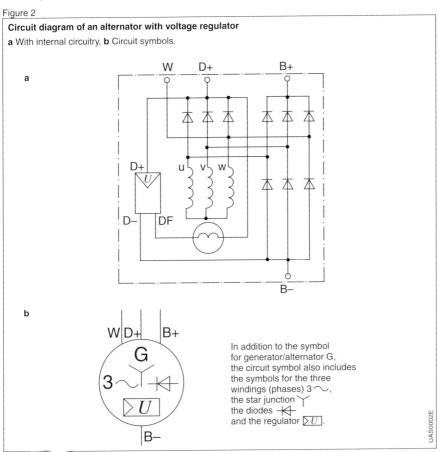

In addition to the symbol for generator/alternator G, the circuit symbol also includes the symbols for the three windings (phases) 3 ⌇, the star junction Y the diodes ⊣◁⊢ and the regulator ▷U⃞.

UAS0002E

Representation

Symbols show circuits in their passive (basic) state, unaffected by physical parameters such as the application of current, voltage or mechanical force. Other operating states, i.e., any condition that varies from the basic status defined above, is indicated by a juxtaposed double arrow (Figure 3).

Symbols and connecting lines (electrical and mechanical linkage elements) feature a single standard line width, with a minimum of 0.25 mm being used to ensure microfilm legibility.

In order to keep the connecting lines as straight as possible and to avoid crossed lines, symbols can be pivoted by 90° increments or shown as mirror images as long as their significance remains unaltered. Tangential connections may be shown as emerging from any convenient point on the symbols; the sole exceptions are resistors (connections only at the narrow ends) and the terminals for electromechanical actuators (connections only at the long sides, Figure 4).

Junctions may be illustrated with or without node dots, no dot being present at intersections (crossings) without an electrical connection (Figure 5). There is no mandatory format for illustrating terminals on electrical devices. Terminals, plugs, sockets and threaded connections are identified by symbols only at those points relevant for installation and removal. All other junctions are represented by dots.

In the case of assembled representation, actuators with a common drive are illustrated as responding to this motive force by all moving in the direction indicated by the dashed line (– – –) which represents the mechanical linkage (Figure 6).

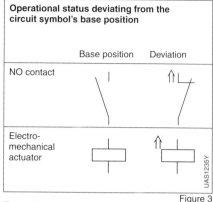

Operational status deviating from the circuit symbol's base position

	Base position	Deviation
NO contact		
Electro-mechanical actuator		

UAS1235Y

Figure 3

Figure 4

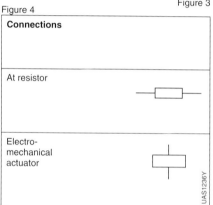

Connections

At resistor

Electro-mechanical actuator

UAS1236Y

Figure 5

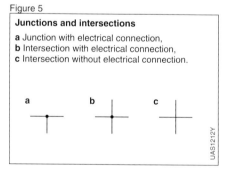

Junctions and intersections

a Junction with electrical connection,
b Intersection with electrical connection,
c Intersection without electrical connection.

a   b   c

UAS1212Y

Figure 6

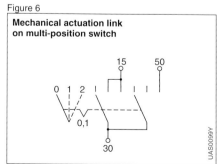

Mechanical actuation link on multi-position switch

UAS0099Y

Selected circuit symbols

(Refer to circuit diagram for additional information)

Connections

Electrical conductor; intersection (with/without connection)

Shielded conductor

Mechanical actuation link; electrical conductor (installed subsequently)

Junctions (with/without electrical connection)

Connection, general; seperable connection (if portrayal is required)

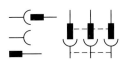

Plug connection; socket; plug; 3-pin plug connection

Ground (housing or vehicle ground)

Mechanical functions

Switch positions (base position: solid line)

0 1 2

0 1 2

Manual activation, with contact element (cam lobe), thermal (bimetallic)

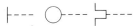

Detent; non-automatic/ automatic return in arrow direction (button)

Activation, general (mech., pneum., hydraul.); piston drive

Actuation at rotation rate n, pressure p, quantity Q, time t, temp. $t°$

$$\boxed{n} -- \boxed{p} -- \boxed{Q} --$$
$$\boxed{t} -- \boxed{t°} --$$

Variability, not intrinsic (external), general

Variability, instrinsic, in response to physical factor, linear/non-linear

Adjustability, general

Switches

Pressure switch, NO/NC contact

Position switch, NO/NC contact

Changeover switch, make before break or break before make

UAS1230E

19

Switches

Various components

Three-position switch with three contact modes (e.g., turn signal)

Make and break contact

Double-make contact

Multiple-position switch

Cam-lobe switch (e.g., ignition points)

Thermal switch

Trigger

Actuators with one winding

Actuator with two windings acting in same direction

Actuator with two opposed windings

Electrothermal actuator (thermal relay)

Electrothermal actuator, tractive solenoid

Solenoid valve (closed)

Relay (actuator and switch), example: NC contact operates without delay, NO contact operates with delay

Resistor

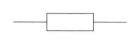

Potentiometer (with three connections)

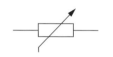

Resistor heater element, glow plug, flame plug, screen defroster

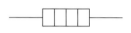

Antenna

Fuse

Permanent magnets

Winding, inductive

UAS1231E

Various components

Devices for automotive applications

Positive Temperature Coefficient (PTC) resistor

Negative Temperature Coefficient (NTC) resistor

Diode, general, current flows toward apex of triangle

PNP transistor
NPN transistor

E = Emitter (arrow indicates flow direction)
C = Collector, positive
B = Base (horizontal), negative

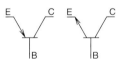

Light Emitting Diode (LED)

Hall generator

Dotted/dashed line used to delineate circuit sections or to indicate logically linked circuit components

Shielded device, frame rectangle connected to ground

Regulator, general

Electronic control units (ECUs)

Display element in general; voltmeter, clock.

Tachometer; temperature gauge; speed display.

Battery

Plug connector

Lamp, headlight

Horn, air horns

Rear screen defroster (heating element in general)

Switch, basic without indicator lamp

Switch, basic with indicator lamp

UAS1232E

Devices for automotive applications

Push-button switch	Spark plug	Motor with blower fan
Relay, general	Ignition coil	Starter motor with solenoid switch (with/without internal circuitry)
Solenoid valve, injection valve (injector), cold-start valve	Ignition distributor, general	
Thermo-time switch	Voltage regulator	Wiper motor (one/two wiper speeds)
Throttle-valve switch	Alternator with voltage regulator (with/without internal circuitry)	
Rotary actuator		Intermittent-wiper relay 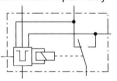
Auxiliary air valve with electrothermal drive	Electric fuel pump, hydraulic pump motor	Car radio

UAS1233E

Devices for automotive applications

Speaker

Piezoelectric sensor

Velocity sensor

Voltage stabilizer

Resistive position indicator

ABS wheel-speed sensor

Inductive sensor, reference-mark controlled

Air-flow sensor

Hall sensor

Flasher, pulse generator, interval relay

Air-mass meter

Converter, transformer (quantity, voltage)

Lambda oxygen sensor (not heated/heated)

Flow sensor, fuel-gauge sensor

Inductive sensor

Temperature switch, temperature sensor

Instrument cluster (dashboard)

N1   P2   P3   P4   P5   H1 H2   H3   H4   H5   H6

Circuit diagrams

Circuit diagrams are idealized representations of electrical devices, rendered in the form of symbols. Such diagrams also include illustrations and simplified design drawings as needed (Figure 1).

The circuit diagram illustrates the functional interrelationships and physical links that connect various devices. It may be supplemented by tables, graphs and descriptions. Circuit diagrams vary according to the intended application (e.g., showing circuit operation) and the selected representation mode.

A "legible" circuit diagram will meet the following requirements:

– The representations must reflect the applicable standards; explanations should be provided for any exceptions.
– Electrical current should be portrayed as flowing from left to right and/or from top to bottom.

In automotive electrical systems, block diagrams are used to provide a quick overview of circuit and device functions. They are usually unipolar and also dispense with representations of internal circuitry components.

The schematic diagrams in their various permutations (as defined by differences in symbol arrangements) provide a detailed diagram of the circuit. Because they illustrate how the circuit operates, they are suitable for use as a reference for repair operations.

The terminal diagram (with equipment connection points) is used by service facilities in replacing defective electrical equipment and when installing supplementary equipment.

Depending upon the type of representation, we distinguish between:

– Unipolar and multipolar representation (according to symbol arrangement),
– Assembled representation, semi-assembled representation, detached representation, and topographical (positionally correct) representation. One circuit diagram may employ all of the above forms of representation.

Fig. 1

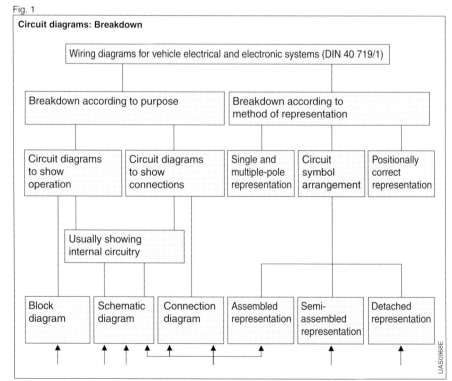

Circuit diagrams: Breakdown

Wiring diagrams for vehicle electrical and electronic systems (DIN 40 719/1)

Breakdown according to purpose

Breakdown according to method of representation

Circuit diagrams to show operation

Circuit diagrams to show connections

Single and multiple-pole representation

Circuit symbol arrangement

Positionally correct representation

Usually showing internal circuitry

Block diagram

Schematic diagram

Connection diagram

Assembled representation

Semi-assembled representation

Detached representation

UAS0968E

Block diagram

The block diagram is a simplified representation of a circuit showing only the most significant elements (Figure 2). It is designed to furnish a rapid overview of function, structure, layout and operation of an electrical system, or part of it. This format also serves as the initial reference for understanding more detailed schematic diagrams.

Squares, rectangles and circles together with attendant symbols are employed to illustrate the components. The basic reference is DIN 40900, Part 2. Wiring is usually shown in single-pole form.

Fig. 2

Motronic ECU block diagram

A1 ECU
B1 Engine-speed sensor
B2 Reference-mark sensor
B3 Air-flow sensor
B4 Intake-air temperature sensor
B5 Engine-temperature sensor
B6 Throttle-valve switch
D1 Microprocessor (CPU)

D2 Address bus
D3 Working memory (RAM)
D4 Program data memory (ROM)
D5 I/O
D6 Data bus
D7 Microcomputer
G1 Battery
K1 Pump relay

M1 Electric fuel pump
N1...N3 Power-output stages
S1 Ignition switch
S2 Program map selector
T1 Ignition coil
U1 and U2 Pulse generators
U3...U6 A/D converters
Y1 Injector

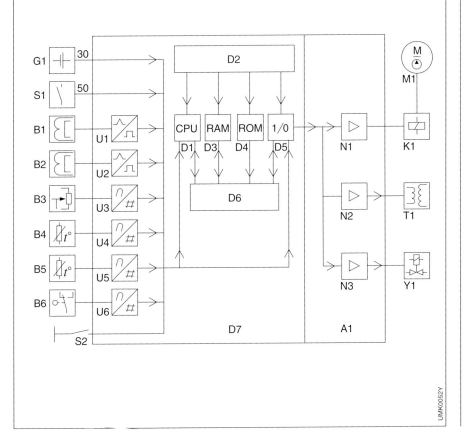

UMK0052Y

Schematic diagram

The schematic diagram shows a circuit and its elements in detail. By clearly depicting individual current paths it also indicates how an electrical circuit operates. In the schematic diagram, presentation of individual circuit components and their spatial relationship to each other must not interfere with the clear, logical and legible presentation of circuit operation. Figure 3 shows the schematic diagram for a starter, in the form of assembled (composite) and detached (exploded) representations.

The schematic diagram must contain the following:
- Wiring and general circuitry,
- Device designations (DIN 40719, Section 2), and
- Terminal designations (DIN 72552, DIN 42400).
- It must be suitable for recording on micro-fiche (minimum line width 0.25 mm).

The schematic diagram may also include:
- Comprehensive illustrations including internal circuitry, to facilitate testing, trouble-shooting, maintenance and replacement (retrofit installation);
- Reference codes to assist in finding symbols and installation locations, especially in detached representation diagrams.

Fig. 3

Two different ways of showing the circuit diagram for a Type KB starter motor for parallel operation

a Assembled representation,
b Detached representation,
K1 Control relay,
K2 Solenoid switch, hold-in winding, pull-in winding,
M1 Starter motor with series and shunt windings.

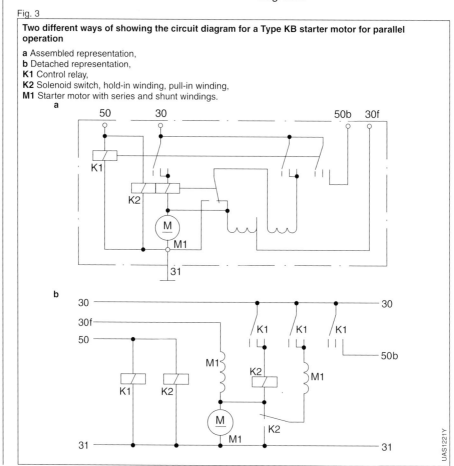

UAS1221Y

Circuit representation
Schematic diagrams usually use multipolar component connections. In accordance with DIN 40719, Section 1, symbols can be represented in the following ways, all of which may be combined within the same circuit diagram:

Assembled (composite) representation
All parts of a device are shown directly next to one another, and mechanical linkage of one part to another is indicated by a double line or broken connecting lines (dashes). This format may be employed to depict simple circuits of relatively limited complexity without unduly hindering clarity (Figure 3a).

Fig. 4

Ground representation

a Individual ground symbols,
b Common ground connection,
c With common ground point.

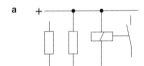

a

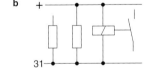

b

31

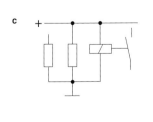

c

UAS1222Y

Detached representation
Symbols for the elements within electrical devices are shown separately in displays designed to show with maximum clarity the various routes taken by the current. No attempt is made to orient the symbols for individual devices and component parts in their actual spatial relationship to each other. Priority is assigned to arranging the symbols so that the individual current paths are as clear and free of crossings as possible.
Primary purpose: To indicate function and operation of a circuit.

A system of symbols defined in DIN 40719, Part 2 can be used to indicate the relationships between the individual components. Each separately illustrated device symbol includes the code for the device. If it is necessary in the interests of clarity and comprehension, a section of the diagram should be set aside for the complete and assembled representation of devices which have otherwise been shown in the detached form (Figure 3b).

Topographical representation
This type of representation places the symbol in a position that completely or partly corresponds to its location within the device or component.

Ground (earth) symbols
For the sake of conceptual simplicity, most vehicles employ a single (hot) conductor layout, relying on the metallic body to conduct the return current. Designers resort to insulated return wiring either when restraints prevent using the body for satisfactory ground connections or when voltages in excess of 42 volts are being handled.
All terminals represented by the ground symbol ($\perp$) are mutually connected electrically through component (housing) or vehicle ground.
All components with a ground symbol must be mounted on the vehicle ground to which they must have a direct electrical connection.
Figure 4 depicts several options for showing connections to ground.

Current paths and conductors (wiring)

Circuits should be arranged to be clear and easy to follow. When possible, individual current paths should indicate signal flow from left to right and/or from top to bottom, as well as being straight and free of intersections and changes in direction. They should also be parallel to the border of the circuit diagram.

When a number of conductors run parallel to each other, they are grouped in sets of three with spaces between the groups.

Lines of demarcation, borders

Dot/dash demarcation or border lines are used to separate individual components within a circuit as an indication of functional and/or structural relationships.

In illustrations of automotive electrical systems, these alternating dots and dashes represent a non-conductive border around a device or circuit component. The line will not always correspond with the component housing and does not indicate ground. In high-tension circuits this outer line is frequently combined with the protective conductor (PE), also represented as a broken dotted line.

Interruptions, codes, destination

For clarity, connecting lines (conductors and lines denoting mechanical linkage) can be interrupted if they would otherwise extend too far within the schematic diagram. Only beginning and end of the connecting line are shown. An unambiguous means - usually based on codes or or designations for the open sections – is required to identify the respective ends of the interrupted circuit section.

The matching codes for the respective ends of the opened circuit section are:

– Terminal designations (DIN 72 552, DIN 42 400), on left of Figure 5,
– Indication of function,
– Identification using alphanumeric symbols.

The destination is given in parentheses to distinguish it from the code; it consists of the section number of the destination (Figure 5 on right).

Fig. 5

Broken-line designation using terminal destinations

a With terminal designations, e.g., Term. 15
b With destination indication, e.g., in sections 8 and 2.

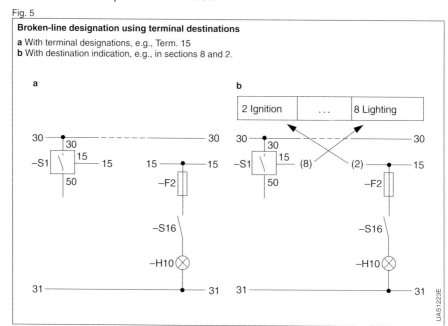

UAS1223E

Section identification

The section-identification code along the upper border of the diagram is used for locating circuit sections. This designation can be in one of three forms.
– Consecutive numbers at equal intervals from left to right (Figure 6a),
– Indication of the content of the circuit sections (Figure 6b),
– Or a combination of the two (Figure 6c).

In the schematic diagrams shown in the manual, section identification is used in accordance with Fig. 6b.

Legends

Devices, parts or symbols are labelled in circuit diagrams with a letter and a cardinal number as defined in DIN 40 719, Part 2. This code is located to the left of or underneath the symbol.

Devices, parts or symbols are labelled in circuit diagrams with a letter and a cardinal number as defined in DIN 40 719, Part 2. This code is located to the left of or underneath the symbol.

Fig. 6

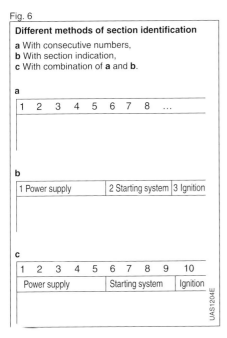

Different methods of section identification

a With consecutive numbers,
b With section indication,
c With combination of **a** and **b**.

a

| 1 | 2 | 3 | 4 | 5 | 6 | 7 | 8 | ... |

b

| 1 Power supply | 2 Starting system | 3 Ignition |

c

| 1 | 2 | 3 | 4 | 5 | 6 | 7 | 8 | 9 | 10 |
| Power supply | | Starting system | | Ignition |

UAS1204E

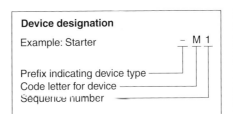

Device designation

Example: Starter   – M 1

Prefix indicating device type
Code letter for device
Sequence number

Terminal designation

Example: Terminal 30   : 30

Terminal prefix
Terminal code

The prefix used to designate the type of device can be omitted if the device is clearly recognizable.

In nested devices, one device is a component part of another, e.g., starter M1 with built-in solenoid switch K6. The designation for the entire device is then: – M1 – K6.

Identification of related symbols in detached views: Each individual symbol is shown separately and all symbols for a particular device are assigned a code which is the same as that used for the device itself.

Terminal designations (such as those defined in DIN 72 552) must be placed outside the symbol and, if boundary lines are present, outside these lines if possible.

For horizontal current paths:
The data applying to the individual symbol are given underneath it. The terminal code is above the connecting line, just outside the symbol proper.

For vertical current paths:
The data applying to the individual symbol are provided to its left. The terminal code is just outside the symbol. It the type is horizontal, the code is provided next to the connecting line on the symbol's right; if the type is vertical, it is on the left.

Terminal diagram

The terminal diagram shows the connection points of electrical devices. It also illustrates the external (and internal as required) connections (lines) at these points.

Representation

Individual electrical devices are illustrated using squares, rectangles, circles, symbols or illustrations, and their locations may correspond to their installed positions. The connections are represented by circles, dots, plug connectors, or simply by the connecting line.

The following conventions govern the methods of representation used in automotive electrical systems:
- Assembled (composite) symbols defined in DIN 40 900 (Fig. 7a),
- Assembled, pictorial representation of the device (Fig. 7b),
- Detached representation of the device, with symbol, connections with destination references; wiring color codes optional (Fig. 8a),
- Detached, pictorial representation, including connections with destination references; wiring color codes optional (Fig. 8b).

Color codes for electrical wiring (as defined in DIN 47 002)

bl	blue	gn	green	sw	black
br	brown	or	orange	tk	turquoise
ge	yellow	rs	pink	vi	purple
gr	gray	rt	red	ws	white

Designation codes

Electrical devices are identified as defined in DIN 40 719, Section 2. Terminals and plug connections are indicated with terminal designations on the device (Fig. 7). Detached views dispense with continuous connecting lines between devices. All conductors leaving a device are provided with a termination code (DIN 40 719, Section 2), consisting of the code for the element at the other end of the wire and the terminal with – if necessary – the wiring color code as specified in DIN 47 002 (Fig. 9).

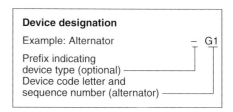

Device designation

Example: Alternator — G1

Prefix indicating device type (optional) ——
Device code letter and sequence number (alternator) ——

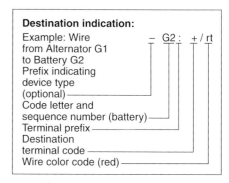

Destination indication:

Example: Wire — G2 : + / rt
from Alternator G1
to Battery G2
Prefix indicating device type (optional) ——
Code letter and sequence number (battery) ——
Terminal prefix ——
Destination terminal code ——
Wire color code (red) ——

Fig. 7

Terminal diagram, assembled representation
a With symbols, **b** With devices.

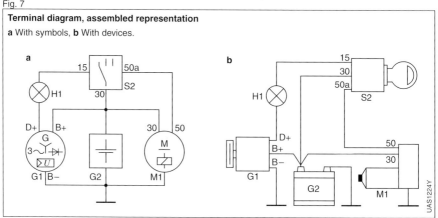

Terminal diagram, detached view

a With symbols and indications of destination, **b** With devices and destinations,
G1 Alternator with voltage regulator,
G2 Battery,
H1 Charge indicator lamp,
M1 Starter motor,
S2 Ignition switch,
XX Device ground on vehicle chassis,
Y Y Terminal for ground connection,
:15 Conductor potential, e.g., Terminal 15

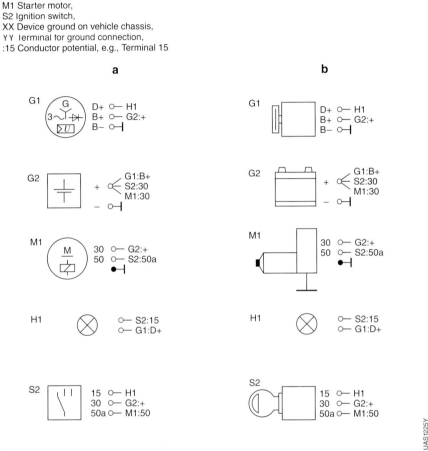

a   **b**

Fig. 8

Fig. 9

Device designation. Example: Alternator

a Device designation (code letter and sequence number)
b Terminal code on device
c Device to ground
d Destination indication (code letter plus sequence number/terminal designation/wire color code)

Device representation   Destination indication

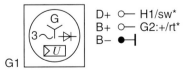

a   **b**   **c**   **d**   *sw: black
                                              rt: red

UAS1055E

Designations for electrical devices

The designations defined in DIN 40717, Section 2 serve as the basis for unambiguous and internationally-recognized labeling conventions for assemblies, components, etc., as represented in circuit diagrams by electrical symbols. The designation as per DIN 40717 is a defined sequence consisting of prefix, letter and numerals. It appears next to the symbol.

Example: Alternator G2, Terminal 15

Prescribed designation code: – G 2:15

Prefix
(may be deleted
if redundant)

Type code letter
(here: alternator)
from Table 1

Sequence number

Connection
(here: Terminal 15)
as standard designation and/
or as indicated on device)

Table 1

Code letter	Type	Examples
A	System, assembly, component group	ABS control units, car radios, two-way radios, mobile phones, alarm systems, equipment assemblies, triggering units, ECUs, cruise control
B	Transducer for converting non-electrical parameters into electrical values and vice versa	Reference-mark sensors, pressure switches, air horns, horns, Lambda oxygen sensors, loudspeakers, air-flow sensors, microphones, oil pressure switches, various sensors, ignition-triggering devices
C	Condenser, capacitor	All types of condensers/capacitors
D	Binary element, data storage	On-board computers, digital equipment, integrated circuits, pulse counters, magnetic-tape recorders
E	Various devices and accessories	Heater systems, air-conditioner, lamps, headlights, spark plugs, ignition distributors
F	Protective equipment	Triggers (bimetallic), reverse-polarity guards, fuses, current protection circuitry
G	Current supply, alternator	Batteries, alternators, battery chargers
H	Monitor, signaling, warning, display device	Audio alarms, display lamps, flasher indicators, turn signals, brake-pad wear indicators, stop lamps, high-beam indicators, charge indicator lamps, indicator lamps, signaling devices, oil-pressure warning lamps, optical indicators, signal lamps, warning buzzers
K	Relay, protective device	Battery relays, turn-signal relays, flasher relays, solenoid relays, starter relays, hazard-warning and turn-signal flashers
L	Inductor	Choke coils, windings

Code letter	Type	Examples
M	Motor	Blower motors, fan motors, pump motors for ABS/TCS/ESP hydraulic modulators, windshield washer/wiper motors, starter motors, step motors
N	Regulator, amplifier	Regulators (electronic or electromechanical), voltage stabilizers
P	Measuring instruments/ Monitoring equipment	Ammeters, diagnosis interfaces, tachometers, manometers, tachographs, test taps/connections, test points, speedometers
R	Resistor	Glow plugs, flame plugs, resistive heater elements, PTC and NTC resistors, potentiometers, regulating resistors, in-line resistors
S	Switch	Switches and contacts of all kinds, ignition points
T	Transformer	Ignition coils, ignition transformers
U	Modulator, converter	DC transformers
V	Semiconductor, electron tube	Darlington transistors, diodes, electron tubes, rectifiers, all semiconductors, varactors, transistors, thyristors, Zener diodes
W	Transmission path, conductor, antenna	Vehicle antennas, shielding, shielded cable, all types of cable, wiring harnesses, (common) ground conductors
X	Terminal, plug, plug-in connection	Terminal studs, all types of electrical connection, spark-plug connectors, terminals, terminal bars, electrical wiring couplings, wiring connectors, plugs, sockets, socket rails, (multiple-pin) plug connections, junction plugs
Y	Electrically operated mechanical devices	Permanent magnets, (solenoid) injection valves, solenoid clutches, electric air valves, electric fuel pumps, solenoids, electric start valves, transmission-shift controls, tractive magnets, kickdown solenoids, headlight leveling controls, ride-height control valves, circuit-control valves, start valves, door locks, central locking systems, auxiliary air devices
Z	Electrical filtering elements	Elements and filters for interference suppression, filter networks, clocks

DIN 72 552 terminal designations

The system of standard terminal designations prescribed for use in automotive applications has been designed to facilitate correct connection of devices and their wiring, with emphasis on repairs and replacement installations.

The terminal codes are not wire designations, as devices with differing terminal codes can be connected to the opposite ends of a single wire. It is therefore not essential that the terminal codes be provided on the wiring.

The DIN 72552 codes may be supplemented by the designations defined in DIN-VDE standards for electrical machinery. Multi-pin plug connections large enough to exhaust the range provided by DIN 72552 are allocated consecutive numbers or letters, avoiding any characters to which the standard has already assigned a specific function.

Terminal	Definition
1	**Ignition coil, distributor** Low-tension circuit
	Ignition distributor with two insulated circuits
1 a	to ignition point set I
1 b	to ignition point set II
2	Short-circuit terminal (magneto ignition)
4	**Ignition coil, distributor** High-tension circuit
	Ignition distributor with two insulated circuits
4 a	Terminal 4, from coil I
4 b	Terminal 4, from coil II
15	Switch-controlled plus downstream from battery (from ignition switch)
15 a	In-line resistor terminal leading to coil and starter
	Glow-plug switch
17	Start
19	Preglow
30	Line from battery positive terminal (direct)
30 a	**Series/parallel battery switch 1 2/24V** Line from battery positive terminal II

Terminal	Definition
31	Return line from battery negative terminal or ground (direct)
31 b	Return line to battery negative terminal or ground via switch or relay (switch-controlled ground)
	Battery changeover relay 12/24 V
31 a	Return line to Battery II negative pole
31 c	Return line to Battery I negative pole
	Electric motors
32	Return line [1]
33	Main connection [1]
33 a	Self-parking switch-off
33 b	Shunt field
33 f	for reduced-rpm operation, speed 2
33 g	for reduced-rpm operation, speed 3
33 h	for reduced-rpm operation, speed 4
33 L	Rotation to left (counterclockwise)
33 R	Rotation to right (clockwise)
	Starter
45	Separate starter relay, output: starter; input: primary current
	Dual starters, parallel activation Relay for pinion-engagement current
45 a	Starter I output Starters I and II input
45 b	Starter II output

[1] Polarity reversal terminal 32/33 possible

Terminal	Definition
48	Terminal on starter and start-repeating relay for monitoring starting process
	Flasher relay (pulse generator)
49	Input
49 a	Output
49 b	Output to second flasher relay
49 c	Output to third flasher relay
	Starter
50	Starter control (direct)
	Battery switching relay
50 a	Output for starter control
	Starter control
50 b	Dual starters in parallel operation with sequential control
	Starting relay for sequential control of engagement current for dual starters in parallel operation
50 c	Starter I input at starter relay
50 d	Starter II input at starter relay
	Start-locking relay
50 e	Input
50 f	Output
	Start repeating relay
50 g	Input
50 h	Output
	AC generator (alternator)
51	DC voltage at rectifier
51 e	DC voltage at rectifier with choke coil for daylight operation
	Trailer signaling devices
52	Supplementary signal transmission from trailer to towing vehicle
53	Wiper motor, input (+)
53 a	Wiper (+), end position
53 b	Wiper (shunt winding)
53 c	Electric windshield-washer pump
53 e	Wiper (brake winding)
53 i	Wiper motor with permanent magnet and third brush (for higher speed)

Terminal	Definition
55	Front fog lamp
56	Headlights
56 a	High-beam with indicator lamp
56 b	Low beam
56 d	Headlight flasher contact
57	Motorcycle/Moped parking lamps (also for passenger cars, trucks, in some export markets)
57 a	Parking lamps
57 L	Parking lamps, left
57 R	Parking lamps, right
58	Side-marker lamps, tail lamps, license-plate and instrument illumination
58 b	Tail light mode selection on single-axle tractors
58 c	Trailer gladhand assembly for single-strand tail light with fuse in trailer
58 d	Rheostatic instrument illumination, tail and side-marker lamps
58 L	left
58 R	right, license-plate lamps
	AC generator (alternator) (magneto generator)
59	AC voltage output, rectifier input
59 a	Charging-armature output
59 b	Tail-lamp armature, output
59 c	Stop-lamp armature, output
61	Charge indicator lamp
	Tone-sequence controller
71	Input
71 a	Output to Horns I and II (bass)
71 b	Output to Horns 1 and 2 (treble)
72	Alarm switch (rotating beacon)
75	Radio, cigarette lighter
76	Speakers

Terminal	Definition
77	Door valve control
	Trailer signaling equipment
54	Trailer gladhand assembly and light combinations Stop lamps
54 g	Pneumatic valve for continuous-duty trailer brake with solenoid control
	Switches, NC contacts and changeover contacts
81	Input
81 a	First output on NC-contact side
81 b	Second output on NC-contact side NO contacts
82	Input
82 a	First output
82 b	Second output
82 z	First input
82 y	Second input Multiple position switch
83	Input
83 a	Output (Pos. 1)
83 b	Output (Pos. 2)
83 L	Output (Pos. left)
83 R	Output (Pos. right)
	Current relay
84	Input: Actuator and relay contacts
84 a	Output: Actuators
84 b	Output: Relay contacts
	Switching relay
85	Output: Actuator (negative winding end or ground)
86	Input: Actuator Start of winding
86 a	Start of winding or first winding coil
86 b	Winding tap or second winding coil

Terminal	Definition
	Relay contact for NC and changeovers contacts
87	Input
87a	First output (NC-contact side)
87 b	Second output
87 c	Third output
87 z	First input
87 y	Second input
87 x	Third input
	Relay contact for NO contact
88	Input
	Relay contact for NO contact and changeover contacts (NO side)
88 a	First output
88 b	Second output
88 c	Third output
	Relay contact for NO contact
88 z	First input
88 y	Second input
88 x	Third input
	Generator/alternator and voltage regulator
B +	Battery positive terminal
B −	Battery negative terminal
D +	Generator positive terminal
D −	Generator negative terminal
DF	Generator field winding
DF 1	Generator field winding 1
DF 2	Generator field winding 2
	Alternator
U, V, W	Three-phase terminals
	Turn signals (turn-signal flasher)
C	Indicator lamp 1
C 0	Main terminal connection for indicator lamp not connected to turn-signal flasher
C 2	Indicator lamp 2
C 3	Indicator lamp 3 (e.g., for dual-trailer operations)
L	Left-side turn signals
R	Right-side turn signals

Circuit diagram for passenger cars (Examples)

Purpose

Circuit diagrams are the only way to provide an overview of complex automotive electrical systems with their numerous terminals and connections (Fig. 1). Modern systems include more than just the lights, with a large number of electric and electronic devices for open and closed-loop engine management as well as numerous accessories for safety, comfort and convenience.

Design

The schematic diagrams in the following section portray various vehicle circuits. They are intended to facilitate understanding of the text; they are not intended for use in manufacture or installation.

Sample designation codes

A1 Device code (DIN 40 719)
15 Terminal code (DIN 72 552)
1 Section identification code (DIN 40 719)

Fig. 2

Section of an automotive wiring harness

UKE0454Y

Fig. 1

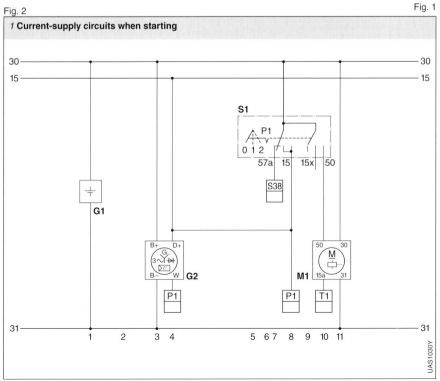

1 **Current-supply circuits when starting**

UAS1030Y

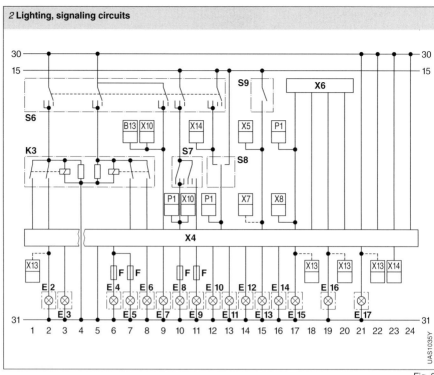

Fig. 3

Fig. 4

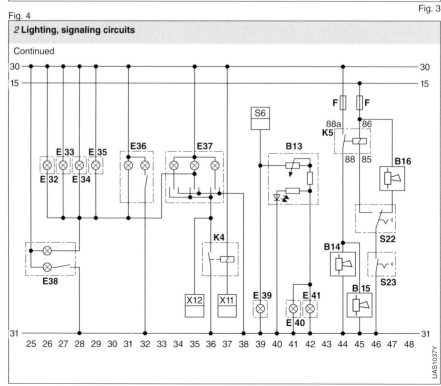

2 Lighting, signaling circuits

Continued

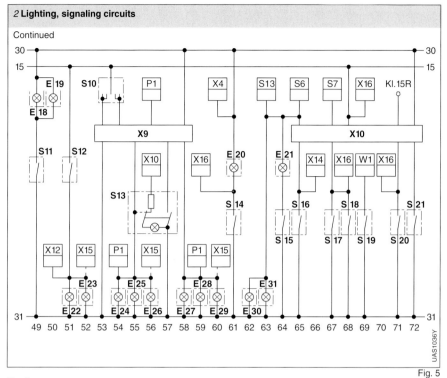

Fig. 5

Fig. 6

3 Radio

4 Indicating instruments (instrument cluster)

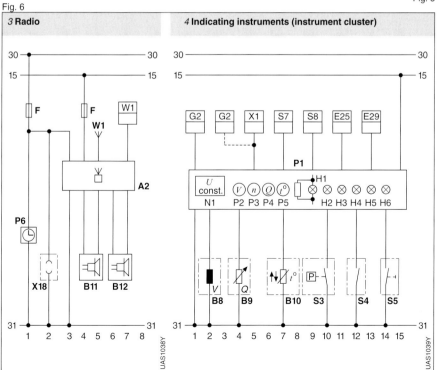

5 ABS with CAN bus

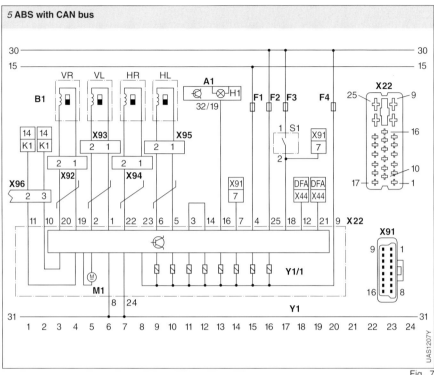

Fig. 7

Fig. 8

6 Motronic M

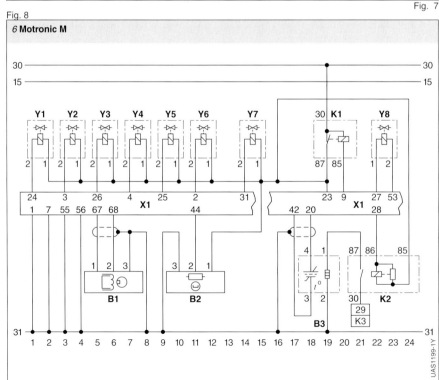

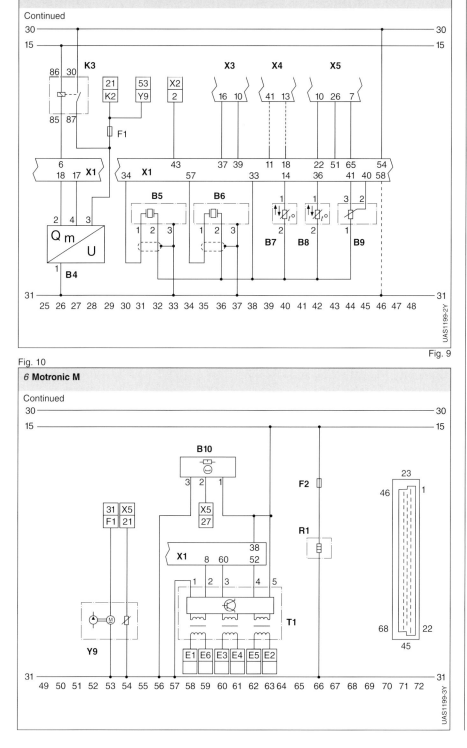

Fig. 10

Fig. 9

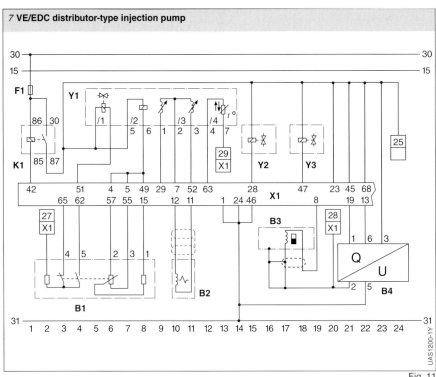

Fig. 11

Fig. 12

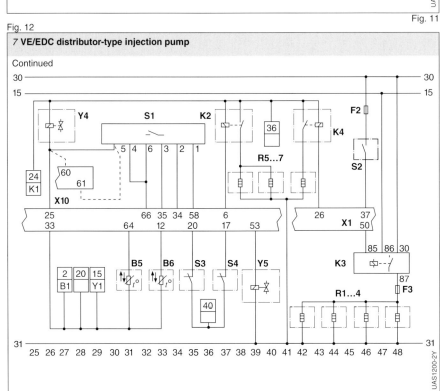

7 VE/EDC distributor-type injection pump

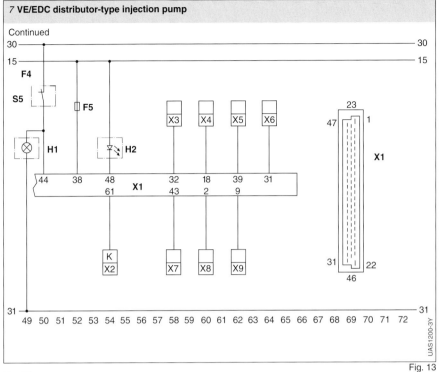

Fig. 13

Fig. 14

8 Diesel preglow circuit

9 Vehicle alarm system

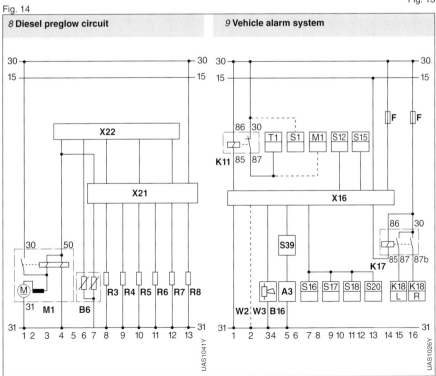

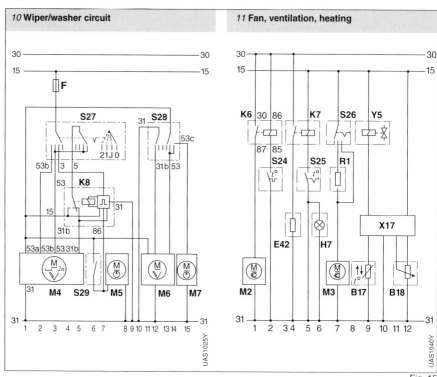

10 Wiper/washer circuit

11 Fan, ventilation, heating

Fig. 16

Fig. 15

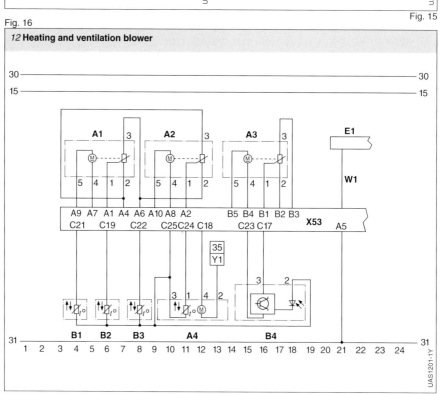

12 Heating and ventilation blower

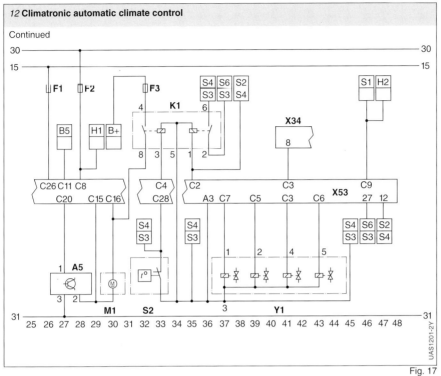

Fig. 17

Fig. 18

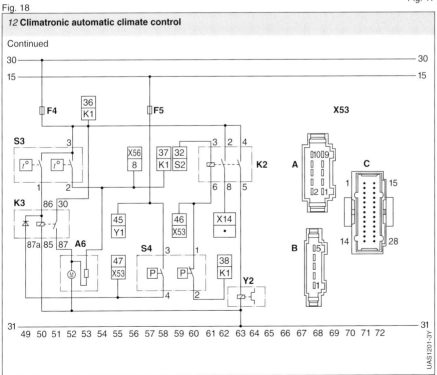

13 AG4 Transmission-shift control

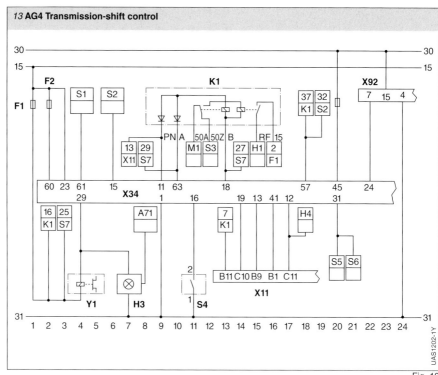

Fig. 19

Fig. 20

13 AG4 Transmission-shift control

Continued

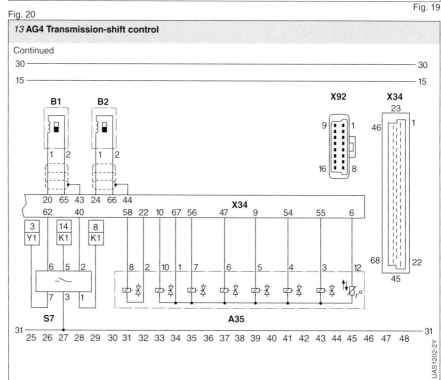

Section designations and device locations

Table 2 contains all of the section designations contained in the section "Circuit diagram for a passenger car with spark-ignition engine." The "Sections in the schematic diagram" (Table 3 below) define specific areas of the schematic diagrams in which a particular system can be found. Table 3 contains devices and their codes along with section numbers for passenger-car circuit diagrams.

Table 2
Sections

Section	System
1	Starting, current supply
2	Lighting, signaling equipment
3	Clock, radio
4	Display assembly (instrument cluster)
5	ABS with CAN bus
6	Motronic M
7	VE/EDC distributor-type injection pump
8	Diesel preglow system
9	Vehicle alarm
10	Wash/wipe system
11	Heater and ventilation blower
12	Climatronic automatic climate control
13	AG4 transmission-shift control

Table 3
Device classification

Code	Device	Section
A1	Warning-lamp display unit	5
A1	Central-flap control motor	12
A2	Radio	3
A2	Dynamic pressure-flap actuator	12
A3	Ignition system with knock control (EZ-K)	9
A3	Temperature-flap actuator motor	12
A4	Dashboard temperature sensor with blower	12
A5	Blower control unit	12
A6	Radiator fan	12

Code	Device	Section
A35	Transmission unit, electric	13
B1	Rotation rate/reference-mark sensor	5, 6
B1	Accelerator-pedal travel sensor	7
B1	Coolant-temperature sensor	12
B1, 2	Vehicle-speed sensor	13, 6
B2	Needle-motion sensor	7
B2	Outside-temperature sensor	12
B2	Transmission-input shaft rpm sensor	13
B3	Lambda oxygen sensor	6
B3	Rotation-rate/ reference-mark sensor	7
B3	Intake-air temperature sensor	12
B4	Air-mass meter	6, 7
B4	Photosensor	12
B5	Knock sensor 1	6
B5, 7	Coolant-temperature sensor	7, 6
B6	Knock sensor 2	6
B6	Fuel-temperature sensor	7, 8
B8	Speed sensor	4
B8	Intake-air temperature sensor	6
B9	Fuel-level (gauge) sensor	4
B9	Throttle-valve-potentiometer sensor	6
B10	Coolant-temperature sensor	4
B10	Cylinder-identification sensor	6
B11,12	Speaker	3
B13	Instrument illumination rheostat	2
B14, 15	Supertone horn	2
B16	Horn 2	2, 9
B17	Interior-temperature sensor	11
B18	Setpoint selector	11
E1	Climatronic display unit	12
E2, 3	Fog-warning lamp, L/R	2
E4, 5	Driving lamp, L/R	2
E6, 7	Front fog, L/R	2
E8, 9	Low-beam headlamp, L/R	2
E10, 11	Side-marker lamp, L/R	2
E12, 13	License-plate lamp, L/R	2
E14, 17	Brake light, L/R	2
E15, 16	Tail lamp, L/R	2
E18	Trunk-lid lamp	2
E19	Luggage-compartment lamp	2
E20	Glove-compartment lamp	2
E21	Engine-compartment lamp	2
E22, 23	Backup lamp, L/R	2
E24, 26	Turn signal, LF, LR	2

Code	Device	Section
E25, 28	Hazard warning flasher L/R	2
E27, 29	Turn signal, RF, RR	2
E30, 31	Ashtray lamp, front and rear	2
E32, 33	Footwell lamp, LR, LF	2
E34, 35	Footwell lamp RF, RR	2
E36, 38	Rear map light, R/L	2
E37	Interior map light	2
E39	Vanity-mirror lamp	2
E40	Instrument illumination	2
E41	Control/dashboard illumination	2
E42	Rear-screen defroster	11
F..	Fuses	
G1	Battery	1
G2	Alternator	1
H1	Charge-indicator lamp	4
H1	ABS warning lamp	5
H1	Preglow indicator lamp	7
H2	Oil-pressure warning lamp	4
H2	Stop lamp	7
H3	Parking-brake indicator lamp	4
H3	Selector-lever illumination	13
H4	Brake-pad wear-indicator lamp	4
H5	High-beam indicator lamp	4
H6	Turn-signal indicator lamp	4
H7	Rear-screen defroster indicator	11
K1	Main relay	6, 7
K1	A/C relay	12
K1	Starter lockout relay	13
K2	Lambda-Oxygen-sensor heater relay	6
K2	Relay for minor heater wire	7
K2	A/C compressor relay	12
K3	Parking-lamp monitor relay	2
K3	Electric-fuel-pump relay	6
K3	Glow-plug relay	7
K3	Radiator-fan starter relay	12
K4	Interior-lamp control relay	2
K4	Heater-line relay	7
K5	Hightone-horn relay	2
K6	Engine-fan relay	11
K7	Rear-screen-defroster relay	11
K8	Intermittent-wiper relay	10
K11	Starter/ignition lockout relay	9
K17	Visual-alarm relay	9
M1	Starter motor	1, 8

Code	Device	Section
M1	Pump motor Hydraulic modulator	5
M1, 3	Fresh-air ventilation fan motor	11, 12
M2	Blower motor	11
M4	Wiper motor	10
M5	Windshield-washer motor	10
M6	Engine-fan relay	10
M7	Rear-screen washer motor	10
N1	Voltage stabilizer	4
P1	Instrument cluster	4
P2	Electric speedometer	4
P3	Tachometer	4
P4	Fuel gauge	4
P5	Engine-temperature gauge	4
P6	Clock	3
R1	Heater-resistance element	6
R1..4	Glow plugs	7
R1	Blower resistor	11
R5..7	Auxiliary heater (with manual transmission)	7
R3..8	Glow plugs	8
S1	Ignition/starter switch	1
S1	Brake-light switch	5
S1	Cruise-control selector unit	7
S1	Light switch	12
S2	A/C switch	7
S2	Evaporator temperature switch	12
S3	Oil-pressure switch	4
S3	Brake-pedal switch	7
S3	Radiator-fan temperature switch	12
S4	Parking-brake switch	4
S4	Clutch-pedal switch	7
S4	A/C system-pressure switch	12
S4	Kickdown switch	13
S5	Brake-pad-wear indicator contact	4
S5	Brake-light switch	7
S6	Light switch	2
S7	Fog-lamp switch	2
S7	Multifunction switch	13
S8	Low-beam switch	2
S9	Brake-light switch	2
S10	Turn-signal switch	2
S11	Trunk-lid lamp switch	2
S12	Backup-lamp switch	2
S13	Hazard-warning-flasher switch	2
S14	Glove-compartment lamp switch	2

Code	Device	Section
S15	Engine-compartment lamp switch	2
S16..18	Door-contact switch, LF, RR, LR	2
S19	Impact switch	2
S20	Door-contact switch, ПП	2
S21	Door-handle switch	2
S22	Horn-selector switch	2
S23	Horn-contact switch	2
S24	Thermal switch	11
S25	Rear-screen defroster switch	11
S26	Blower switch	11
S27	Wiper switch	10
S28	Rear-screen wiper washer switch	10
S29	Washer switch	10
S39	Alarm-system code-entry switch	9
T1	Ignition coil	6
W1	Car antenna	3
W1	Socket connector for 16-pin flat cable	12
W2,3	Code entry line	9
X1	Motronic/VE/EDC control-unit plug	6, 7
X3	A/C control-unit plug	6
X4	Lamp-control-module plug	2
X4	Transmission-shift control ECU plug	6
X5	Instrument-cluster connection plug	6
X6	Check Control plug	2
X9	Hazard-warning-flasher relay socket	2
X10	Basic module plug for central bodywork electronics	2, 7
X11	Engine-management-ECU plug	13
X16	Alarm-system-ECU plug	9
X17	A/C & heater-control ECU plug	11
X18	Diagnosis interface	3
X21	Glow-control-unit ECU plug	8
X22	ABS/ABD ECU plug	5
X22	Diagnosis plug	8
X34	Transmission-shift-control ECU plug	12, 13
X44	Navigation-system plug	5
X53	Automatic climate-control plug	12
X91, 92	Diagnosis interface	5, 13
Y1	Hydraulic unit	5
Y1	Injection valve 1	6, 7

Code	Device	Section
Y1	Fuel rail	12
Y1	Shift-lockout solenoid	13
Y2	A/C performance control	7
Y2	A/C solenoid clutch	12
Y2..5	Injection valves	6, 7
Y5	Hot-water valve	11
Y6	Injection valve 6	6
Y7	Canister-purge valve	6
Y8	Idle actuator	6
Y9	Electric fuel pump	6

Circuit diagrams

49

Schematic diagram

Bosch has responded to the requirements associated with trouble-shooting on complex, networked systems by developing system-specific circuit diagrams. Bosch makes schematic diagrams for numerous vehicles available on its "P" CD-ROM, which is an integral element within the Bosch ESI Electronic Service Information system. This reference source furnishes vehicle service operations with valuable assistance in localizing defects, while also helping personnel install auxiliary equipment with maximum efficiency. Figure 2 illustrates the schematic diagram for a door-locking system.

The representations in the schematic diagrams diverge from those in the standard circuit diagrams by relying on US symbols with supplementary legends (Figure 1). These legends include component codes (for instance: "A28" for anti-theft system) as well as the color codes for wiring (Table 2). Both tables can be accessed with the "P" CD-ROM.

Schematic diagrams are classified according to system circuits, with further

Figure 1

Supplementary data in the schematic diagram

1 Wire color, **2** Connector number,
3 PIN number (dashes between PINs indicate that all PINs are part of the same plug).

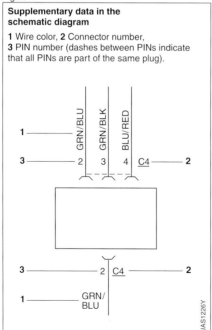

Table 1

Explanation of component codes.

Position	Description
A1865	Electric power-seat system
A28	Theft-deterrent system
A750	Fuse/relay module
F53	Fuse C
F70	Fuse A
M334	Supply pump
S1178	Warning-buzzer switch
Y157	Vacuum actuator
Y360	Actuator, door, right front
Y361	Actuator, door, left front
Y364	Actuator, door, right rear
Y365	Actuator, door, left rear
Y366	Fuel filler-flap actuator
Y367	Actuator, lock, luggage compartment, trunk lid

Table 2

Wire color code chart.

Position	Description
BLK	Black
BLU	Blue
BRN	Brown
CLR	Transparent
DK BLU	Dark blue
DK GRN	Dark green
GRN	Green
GRY	Gray
LT BLU	Light blue
LT GRN	Light green
NCA	No color assignment
ORG	Orange
PNK	Pink
PPL	Purple
RED	Red
TAN	Tan
VIO	Violet
WHT	White
YEL	Yellow

UAS1226Y

Figure 2

Schematic diagram for a door-locking system (example)

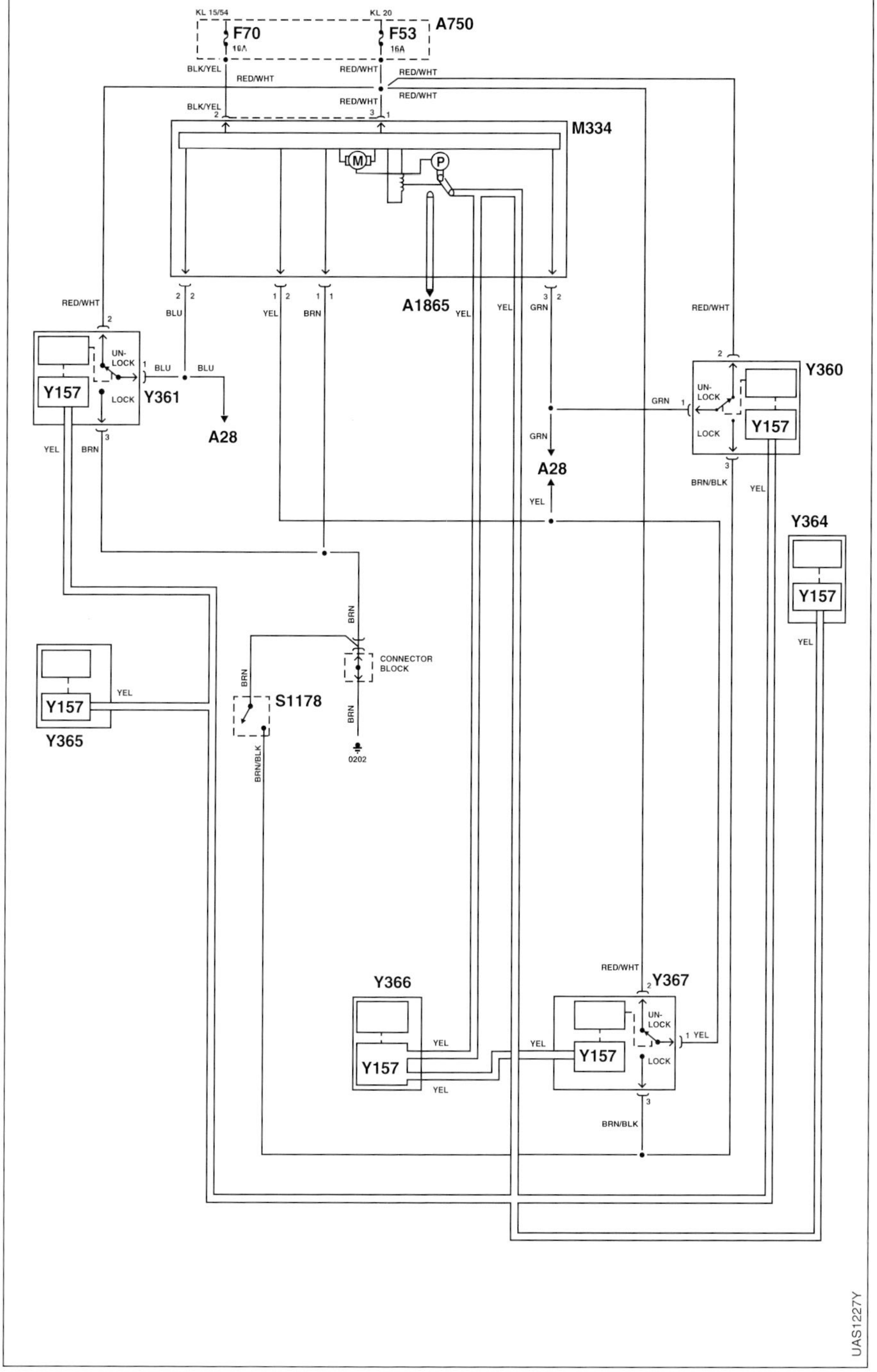

KL 15/54   KL 20   **A750**

F70   F53
16A   16A

BLK/YEL   RED/WHT   RED/WHT
RED/WHT   RED/WHT

BLK/YEL   RED/WHT
2   3 1

M334

(M)   (P)

RED/WHT   2 2   1 2 1 1   **A1865**   YEL   3 2   RED/WHT
BLU   YEL   BRN   YEL   YEL   GRN

UN-
LOCK   1 BLU   BLU   **Y360**

Y157   LOCK   **Y361**   UN-
LOCK

A28   GRN 1   **Y157**

GRN   LOCK

YEL   BRN   **A28**   3   **Y364**
YEL   BRN/BLK   YEL

Y157
YEL

Y157   YEL
BRN

Y365   BRN   CONNECTOR
BLOCK

BRN   **S1178**   BRN

BRN/BLK   0202

RED/WHT   **Y367**
Y366   2   UN-
LOCK
YEL   YEL   1 YEL
Y157   **Y157**   LOCK
YEL

3

BRN/BLK

UAS1227Y

51

divisions by subsystem as indicated (Table 3).

Classification of system circuits reflects the standard ESI practice as used for other systems by employing assignments to one of four assembly groups:
- Engine,
- Bodywork,
- Suspension, and
- Drivetrain.

Table 3

System sub-circuits

1	Engine management
2	Starter/charging circuit
3	Heating and air conditioning
4	Blower fan
5	ABS
6	Cruise control
7	Power windows
8	Central locking system
9	Instrument panel
10	Wiper/washer system
11	Headlights
12	External lighting
13	Current supply
14	Ground assignments
15	Data cable
16	Shift lockout
17	Theft-deterrent system
18	Passive safety system
19	Electric antenna
20	Alarm system
21	Screen/mirror defroster
22	Supplementary safety systems
23	Interior lighting
24	Power steering
25	Adjustable mirrors
26	Power convertible top
27	Horn
28	Luggage compartment, trunk lid
29	Power seats
30	Electronic damping
31	Cigarette lighter, socket
32	Navigation
33	Transmission
34	Active bodywork components
35	Vibration damping
36	Mobile phone
37	Radio/sound system
38	Vehicle immobilizer

Because ground-point identification is always important, and absolutely vital for installations of supplementary equipment, CD-ROM "P" supplements the schematic diagrams for specific vehicles with individual diagrams showing ground locations (Figure 3).

Figure 3

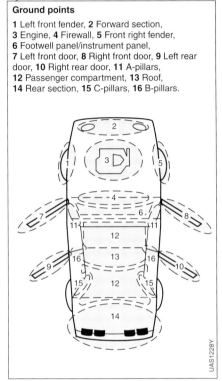

Ground points

1 Left front fender, **2** Forward section, **3** Engine, **4** Firewall, **5** Front right fender, **6** Footwell panel/instrument panel, **7** Left front door, **8** Right front door, **9** Left rear door, **10** Right rear door, **11** A-pillars, **12** Passenger compartment, **13** Roof, **14** Rear section, **15** C-pillars, **16** B-pillars.

UAS1228Y

The schematic diagrams rely on US symbols that differ from the DIN and IEC standards. Samples of these American symbols are provided in Figure 4.

Figure 4

Sample US symbols

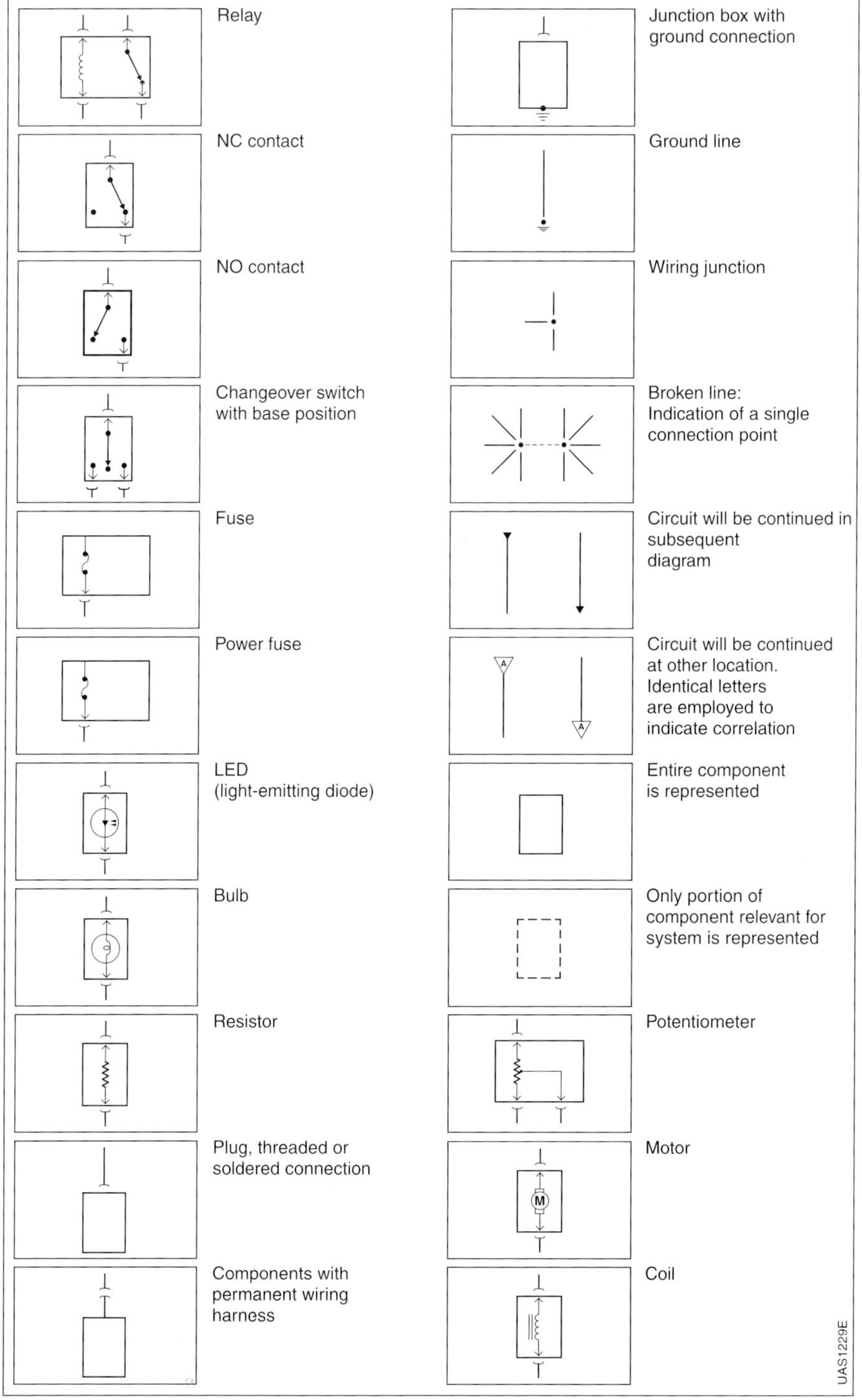

Relay	Junction box with ground connection
NC contact	Ground line
NO contact	Wiring junction
Changeover switch with base position	Broken line: Indication of a single connection point
Fuse	Circuit will be continued in subsequent diagram
Power fuse	Circuit will be continued at other location. Identical letters are employed to indicate correlation
LED (light-emitting diode)	Entire component is represented
Bulb	Only portion of component relevant for system is represented
Resistor	Potentiometer
Plug, threaded or soldered connection	Motor
Components with permanent wiring harness	Coil

UAS1229E

Electromagnetic compatibility (EMC) and interference suppression

Electromagnetic compatibility consists of two elements. One is understood as the ability of a device to continue providing reliable service when exposed to electro-magnetism from external sources. The second aspect focuses on electro-magnetic fields generated by the same device; these should remain minimal in order to avoid creating interference that would impinge upon the quality of radio reception, etc. in the vicinity.

Modern-day vehicles are equipped with a wide range of systems that rely on electrical or electronic componentry to perform an array of functions that were either non-existent or purely mechanical in earlier automotive applications. The continuing proliferation of electrical and electronic devices within the modern automotive environment has been marked by a proportionate increase in the relative significance of electromagnetic compatibility.

With the exception of two-way transceivers used in special-purpose applications, a car's radio was once the only device in which signal reception was of any importance. In contrast, today's vehicles feature a host of devices that rely on the reception of electromagnetic radio waves. Mobile phones, navigation systems, theft-deterrent systems with remote radio control, and integrated fax and PC units are now being installed and used in vehicles. This trend has led to a commensurate increase in the importance of suppressing interference and ensuring reception of the operationally vital radio waves.

EMC ranges

The design of electrical and electronic

systems for automotive applications must focus on three main priority areas.

Transmitter and receiver

All vehicle systems must remain impervious to electromagnetic radiation emitted from such external sources as extremely powerful radio transmitters. In other words, there must be no threat to the vehicle's operational integrity, and functional irregularities representing a potential source of driver irritation are inacceptable. Another consideration is that stationary receivers should remain unaffected by passing traffic. Both considerations are governed by national and international codes (EC ordinances, German StVZO).

Electrical and electronic components

Vehicles contain an extensive array of electrical and electronic components including servo and fan motors, solenoid valves, electronic sensors and ECU's with microprocessors. These devices must rely for their power supply on a single on-board network. They must all function simultaneously in an environment characterized by lack of space and close proximity between units. It is thus vital to avoid mutual interference and feedback phenomena generated by one or several systems so that these do not cause malfunction.

On-board electronic systems

Mobile communications equipment – such as the radio – also exists within an interlinked environment including all of the vehicle's electronic systems. Every device is powered via the same on-board electrical system, and its reception antenna is located in the immediate vicinity of potential interference sources. These considerations make it imperative that strict limits are imposed on the levels of interference emitted by on-board electronic systems. Compliance with official regulations is essential, and it is important to maintain interference-free on-board reception even when conditions are well below optimal.

EMC between various vehicular systems

Shared on-board power supply

All of the motor vehicle's electrical systems rely on a single shared on-board power-supply network. Because the wires and cables leading to the individual systems are frequently combined within a single wiring harness, feedback pulses can easily travel from one system to the I/O ports of its neighbor (Figure 1).

This transfer of interference can be in the form of signal pulses (abrupt, steep jumps in current and voltage) generated during the switching on and off of various electrical components such as motors and solenoid valves. Yet another source is the ignition system's high-tension circuit. Similar to other interference signals (for instance the ripple on the power supply), these signal pulses can propagate through the wiring harness. These interference pulses then proceed to the I/O ports of adjacent systems either directly, through shared conductors such as the power supply (galvanic coupling), or indirectly, through capacitive and inductive coupling stemming from electromagnetic emissions.

Galvanic coupling

Currents for two different circuits (such as a solenoid valve's trigger loop and the circuit for assessing sensor data) flowing through a single conductive path (common ground through the vehicle chassis, etc.) will both generate a voltage owing to the consistent resistance in the shared conductor (Figure 2a). Continuing with this sample scenario, the voltage produced by interference source U1 has the effect of a supplementary signal voltage in signal circuit 2, and could lead to erroneous interpretation of the sensor signal. One remedy is to use separate return lines for each circuit (Figure 2b).

Figure 1

Mutual interference between two systems as transmitted through the shared vehicle power supply (A) and wiring harnesses (B and C)

System I: **1** ECU, **2** Actuator, **3** Sensor. System II: **4** ECU, **5** Actuator, **6** Sensor.

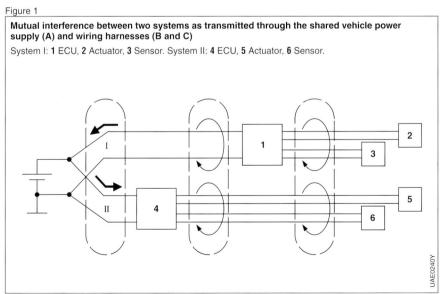

Capacitive coupling

Capacitive coupling allows variable-periodicity signals such as pulse voltages and sinusoidal DC voltage to produce interference and crosstalk in adjacent circuits, even without the existence of a direct physical link (Figure 3). The potential level of capacitive (interference) voltage is proportional to such factors as the closeness of the neighboring conductor paths and the rise rate of the pulse-shaped voltage shifts (or the frequency of the AC voltage).

The first step is to separate the conductive paths while at the same time extending the ramp periods during which signals rise and fall (or to limit the frequency of the AC voltages to the absolute minimum required for the function).

Inductive coupling

Currents recurring with variable periodicity in one conductor can induce voltage pulses in adjoining circuits. These voltage pulses then generate current in the secondary circuit (Figure 4). This is the inductive principle exploited in transformer design. One prime factor defining susceptibility to overcoupling is the signal's rise and fall time (or AC voltage frequency), reflecting the situation encountered with capacitive coupling. Also significant is the effective mutual inductance, as determined by such factors as the size of the wires and their relative routing. Strategies for avoiding inductive coupling include minimizing the dimensions of circuit wires, keeping critical circuits as far apart as possible from each other, and the avoidance of parallel con-

Figure 2

Galvanic coupling of interference signals

a With shared return conductor path,
b With separate return lines.
u_1, u_2 : Voltage source,
Z_i: Internal resistor,
Z_a: Terminal resistor.

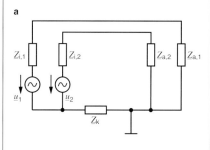

a

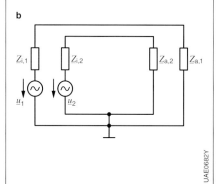

b

Figure 3

Capacitive coupling of interference signals

1 Circuit 1, **2** Circuit 2.
u_1: Voltage source, Z_i: Internal resistor,
R_E: Input resistance, C_E: Input capacitance,
$C_{1,2}$: Capacitance between two conductor paths,
u_s: Interference voltage.

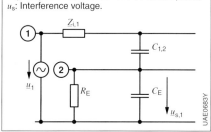

Figure 4

Inductive coupling of interference signals

1 Circuit 1, **2** Circuit 2.
u_1: Voltage source, u_2: Voltage source,
Z_i: Internal resistance, Z_a: Terminal resistance,
L_1, L_2: Inductance of conductors,
$M_{1,2}$: Inductive coupling, u_s: Interference voltage.

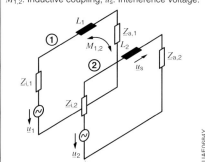

ductive paths. The tendency toward inductive interference is especially pronounced in circuits carrying low-frequency signals (e.g., coupling into loudspeaker wiring).

Pulsation in the vehicle electrical system

One strategy for dealing with interference pulses in the vehicle's electrical system entails limiting the amplitude of the interference emanating from the source. On the other hand, the affected electronic components are designed for insensitivity to pulses of specific shapes and amplitudes. The initial step was to list and classify the pulses encountered within vehicular electrical systems (Table 1). Special-purpose pulse generators can be used to generate the test patterns defined in Table 1 as a basis for assessing resistance to these interference waves in exposed devices. Both the test pulses and the examination procedures are codified in standards (DIN 40839,

Section 1; ISO 7637, Section 1) including definitions of the measurement technology for evaluating emissions of pulse-pattern interference. Classifications based on pulse amplitude levels facilitate effective definition of interference sources and the susceptible devices (interference receptors) within each vehicle. It would thus be possible to specify Class II for all of the interference sources within a vehicle while designing all the susceptible devices interference receptors (such as ECU's) to comply with and exceed – by a certain safety margin – Class III. If suppressing interference at the source proves to be cheaper or to involve less technical complexity than reducing sensitivity at the receptors, the logical response might be to shift the definitions to Classes I/II. If the scenario is inverted, with the shielding of potential receptors as the cheaper and simpler solution, then a move to Classes III/IV is warranted.

Because numerous wires are combined within a single wiring harness, each

Table 1

Mutual interference within voltage supply

Test pulses as defined in DIN 40 389, Section 1			Max. pulse amplitude classes			
Pulse pattern	Internal resistance	Pulse duration	I	II	III	IV
1	10 Ω	2 ms	−25 V	−50 V	−75 V	−100 V
2	10 Ω	50 µs	+25 V	+50 V	+75 V	+100 V
3a	50 Ω	0.1 µs	−40 V	−75 V	−110 V	−150 V
3b			+25 V	+50 V	+75 V	+100 V
4	10 Ω	up to 20 s	12 V	12 V	12 V	12 V
		−3 V	−5 V	−6 V	−7 V	
5	1 Ω	up to 400 ms	+35 V	+50 V	+80 V	+120 V

UAE0630E

individual conductor is potentially susceptible to inductive and capacitive interference. Although reduced in intensity, the resulting voltage pulses in adjacent wiring can then appear as spurious signals at the input ports and control outputs in neighboring systems. The test procedure for simulating crosstalk interference within wiring harnesses (as defined in DIN 40839, Section 3 and ISO 7637, Section 3) uses a standardized substitute wiring layout (capacitive clip) with a defined wiring capacitance. Test pulses are fed into this layout and through the specimen's wiring harness to produce overcoupling in the signal and control lines. The effects of low-frequency oscillations within vehicular electrical systems can be simulated by producing the desired signals with a signal generator and projecting these into the wiring harness through an inductive clamp. This process reflects the procedure described above by serving as the basis for the correct balance between the amplitude of radiated interference pulses and the resistance to interference of the potentially susceptible devices (receptors).

Figure 5

Voltage amplitude

a As a function of time, b As a function of frequency.
T: Period, T_r: Rise time, T_i: Pulse duration, $f_0 = T^{-1}$: Fundamental wave, f_g: Corner frequencies, f_{min}: Periodic minima, H: Envelope curve.

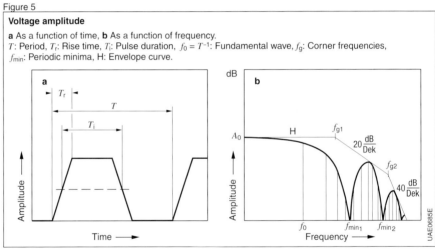

Figure 6

Interference-signal spectrum

a Wide-band interference, b Narrow-band interference.

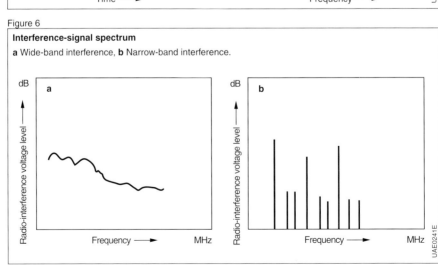

Effects on mobile radio reception of high-frequency signal feedback in the electrical system

Along with the pulses and other interference signals described above, high-frequency signals constitute yet another source of undesired interference within the vehicle's electrical system. These signals can stem from periodic switching operations, as found in high-intensity ignition systems, at the commutators in DC motors and from the CPU clock signals that are generated in microprocessor-equipped ECU's. The interference these signals can induce in mobile communications receiver equipment may impair reception or even render it impossible.

Spectrum

Pulsation within on-board electrical systems is usually based on the analysis of current and voltage progressions (Figure 5a) as function of time. Amplitudes at specific frequencies are generally viewed as the prime criterion for evaluating interference signals affecting radio reception (Figure 5b). Under standard conditions, the interference signals encountered in the automotive environment are rarely isolated sinusoidal waves with single amplitudes recurring at consistent frequencies. Much more common are superimpositions consisting of numerous oscillation components with a variety of amplitudes and frequencies. The "spectrum" for an interference signal is a portrayal of amplitude versus frequency designed to facilitate evaluation of interference potential in individual wavebands (Figure 6a and 6b). Table 3 (refer to section on interference classes) lists the most significant wavebands for automotive applications.

Interference test technology divides interference emissions into wide-band and narrow-band signals. A spectrum with a continuous, progressive curve (Figure 6a) is indicative of wide-band interference, and identifies the corresponding source as a wide-band interference source or emitter. A contrasting pattern composed of isolated spikes forming a so-called line spectrum indicates narrow-band interference stemming from narrow-band emitters (Figure 6b). Initially, the classifications in either category are arbitrary: Among the factors that determine whether interference is emanating from a "wide-band" or "narrow-band" source are the reception properties of the potentially susceptible devices as reflected in the characteristics of the test equipment employed to monitor the emissions. In the case of radio emissions this equipment is in the form of a selective laboratory receiver or spectral analysis unit. Similar to a radio receiver therefore, the tester thus measures the signal amplitude only within a specific narrow frequency band (receiver bandwidth). The test progresses through the entire relevant frequency range using a procedure analogous to the station search function in standard radios: while maintaining the initial bandwidth, the laboratory receiver proceeds through a range of frequencies, either at graduated intervals or in a continuous progression. Interference signals recurring at frequencies lower than the test bandwidth produce the continuous signal pattern indicative of wide-band interference. If the frequency is higher than the test bandwidth, the test monitor will also pick up gaps in the spectrum, and the result will be the line spectrum that indicates narrow-band interference. Electric motors are a typical source of wideband interference.

Commutation frequencies in electric motors are located around just a few 100 Hz, with the exact figure depending on the motor's number of pole's and its rotation rate. At a test bandwidth of 120 kHz (bandwidth corresponding to an FM receiver bandwidth) this produces a continuous spectrum. At the same test bandwidth, a 2 MHz clock signal (of the kind that might be encountered in a microprocessor-equipped ECU) will generate a completely different spectrum, this time with the spiked line spectrum typical of narrow-band interference (interference signal recurs at frequency greater than test bandwidth).

While all electric motors – fans, windshield wipers, servo units and fuel pump, etc. – join the alternator as typical emitters of wide-band interference, yet another potential emitter is the high-tension ignition circuit. In addition, low-frequency clock signals stemming from devices such as switching elements can produce wide-band interference. The list of narrow-band interference emitters includes the microprocesors in ECUs as well as all other devices using high-frequency control signals.

Interference signals picked up by radio receivers may also be of the so-called conducted interference type, that is they may propagate through wiring and conductors (such as the radio's power supply) or they can enter the receiver's I/O ports via inductive and capacitive radiation in the wiring harness. However, interference usually enters through the antenna, either through coupling in the antenna cable, or because the antenna picks up the electromagnetic fields generated by interference sources. One particularly effective transmission antenna is the wiring harness. Other factors influencing the amplitude of the interference signals picked-up by the receiver are the vehicle-body structure and the type and location of the antenna.

Measuring incident interference
Test techniques (DIN 57 879/VDE 0879, Sections 2 and 3; CISPR 25) are stipulated for monitoring interference transmitted through wiring and antennas. Research on individual components is carried out using laboratory setups and standardized test setups in shielded chambers. Interference is monitored with a laboratory receiver. Precise compliance with specified test arrangements, including specifications for wire lengths and other geometrical dimensions, is essential for obtaining reproducible results. Voltage sources are also precisely defined, with laboratory specimens being powered from simulated vehicle electrical systems (Figure 7).

Figure 7

Schematic diagram of simulated automotive electrical circuit

Connections: P-B Test specimen, A-B Power supply, M-B Radio-interference emitter, S Switch, B Reference ground (metal sheet, shielding for simulated circuit).

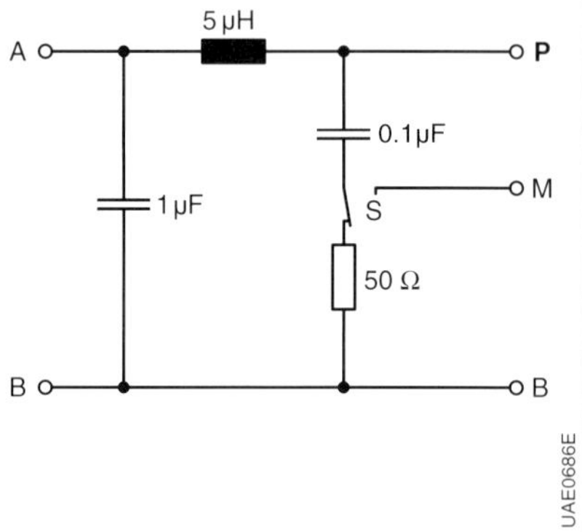

Table 2

Interference-suppression levels: Radio-interference voltage limits for individual frequency ranges in dBµV for wide-band (B) and narrow-band (S) interference as defined in CISPR 25

Interference-suppression levels	0.15... 0.3 MHz (LW)		0.53 ... 2.0 MHz (AM)		5.9 ... 6.2 MHz (SW)		30 ... 54 MHz		70 ... 108 MHz (FM)	
	B	S	B	S	B	S	B	S	B	S
1	100	90	82	66	64	57	64	52	48	42
2	90	80	74	58	58	51	58	46	42	36
3	80	70	66	50	52	45	52	40	36	30
4	70	60	58	42	46	39	46	34	30	24
5	60	50	50	34	40	33	40	28	24	18

Interference-suppression classes

Similar to pulsation in the vehicular electrical system, narrow and wide-band interference factors are classified in order to facilitate selection and design for specific applications. Within this classification system, demands on sporadically active interference generators are less stringent than the requirements placed on components in continuous operation, such as the alternator. The permissible CISPR 25 radio interference levels are listed in Table 2. CISPR 25 also contains tables defining maximum interference field strengths for radiation measurements using antennas and other test procedures.

Narrow-band interference of the kind generated by CPU clock signals in ECUs poses an especially acute problem for radio reception. These interference signals are always present (the ECUs are usually switched on along with the ignition). Also of significance is the fact that the radio receiver cannot distinguish these spurious signals from actual broadcast signals, making reception of weak transmissions impossible. This situation is reflected in the categories defined for interference levels. In a given interference-suppression class, narrow-band emissions are assigned lower levels than wide-band emissions.

Because individual vehicle configuration also has a substantial effect on the quality of broadcast reception, the interference-suppression data derived from laboratory testing must be confirmed in a practical automotive environment. The test involves measuring antenna voltage at the end of the antenna cable to which the radio receiver will subsequently be connected. CISPR 25 also prescribes limits for interference voltage as determined using this procedure (Table 3). It contains voltage levels defined to reflect the unfavorable reception conditions as encountered in motor vehicles, where effective signal strengths not only fail to exceed just a few µV, but also suffer from substantial fluctuation owing to vehicle motion and

Table 3

Limits defined for interference voltage at vehicle antenna in dBµV						
QP-B: Quasi-peak detector relays aural impression produced by interference,						
B: Wide-band interference emitter with peak detector indicates maximum level,						
S: Narrow-band interference emitter with peak detector indicates maximum level,						
* Ignition system high-tension voltage limits.						

Frequency-range	Frequency	Continuous broad-band interference		Sporadic broad-band interference		Narrow-band interference
	MHz	QP-B	B	QP-B	B	S
LW	0.14 ... 0.30	9	22	15	28	6
AM	0.53 ... 2.0	6	19	15	28	0
SW	5.9 ... 6.2	6	19	6	19	0
2-way transceivers	30 ... 54	6 (15*)	28	15	28	0
2-way transceivers	70 ... 87	6 (15*)	28	15	28	0
FM	87 ... 108	6 (15*)	28	15	28	6
2-way transceivers	144 ... 172	6 (15*)	28	15	28	0
C Network mobile phones	420 ... 512	6 (15*)	28	15	28	0
D Network mobile phones	800 ...1,000	6 (15*)	28	15	28	0

the multiple reception paths generated by signal reflections.

Interference from electrostatic charges

ESD, or electrostatic discharge, can damage or even destroy electronic components. As this kind of discharge can generate extremely high pulse-shaped currents with magnitudes of several thousand volts, preventative action to avoid static electricity's destructive effects – or, better still, preemptive measures to inhibit its generation in the first place – are essential. Electronic components that come into contact with vehicle passengers are particularly endangered.

ISO TR 10605 defines test procedures for examining the effects of electrostatic discharge and the immunity to interference that components exhibit both in the laboratory and when installed in the vehicle. The test is conducted with a suitable ESD test-pulse generator, usually in the shape of a pistol. The generator produces the pulses that are then coupled into the test specimen.

EMC between the vehicle and its surroundings

Since early 1996, an official regulation has placed mandatory limits on radiation by the vehicle of interference affecting stationary radio reception, while at the same time defining suppression requirements during exposure to external electromagnetic fields. This ordinance (European Directive 95/54/EC) superseded an earlier directive of more limited scope, which only limited interference emissions while defining the homologation certification procedure governing EMC in motor vehicles.

Emitted interference

Electromagnetic emissions radiated by motor vehicles must remain within the limits for narrow and wide-band signals (Figure 8) to ensure that they do not interfere with radio and television broadcasts or private transmissions. These limits are specified in the Directive 95/54/EU cited above as well as in VDE 0879, Section 1 and the CISPR 12 standard. These regulations contain detailed descriptions of the test procedure, which employs antennas set

Figure 8

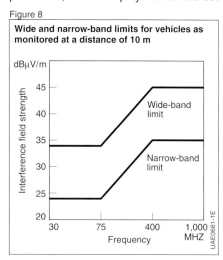

Wide and narrow-band limits for vehicles as monitored at a distance of 10 m

up at defined distances from the vehicle (10 meters, 3 meters). The ignition system usually turns out to be the primary source of interference emissions, but the extensive range of suppression measures implemented to maintain satisfactory on-board radio reception generally reduces interference radiation to levels well below the legal maximum.

Incident radiation

As a vehicle travels through the high-intensity electromagnetic field (near field) immediately surrounding a high-intensity emissions source, radiation penetrates through gaps and apertures in the vehicle's bodywork and interferes with on-board electrical systems. The intensity of this effect varies according to component locations, bodywork configuration and the wiring harness.

Vehicle measurements

At one time the procedure for verifying that various electronic systems would continue to provide satisfactory per-

Figure 9:
Measuring resistance to incident radiation of automotive electrical systems in the EMC anechoic chamber

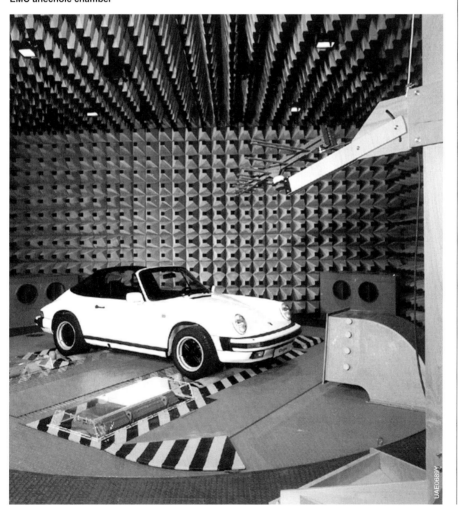

formance under exposure to high-intensity fields involved driving the vehicle to a number of different broadcast transmitter locations. Now, special test chambers are available for this purpose.

These chambers must be surrounded (shielded) with a metallic casing to prevent the electromagnetism generated during testing from radiating to the outside. In addition, the chambers must be equipped with absorption materials to inhibit formation of standing waves (nodes and antinodes), which would lead to major variations in field strength between the various monitoring locations.

The anechoic (absorption chamber) is used for investigating the overall performance of vehicular electrical systems under real-world operating conditions. Figure 9 shows the Bosch EMC anechoic chamber as an example of a typical layout and as an aid to understanding the test procedures.

The Bosch anechoic chamber is equipped with devices capable of generating high-frequency fields within a range extending from 10 kHz to 18 GHz, with maximum field strength $E_{max} = 200$ V/m. Since fields of this intensity represent a health hazard, the test vehicle is operated via remote control from within a shielded control room, while testing is monitored by video cameras. The chamber is screened with metalic sheeting, and non-conductive materials (wood and plastic) line the interior, as metallic substances could foster spurious readings. To prevent reflections and inhibit the formation of standing waves, walls and ceiling are covered by pyramid-shaped absorption elements made of graphite-filled polyurethane foam.

Vehicle testing proceeds on a chassis dynamometer (rolling road) capable of accomodating simulated speeds of up to 200 km/h. A fan is also present to direct up to 40,000 m^3/h (corresponding to a wind speed of 80 km/h) of air over the vehicle.

Among the advantages that distinguish in-door testing from the old open-air tests near transmitters is that the former method allows latitude for considerable variations in both frequency and field strength. This facilitates evaluation of vehicular resistance to incident radiation under a wide range of conditions, and not at just a few available points. The ability to gradually increase field strengths right up to each electronic device's operational limits furnishes information on safety margins. Specifics defining the test procedures used in determining radiation resistance in the overall vehicle are described in DIN 40 839, Section 4. ISO 11 452 Sections 1–4 contain similar descriptions as elements within a more extensive compilation of special-purpose test procedures.

Laboratory measurements
Although the data garnered from assessments of incident radiation on the vehicle as a whole are invaluable, the disadvantage of this type of testing is that it can only be performed at an advanced stage in the vehicle and electronic system design process. Only very limited latitude is available for responding to indications of inadequate radiation resistance uncovered at this stage.

This explains the necessity to determine at an early stage how an electronic system will operate during subsequent use in its automotive environment: prompt information allows effective recourse to remedial action when needed. Various test procedures have emerged for obtaining the required information.

In the first of the three tests described below, interference waves propagating through a conductor are coupled into the wiring harness of the system under investigation. As the suitability of these test arrangements for evaluating the frequency range beyond 400 MHz is restricted, a different procedure is used for this range. The alternate procedure relies on antennas to project electromagnetic fields into standardized bench setups.

The specifics of the methods outlined below are defined in DIN 40839, Section 4, and ISO 11452, Sections 1–7 (which also contains additional process descriptions detailing less widespread methods).

All these methods furnish a precise picture of a system's resistance to incident radiation. This can then be used to implement improvements while the development phase is still in progress. Thanks to this inestimable advantage it is now impossible to imagine development projects without these laboratory measurements as flanking measures.

It must be taken into consideration through that the design of the electrical system in itself is not the only important factor. Because installation conditions in the vehicle and the routing of the wiring harness can also exercise a decisive influence on ultimate levels of interference resistance, all earlier results still need to be confirmed on a production vehicle in the anechoic chamber.

Stripline method (Figs. 10 and 11)
The "stripline" designation is derived from the 4.1 meter long and 0.74 m wide strip conductor suspended at a height of 0.15 meters above a conductive sheet (counterelectrode). A high-frequency generator serves as the source of a transverse magnetic wave generated between the stripline and the counterelectrode, continuing until it reaches the terminal resistor. The stripline dimensions have been selected to minimize the likelihood of reflections occurring during wave propagation, thus ensuring constant field-strength amplitudes relative to frequency.

A typical system setup might include an ECU, wiring harness and peripheral devices (sensors and actuators). These are set up at a height half way between the base plate and the stripline, with the wiring harness in alignment with the waves' direction of propagation. The field

Stripline method

1 High-frequency generator, 2 Resistor,
3 Stripline, 4 Counterelectrode (conductive sheet or cell), 5 System under test, 6 Wiring harness,
7 Peripherals (sensors, actuators).

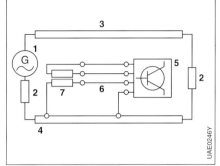

Figure 10

Figure 11

Resistance to incident radiation

Determined using stripline, BCI or TEM cell method.

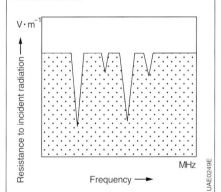

Figure 12

BCI Method

1 High-frequency generator, 2 Resistor,
3 Counterelectrode (conductive sheet or cell),
4 System under test, 5 Wiring harness,
6 Peripherals (sensors, actuators),
7 Current clamp.

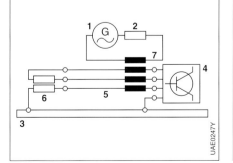

strength between the planes at a given frequency is then increased until the system malfunctions or until a maximum specified level is reached. If the increments are small enough, repeating this process at progressively altered frequencies will produce a detailed diagram portraying resistance to incident radiation as a function of frequency.

Bulk current injection method (Figure 12)
With the BCI method, the system being tested is located above a conductive plate (counterelectrode) similar to that used in the stripline procedure. A clamp attached to the wiring harness injects transformer current into its individual wires. The vectorial sum of these currents corresponds to the total current radiated by the clamp. While testing using the stripline method proceeds through a range of increasing field strengths, BCI relies on progressively higher current flows as it continues until the system malfunctions or until current reaches the specified terminal value.

TEM cell (Figure 13)
As with the stripline method, a transverse electromagnetic (TEM) field is generated here between a strip conductor and a counterelectrode. The TEM counter-electrode, however, is an enclosed housing instead of a metal plate. This leads to another distinguishing feature: TEM test benches do not need to be set up within specially shielded chambers.

A further difference lies in the fact that only the test specimen per se (for example: an ECU) is subject to electromagnetic exposure. The peripheral equipment is located outside the TEM cell and is connected to the test specimen via a short wiring harness arranged at right angles to the propagation direction of the electromagnetic waves. Measurement proceeds as in the stripline process, with field strength being raised until the system malfunctions or until field strength reaches the maximum value specified for testing.

Incident radiation via antenna
Once again similarities with the stripline method are encountered, as the test specimen, including ECU, wiring harness and peripherals, is set up on a baseplate. The wiring harness is laid out at a defined distance from this baseplate. An antenna located at a specified distance from this

Figure 13

TEM-cell test setup

1 High-frequency generator, **2** Resistor,
3 Stripline, **4** Counterelectrode (conductive sheet or cell),
5 System under test, **6** Peripherals (sensors, actuators).

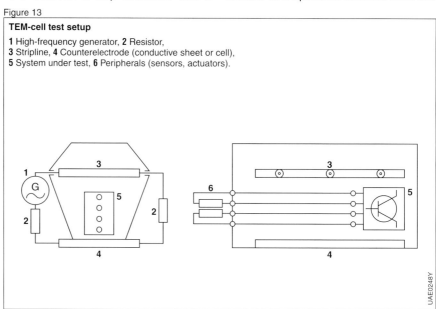

UAE0248Y

assembly generates an electromagnetic field for projection into the entire test setup. Again, field strength is raised until the system malfunctions or until field strength reaches the maximum value specified for testing.

Ensuring interference suppression and immunity to interference

As early as the planning and conception phases, EMC requirements for immunity to and suppression of radio interference have to be considered as part of every development project focusing on any electronic system or component. The finished products must incorporate EMC measures as an integral design element.

EMC in the electronic control unit (ECU)

The first EMC requirement reflected in any ECU is to select microprocessor CPU clock frequencies that are as low as possible while simultaneously making every effort to keep the signal transit curves as flat as possible. EMC characteristics join standard operational aspects as an essential criterion governing the selection of components (integrated circuits). Whereas the components' resistance to interference must be maximal, its interference emissions should be minimal. On printed-circuit boards designed for optimal EMC performance, all components with pronounced sensitivity to interference, and all of those with the potential to cause it, are decoupled from the connected wiring harness. This is done by locating the components as far as possible from the plug connections.

Suppression components, usually capacitors designed for high-frequency applications, place the required restrictions on potentially disruptive interference. These interference-suppression capacitors are located either directly within the integrated circuits or in the plug-connector area. The suppression elements in the plug-connector area combine with housings featuring optimal conductive properties (shielded casing) to prevent high-frequency radiation in the immediate vicinity from reaching the inside of the ECU. This prevents signals outside the housing from inducing malfunctions within the unit, while also preventing the high-frequency signals generated within the housing from causing unwanted interference in the unit's surroundings.

Electric motors and other electromechanical components

Standard practice in the design of electric motors reflects the procedures used for ECUs and sensors by incorporating interference suppression in the design process right from the initial development stages. A sample problem is the interference produced by the brush sparks during current reversal in motors with commutator control. This can severely disrupt radio reception. Suitable suppression elements (capacitors and throttles) limit the interference. Motors are designed with optimal operation of these interference suppressors in mind.

The voltage pulses produced during switching operations in electromagnetic actuators are kept to an acceptable level by suitable elements, such as quench resistors, in the circuit.

High-voltage ignition

The ignition system's high-voltage circuit represents a potential source of substantial interference to radio reception. One standard solution is to use spark plugs and cable sockets with suppression resistors. While the resistor element is usually located at the end of the ignition cables, in modern (multiple-coil) systems it can be integrated within the dedicated ignition coils plugged directly onto the spark plugs for the individual cylinders. In using resistors, the object is to find a balance between the demand for ignition energy and the required level of interference suppression.

Retrofit interference suppression

As made clear in the above sections, EMC design considerations must be carefully aligned and balanced against operational requirements. Subsequent remedial measures superimposed on the original design at a later date usually entail substantial effort, and should be avoided in production vehicles. Should the interference-suppression measures incorporated in electrical components prove inadequate for dealing with the conditions encountered in specific applications (in special-purpose emergency vehicles, etc.), supplementary action can be implemented to deal with the problem. Options include installation of filters as well as supplementary shielding for components and wiring.

These options should be exercised with great care, as field modifications to electronic components can lead to operational malfunctions. As is the case with supplementary electrical and electronic systems, under certain circumstances such modifications can pose a critical threat to safety.

National and international EMC standards for motor vehicles

Radio-interference suppression

DIN 57879/VDE 0879, Section 1
Radio-interference suppression in vehicles, in vehicle equipment and in internal-combustion engines – Remote interference suppression in vehicles; long-distance interference suppression of assemblies equipped with internal-combustion engines.

DIN 57879/VDE 0879, Section 2
Radio-interference suppression in vehicles, in vehicle equipment and in internal-combustion engines – interference-suppression using on-board elements.

DIN 57879/VDE 0879, Section 3
Radio-interference suppression in vehicles, in vehicle equipment and in internal-combustion engines – interference-suppression using on-board elements: testing of vehicle equipment.

CISPR 12
Test limits and methods for measuring radio interference in motor vehicles, motor boats, and devices powered by spark-ignition engines.

CISPR 25
Test limits and methods for measuring radio interference to protect receivers used on board motor vehicles.
Part 1: Measurement of emissions received by an antenna on the same vehicle.
Part 2: Measurement on vehicle components and modules.

EMC

DIN 40 839, Section 1
Electromagnetic compatibility (EMC) in road vehicles – conductor-borne interference pulses in supply lines in 12 V and 24 V vehicle electrical systems.

DIN 40 839, Section 3
Electromagnetic compatibility (EMC) in motor vehicles – capacitive and inductive interference in sensor and signal wiring.

DIN 40 839, Section 4
Electromagnetic compatibility (EMC) in motor vehicles – Incident radiation factors.

ISO 7637-0
Road vehicles – Electrical interference from conduction and coupling.
Part 0: Definitions and general information.

ISO 7637-1
Road vehicles – Electrical interference from conduction and coupling.
Part 1: Passenger cars and light commercial vehicles with electrical systems rated at 12 V – Electrical transient conduction along supply lines only.

ISO 7637-2
Road vehicles – Electrical interference from conduction and coupling.
Part 2: Commercial vehicles with electrical systems rated at 24 V – Electrical transient conduction along supply lines only.

ISO 7637-3
Road vehicles – Electrical interference from conduction and coupling.
Part 3: Vehicles with electrical systems rated at 12 V or 24 V – Electrical transient transmission by capacitive and inductive coupling via lines other than the supply.

ISO 11 451
Road vehicles – Electrical interference from narrow-band electromagnetic radiation – Vehicle test methods.

ISO 11 451-1
Part 1: Definitions and general information.

ISO 11 451-2
Part 2: Remote radiation sources.

ISO 11 451-3
Part 3: On-board transmitter simulation.

ISO 11 451-4
Part 4: Bulk Current Injection (BCI).

ISO 11 452
Road vehicles – Electrical interference from narrow-band electromagnetic radiation – Component test methods.

ISO 11 452-1
Part 1: Definitions and general information.

ISO 11 452-2
Part 2: Absorber-lined chamber.

ISO 11 452-3
Part 3: Transverse Electromagnetic Mode (TEM) cell.

ISO 11 452-4
Part 4: Bulk Current Injection (BCI).

ISO 11 452-5
Part 5: Stripline.

ISO 11 452-6
Part 6: Parallel plate antenna.

ISO 11 452-7
Part 7: Direct radio frequency (RF) power injection.

ISO/TR 10 605
Road vehicles – Electrical interference from electrostatic discharge.

National and international EMC standards for motor vehicles

Starter batteries

Battery design

Basically, the vehicle electrical system comprises the alternator as energy generator, the battery as energy store, and the starter and other loads as consumers. All these elements must be perfectly matched to each other. In other words, power generation and power consumption must be in a well-balanced relationship. The following quantities have an effect upon the special relationship between battery, alternator, and starter:
– Electrical load requirements,
– Alternator current output,
– Engine speed when the vehicle is actually being driven,
– Charging voltage, and
– Starting temperature.

Electrical loads

In past years, due to the increasingly extensive and complex equipment fitted in the vehicle, the number of electrical loads has increased sharply. This means that the total power needed by the loads and therefore the required alternator output (Fig. 1) has increased accordingly. Formerly, it was mainly the starter, the ignition, and the lighting system which were the major loads. During the course of the past years though, there has been no let up in the expansion of the vehicle's basic equipment by the addition of electronically controlled ignition and fuel-injection systems, comfort and convenience systems using a variety of different drive motors, and safety and security systems (Fig. 2). These loads have a continually increasing share of the power consumed in the vehicle's electrical system.

Starting systems

The starter must crank the IC engine at a given minimum speed and, after the first ignitions, bring the engine up to the minimum self-sustaining speed. In the process, considerable resistances due to compression, piston friction, and bearing friction must be overcome. These are strongly dependent upon engine type and number of cylinders, as well as upon lubricant characteristics and engine temperature. The friction resistances are at a maximum at low temperatures. Even though the starter is only switched on for a brief period, it has the highest current consumption of all the loads.

Further electrical loads

The vehicle's electrical loads feature different powers and operating times. One differentiates between:
– Permanent loads (e.g. ignition, fuel injection and/or engine management),
– Long-time loads (e.g. lighting, heated rear window), and

Fig. 1

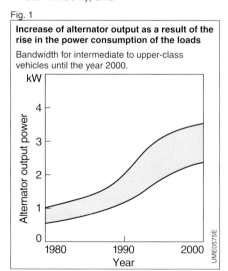

Increase of alternator output as a result of the rise in the power consumption of the loads

Bandwidth for intermediate to upper-class vehicles until the year 2000.

Fig. 2

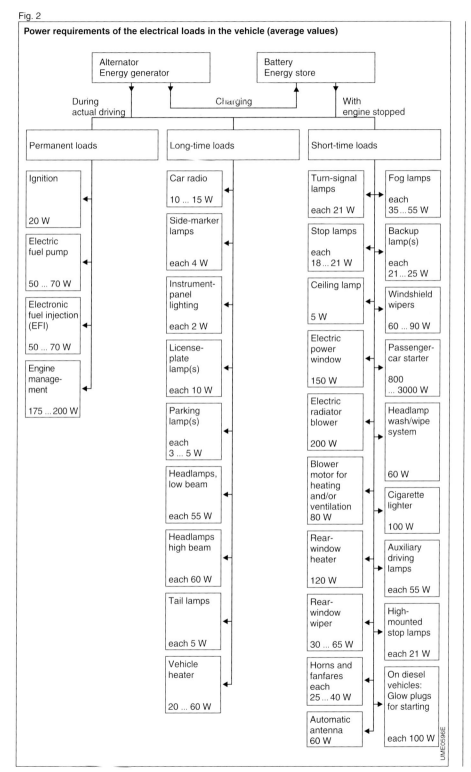

Power requirements of the electrical loads in the vehicle (average values)

Alternator
Energy generator

Battery
Energy store

During actual driving — Charging — With engine stopped

Permanent loads	Long-time loads	Short-time loads	
Ignition 20 W	Car radio 10 ... 15 W	Turn-signal lamps each 21 W	Fog lamps each 35 ... 55 W
Electric fuel pump 50 ... 70 W	Side-marker lamps each 4 W	Stop lamps each 18 ... 21 W	Backup lamp(s) each 21 ... 25 W
Electronic fuel injection (EFI) 50 ... 70 W	Instrument-panel lighting each 2 W	Ceiling lamp 5 W	Windshield wipers 60 ... 90 W
Engine management 175 ... 200 W	License-plate lamp(s) each 10 W	Electric power window 150 W	Passenger-car starter 800 ... 3000 W
	Parking lamp(s) each 3 ... 5 W	Electric radiator blower 200 W	Headlamp wash/wipe system 60 W
	Headlamps, low beam each 55 W	Blower motor for heating and/or ventilation 80 W	Cigarette lighter 100 W
	Headlamps high beam each 60 W	Rear-window heater 120 W	Auxiliary driving lamps each 55 W
	Tail lamps each 5 W	Rear-window wiper 30 ... 65 W	High-mounted stop lamps each 21 W
	Vehicle heater 20 ... 60 W	Horns and fanfares each 25 ... 40 W	On diesel vehicles: Glow plugs for starting each 100 W
		Automatic antenna 60 W	

UME0596E

71

– Short-time loads (e.g. turn signals, stop lamps).

A number of the electrical loads are only seasonal (for instance, air-conditioner in summer, seat heating in the winter months). And the frequency with which electric radiator fans are switched on, for instance, depends upon temperature and upon operating conditions. In the winter months it is usually necessary to have the lights on in commuter traffic.

During a given driving period, the power required by the loads varies. Particularly in the first few minutues after driving off, power requirements are very high (for "heating" and "cooling"), and then sink rapidly (Fig. 3). Examples are:

1. Electrically heated catalytic converters will in future need an extra 1...4 kW in order to achieve the > 300 °C temperature within 10...30 s after start (power requirements depend upon engine size and type of exhaust system).

2. The secondary-air pump, which pumps extra air directly into the combustion chamber for exhaust-gas afterburning, runs at least 200 s after the engine has started.

3. Depending upon the particular situation, further loads such as heating (also including rear-window heating), fan and lighting, are switched on for longer or shorter periods. The engine management though is permanently in operation.

Operating conditions

In order that the series-production vehicles which finally develop from the test vehicles can be equipped with suitable components to cope with everyday energy requirements, the test vehicles are operated under extreme conditions. The demands made upon battery, alternator, and vehicle electrical system depend to a great extent upon operating conditions and the type of vehicle in which they are installed. In addition to the battery characteristics which are matched to the electrical system of the particular vehicle, such as starting power, Ah capacity, and charge-current input within a temperature range of approx. −30...+70 °C, there are a number of other battery specifications which must also be complied with under certain operating conditions. These include, for instance, freedom from maintenance, vibration-resistance, and resistance to deep cycling.

Starting temperature

Among other things, the temperature at which the engine can still be started is dependent upon battery and starter (size, with/without reduction gear, electrically excited or permanent-magnet excited). If the engine is to be started for instance at temperatures as low as −20 °C, it is imperative that the battery has a minimum charge state. With a larger battery (high Ah capacity), minimum charge state can be less than that with a smaller battery (low Ah capacity). In Europe, for instance, the following minimum start temperatures are stipulated:

Passenger cars	−18...−25 °C,
Trucks and buses	−15...−20 °C,
Tractors	−12...−15 °C.

External loading

Vehicles (cars and light commercial vehicles) which are operated normally on asphalt roads in mixed town and country

Fig. 3

Power consumption of the loads as a function of the driving time (Example)

Loads when starting: **A** Starter, catalytic-converter heating, secondary-air pump, heating, fan, headlights, auxiliary lamps, engine management.
B Starter switch-off; **C** Switch-off of catalytic-converter heating; **D** Switch-off of the secondary-air pump.

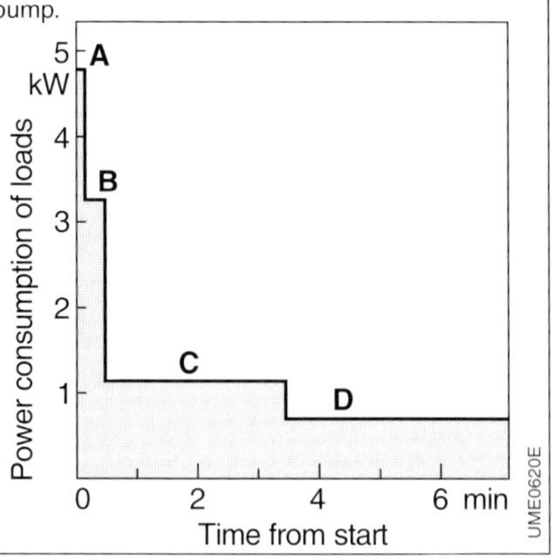

traffic are not subjected to special external mechanical or cyclical loading.

On commercial and industrial vehicles (passenger cars and trucks), and for special off-road applications such as ski-slope preparation etc., the short distances involved, together with the effects of vibration and shock, often result in very high external loading (see Extreme Operating Conditions).

Climatic loading
Generally speaking, batteries are exposed to the effects of damp, dirt, oil, temperature etc. and must also be able to cope with the specific climatic demands of the application in question.

Installation point
Regarding the installation point, the following critera are important:
– Easy access for battery installation,
– Protection against excessive battery heat-up,
– Protection against excessive battery cooling,
– Protection against damp,
– Protection against mechanical damage (e.g. due to excessive vibration). The battery must be securely fastened and not subjected to vibration.
– Protection against oil and fuels etc.

Regarding battery installation, the engine compartment usually provides better accessibility than other points in the vehicle. Furthermore, in the engine compartment short lines can be used between battery, alternator, and starter so that voltage losses are kept to a minimum.

To prevent heat accumulation in or around the battery which could accelerate unwanted chemical reactions inside it, the installation position should not be in the vicinity of engine block or exhaust manifold. The battery should not be exposed permanently to temperatures above 50 °C, or its service life will be reduced (high levels of self-discharge).

When the battery cannot be fitted in the engine compartment, it is installed under the driver's seat or under the rear seat.

Since on batteries with vent plugs the "acid fog" which escapes through the vent holes can attack the metal components in the vicinity, adequate ventilation must be provided around the battery. Even a very slight movement of air eliminates this danger completely.

Formerly, regular servicing, and topping-up with distilled water, was decisive for the battery's operative lifetime. The demand for easy accessibility for the workshops or for the driver himself were therefore more important than they are today. In the case of the completely maintenance-free battery in fact, this form of servicing has become a thing of the past.

Charge balance
Calculation of the charge balance serves to optimize the vehicle's electrical system. It provides information on the battery's behavior in the electrical system following a specified driving cycle with defined load power consumption. It is performed by the vehicle manufacturer. A typical passenger-car cycle consists of vehicle operation in commuter traffic (low engine speeds) combined with winter operation (low charge-current input to the battery). The charge-balance calculation takes into account the frequency of occurrence of certain engine speeds during one cycle, as well as the characteristic curves of alternator, voltage regulator, and battery, as well as the power consumption of all loads connected into the vehicle electrical system.

Engine speed during driving
The rotational speed transferred from the engine to the alternator depends upon the particular vehicle usage. The alternator's charge voltage increases slightly along with its rotational speed. During commuter traffic, due to traffic-light halts and traffic jams, the passenger-car engine runs for a considerable part of the time at idle, in other words at low speeds (with low charge currents). In contrast, when driven on the turnpike or motorway, the engine usually runs in the medium or high-speed range and the charge current is correspondingly higher.

Loads which remain on when the engine is switched off have a negative effect on the battery's state of charge. In municipal bus services, the buses spend more time at idle due to their repeated halts at the bus stopping points. Long-distance coaches on the other hand usually spend very little time at idle.

Power demand

Standard version

The power demand which for a given vehicle results from the power consumption of the loads (see Tables 1 and 2) is determined in accordance with the vehicle operating conditions. It is decisive not only for the dimensioning of the battery but also of the alternator. Under the following test conditions, the original-equipment battery, as specified and installed by the vehicle manufacturer and

Bosch, covers the standard power demand of the starter and the loads installed in the vehicle electrical system:
– Winter commuter traffic with day and night driving for 2 weeks, of which 1 week is at 0 °C and the other at –20 °C,
– The battery's residual capacity is to be at least 50% after completing these driving schedules.
The following must still function:
– Starting at –20 °C,
– The parking lamps for 12 hours,
– The hazard-warning and turn-signal system must operate for 3 hours,
– All loads which operate with the key removed must run for 2 hours. It must then be possible to start the engine.
In addition, the standard values for battery specifications apply.

Auxiliary equipment

If auxiliary equipment is chosen for the vehicle, this can lead to considerable increases in power demand. Such equipment includes comfort and convenience systems with additional servomotors for roof and power-window drives, seat and steering-wheel adjustment; as well as seat heating, air-conditioner, cooler unit or similar. The automaker takes this additional power demand into account when dimensioning the electrical com-

Table 1: Starter power input.

Application	Power range
Passenger car with SI engine	0.7...2.0 kW
Passenger car with diesel engine	1.4...2.6 kW
Buses, trucks, tractors	2.3...9.0 kW

Table 2: Power input of installed equipment as a function of average duty cycle.

Loads	Power input	Mean power input
Motronic, electric fuel pump	250 W	250 W
Radio	20 W	20 W
Side-marker lamps	8 W	7 W
Low-beam headlamps	110 W	90 W
License-plate lamps, tail lamps	30 W	25 W
Indicator lamps, instrument-panel lamps	22 W	20 W
Heated rear window	200 W	60 W
Heating system, blower	120 W	50 W
Electric radiator fan	120 W	30 W
Windshield wipers	50 W	10 W
Stop lamps	42 W	11 W
Turn-signal indicators	42 W	5 W
Front fog lamps	110 W	20 W
Rear fog lamps	21 W	2 W
Sum		
Installed loads	1,145 W	
Average power consumed by loads		600 W

ponents for the vehicle. This means that a vehicle equipped with such auxiliary equipment is delivered with a larger battery (e.g. VW Golf III up to 40% higher Ah capacity), and in some cases with a more powerful alternator. Similarly, specific mechanical, cyclical, or climatic loading may have to be taken into account depending upon the vehicle's application.

Retrofit equipment

Auxiliary loads
Auxiliary loads which are retrofitted at a later date, or connected from time to time, may necessitate the standard battery being replaced by a more powerful one. Examples are as follows:
- Sophisticated, high-performance car radio and audio systems with high power demands,
- Auxiliary lamps and high-mounted stop lamps, fanfare horns, floodlamps and spot lamps, alarm system,
- Auxiliary heating system,
- Connection of equipment powered by the vehicle's battery through the cigarette lighter for instance. These include small-power compressors, small lamps, and floodlamps,
- Trailers and caravans connected to the vehicle's electrical system.

Caravans and mobile homes are often equipped with such electrical appliances as lighting, refrigerator, heating, radio and TV. Here, it is common to fit extra batteries with a separate circuit of their own.

Increased power demand
The above-mentioned auxiliary loads cause an increase in power demand. The existing starter battery though has been carefully matched to the remainder of the vehicle's electrical system, including the alternator, regulator, and starter. Bearing this fact in mind, in order to power the above-mentioned auxiliary loads the original starter battery cannot simply be lifted out and replaced by a battery with a higher Ah capacity. In such cases, the expert from the Bosch Service Agent should be consulted so that the danger of making the wrong choice is ruled out from the very beginning. For such replacements, the Bosch aftermarket program contains cross-reference lists with interchangeable batteries which not only have the same physical dimensions, but which also comply with the particular demands.

Extreme operating conditions

It is impossible to completely cover the wide range of different operating conditions which can be encountered in the field with one single standard battery. Such a battery would be far too large for normal operation and far too expensive.

Outside temperatures
Batteries with higher starting power are needed for cold countries with very low temperatures, where starting must often take place at below $-20\,°C$. Such batteries feature an increased number of thinner plates and separators. In temperate and cold zones there is no need to change the electrolyte's density with its fully-charged freezing limit of $-68\,°C$. In tropical regions though the density must be reduced.

Mechanical and cyclical loading of the battery
In the industrial and commercial sectors (bus, taxi, ambulance, delivery van etc.), the fact that the vehicle is repeatedly driven only short distances means that the current taken from the battery is correspondingly higher. This leads to severe cyclical loading of the battery which is supplemented by further cyclical loading due to high power demands with the vehicle stationary. Such loads include the air-conditioner, the electrohydraulic liftgate, auxiliary heating, and refrigeration unit etc. In addition to the above cyclical loading, batteries in off-road vehicles, commercial vehicles, construction machines, and tractor vehicles, and those in agriculture and forestry applications, must withstand high vibration and impact stresses during off-road operation and on building and construction sites.

Method of operation

Within the vehicle's electrical system, the battery assumes the role of a chemical storage unit for the electrical energy (which is generated by the alternator when the vehicle is being driven). This energy must be made available to start the engine again after it has been switched off. This is one of the reasons for the battery also being known as the "starter battery". Alternator and starter battery must therefore be correctly matched to each other.

On the one hand, with the engine stopped (and therefore also the alternator), the battery must be able to deliver a high current for a brief period in order to start the engine (this is especially critical at low temperatures). And on the other, when the engine is running at idle or is switched off, it must for limited periods be able to supply some or all of the electrical energy to other important components in the vehicle's electrical system.

The battery also absorbs voltage peaks in the vehicle's electrical system so that these do not damage sensitive electronic components.

Generally speaking, the lead-acid storage battery suffices for meeting these demands, as well as at present still being the most cost-effective energy-storage medium for such assignments.

Typical system voltages are 12 V for passenger cars, and 24 V for commercial vehicles (using series-connection of two 12 V batteries).

Electrochemical processes in the lead storage cell

Generation of the cell voltage

If a lead electrode is immersed in dilute sulphuric acid (the electrolyte), positive ions are transferred from the electrode and into the electrolyte due to the effects of the so-called "solution pressure". The transfer of the positive lead ions to the electrolyte means that negative charges (electrons) remain on the lead electrode. In other words, the lead electrode is no longer electrically neutral, but has a negative potential referred to the electrolyte.

If 2 electrodes of different materials (for instance lead [Pb] and lead dioxide [PbO_2]) are immersed in a common electrolyte, different potentials develop at the individual electrodes with respect to the electrolyte (Fig. 1). The difference in potential between the electrodes themselves is the cell voltage.

Since the negative charges remaining on the electrode exert a force of attraction (return force) on the positive ions which have entered the electrolyte, the phenomenon described above, in which charged particles (here the lead ions) are released into the electrolyte, very quickly results in a condition of equilibrium. After a certain period, this return force exactly equals the solution pressure.

If an external voltage is applied across the electrodes, depending upon the direction of current further electrical particles can be released into the electrolyte or they can return from the electrolyte to the lead electrode. It is this fact which makes it possible to recharge the lead-acid storage battery (secondary cell).

In a charged lead cell, the positive electrode consists essentially of lead dioxide, and the negative electrode of pure lead. With the electrolyte (diluted sulphuric acid) used in lead cells, current transportation is via ionic conduction. In the aqueous solution, the sulphuric-acid molecules split into positively charged hydrogen ions and negatively charged residual acid ions. The splitting of the sulphuric-acid molecules is the prerequisite for the electrolyte's conductivity and therefore for the flow of charge or discharge current. When discharge current flows, the positive electrode's lead dioxide and the negative electrode's lead are converted to lead sulphate ($PbSO_4$). In a discharged lead cell (Fig. 1), both electrodes consist of lead sulphate.

The electrolyte is dilute sulphuric acid (17% pure sulphuric acid [H_2SO_4] and 83% water [H_2O]). The sulphuric-acid component means that the pure water becomes conductive so that it can be used as an electrolyte.

These particle transfers involved in the charge and discharge of the lead cell are described in the two paragraphs below in more detail.

Charging

To charge the lead storage cell, its positive electrode is connected to the positive pole of a DC source, and its negative electrode to the negative pole. Contrary to the discharging process described later on, electrical energy must be introduced here in order to force the cell to charge so that it has a higher energy level at the completion of charging than at the beginning. Figs. 1 to 3 demonstrate in schematic form the processes which take place between the individual particles in the electrode mass and the electrolyte.

The source of charging current draws electrons from the positive electrode and forces them to the negative electrode. Due to the electrons which have been forced to the negative electrode by the source of charging current, zerovalent (metallic) lead is formed at this electrode from the <u>bivalent</u> positive lead atoms, whereby the ($PbSO_4$) lead-sulphate molecules are broken down.

At the same time, the negatively charged residual-oxygen ions (SO_4^{--}) are released from the negative electrode into the electrolyte.

At the positive electrode <u>bivalent</u> positive lead (Fig. 2) is transformed into <u>tetravalent</u> positive lead due to the removal of electrons, whereby the lead compound $PbSO_4$ (lead sulphate) is split electrochemically by the applied charging voltage.

The tetravalent positive lead combines with the oxygen removed from the water (H_2O) to form lead peroxide (PbO_2).

At the same time, the sulphate ions released at the positive electrode during this oxidation process (from the lead compound $PbSO_4$), and the hydrogen ions (from the water), pass into the electrolyte. As mentioned above, sulphate ions also enter the electrolyte from the negative electrode. As a result of the charging process therefore, the number of hydrogen ions and sulphate ions in the electrolyte is increased. In other words, fresh sulphuric acid (H_2SO_4) is formed and the density ρ of the electrolyte increases. In a charged cell, this is normally $\rho \approx 1.28$ kg/l corresponding to an electrolyte comprising about 37% sulphuric acid and about 63% water.

Fig. 1

Charging (cell discharged)

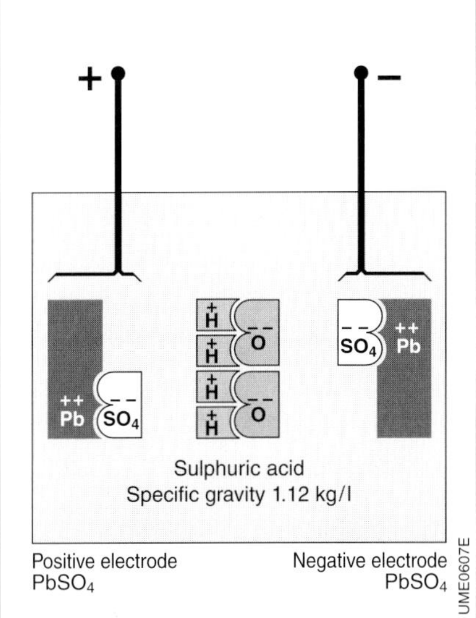

+ •   • −

Sulphuric acid
Specific gravity 1.12 kg/l

Positive electrode	Negative electrode
$PbSO_4$	$PbSO_4$

UME0607E

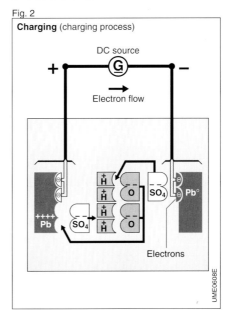

Fig. 2

Charging (charging process)

DC source

+ • — Ⓖ — • −

Electron flow

Electrons

UME0608E

This means that the state of charge can be determined by measuring the electrolyte's specific gravity.

Charging is complete when:
– The lead sulphate at the positive electrode has converted to lead peroxide (PbO_2), and
– The lead sulphate at the negative electrode has converted to metallic lead (Pb), and when
– The charge voltage and the electrolyte's specific gravity remain constant even though charging continues.

If charging is continued although it is in fact finished, all that occurs is electrolytic decomposition of the water. Oxygen is formed at the positive plate and hydrogen at the negative plate. The battery is said to be "gassing".

Once it has been charged, the battery can now be separated again from the source of charging current. In the charging process, the electrical energy applied to the cell has been converted and stored there in chemical form (Fig. 3).

Discharge
(current drain from the battery)
If a load (e.g. a lamp bulb) is connected between the poles of a lead cell, due to the potential difference between the poles (cell voltage) electrons flow from the negative pole through the load to the positive pole. Compared to the battery-charging process, the direction of current flow and the electrochemical process are reversed when discharge takes place.

This flow of electrons leads to the tetravalent lead at the positive electrode changing to bivalent positive lead, and to the breakage of the bond between previously tetravalent lead and the oxygen atoms. The oxygen atoms thus released combine with the hydrogen ions from the sulphuric acid and form water. At the negative electrode, bivalent positive lead also forms as a result of electrons moving from the metallic lead to the positive electrode (Fig. 4). The bivalent negative sulphate ions (SO_4^{--} from the sulphuric acid) combine with the bivalent positive lead at both electrodes so that lead sulphate ($PbSO_4$) is formed as a discharge product at each electrode.

Both electrodes have now returned to their initial condition: the chemical energy stored in the cell has been transformed back into electrical energy by the discharge process.

Fig. 3

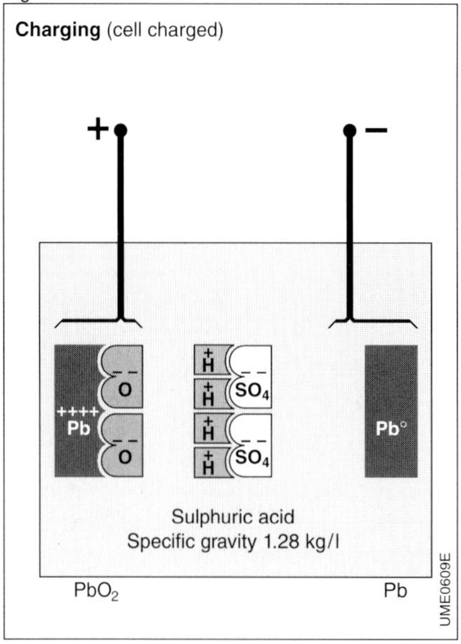

Charging (cell charged)

Sulphuric acid
Specific gravity 1.28 kg/l

PbO_2 Pb

UME0609E

Fig. 4

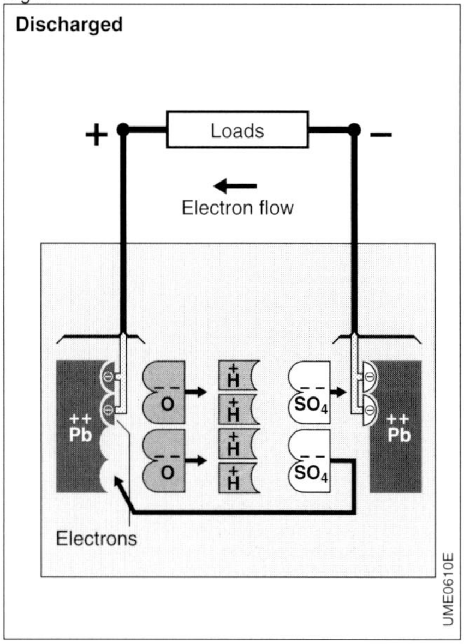

Discharged

Loads

Electron flow

Electrons

UME0610E

Table 1: Overview of the discharge processes.

	Positive electrode	Electrolyte	Negative electrode
Lead cell charged	Active mass: Lead peroxide (PbO_2, brown)	Sulphuric acid (H_2SO_4) high specific gravity	Active mass: Lead (Pb), metallic grey
Discharge Electron flow from the negative electrode through the electrical load to the positive electrode.	The taking on of two electrons reduces the tetravalent lead peroxide ($Pb^{4+}O_2$) to bivalent lead ions (Pb^{++}), which combine with the residual bivalent negative sulphate ions (SO_4^{--}) to form the light-colored lead sulphate ($PbSO_4$).	Oxygen (O_2) in the lead peroxide (PbO_2) at the positive electrode forms water with the released positively charged hydrogen ions (H^+, H_3O^+) from the sulphuric acid and dilutes the electrolyte.	Electron migration results in the generation of bivalent positive lead ions (Pb^{++}) from the zero-valent metallic lead (Pb). These combine with the sulphate ions SO_4^{--} to form the light-colored lead sulphate ($PbSO_4$)
Lead cell discharged	Lead sulphate ($PbSO_4$) from the ions $Pb^{++}+SO_4^{--}$	Sulphuric acid with low specific gravity	Lead sulphate ($PbSO_4$) from the ions $Pb^{++}+SO_4^{--}$

Table 1 provides an overview of the processes concerned in the discharge of a battery. When charging, the processes are reversed.

Parameters

Overview
The European standards EN 60095-1 and national standards define the specifications and test methods for starter batteries. Although these tests are suitable for evaluating and monitoring the quality of new batteries, they make no claims to complete coverage of all aspects of battery performance as encountered in the field.

One characteristic of the chemical current store is that the quantity of electricity (capacity) depends upon the magnitude of the discharge current I_E. Presuming a defined final voltage, therefore, this means that the higher the current withdrawn from the battery the less is its available capacity.

In order to be able make a comparison between starter batteries at all, their capacity is referred to the current which the battery can deliver within 20 hours at constant discharge current down to a defined cutoff voltage per cell (nominal capacity K_{20}).

Cell voltage
The cell voltage U_Z is the difference between the potentials which are generated between the positive and negative plates in the electrolyte. These potentials depend upon the plate materials and upon the electrolyte and its concentration. The cell voltage is not a non-variable figure but depends upon the state of charge (electrolyte specific gravity) and the electrolyte temperature.

Nominal voltage
For lead storage batteries, standards (DIN 40 729) define the nominal (theoretical) voltage U_N of a single cell as 2 V. The nominal voltage of the complete battery results from multiplying the individual cell voltages by the number of cells connected in series. According to the Standard EN 60095-1, the nominal voltage for starter batteries is 12 V. The 24 V required for truck electrical systems is provided by connecting two 12 V batteries in series with each other.

Off-load and steady-state voltage
The off-load voltage is the voltage across the unloaded battery. Following completion of the charging and discharging processes, due to polarization and dif-

fusion it changes to become a final value which is referred to as the steady-state voltage U_0 (Fig. 5).

The steady-state voltage U_0 is the actual voltage obtained by multiplying the number of series-connected cells by the cell voltage U_Z.

For six cells, the following applies:

$$U_0 = U_{Z1} + \ldots + U_{Z6} = 6 \cdot U_{Zi}$$

Similar to the cell voltage, the steady-state voltage is also dependent upon the state of charge and the electrolyte temperature.

Internal resistance

The internal resistance R_i of a cell is composed of a number of individual resistances together. Basically speaking, these consist of R_{i1} the contact resistance between the electrodes and the electrolyte (polarization resistance), together with the resistance R_{i2} presented by the electrodes (plates with separators) to the flow of electrons, and R_{i3} the electrolyte's resistance to the flow of ions. And when a number of cells are connected in series, the resistances of the individual cell connectors R_{i4} must also be added. The following thus applies:

$$R_i = R_{i1} + R_{i2} + R_{i3} + R_{i4}.$$

As the number of plates increases (and therefore their total surface area as well), the cell's internal resistance decreases. In other words: the higher the A·h capacity, the lower the internal resistance (presuming identical plate thickness). On the other hand, the more the battery discharges, and at low temperatures (the sulphuric acid becomes more viscous), the internal resistance increases again. The internal resistance R_i of a 12 V starter battery is comprised of the series connection of the internal resistances of the individual cells, together with the resistances presented by the internal connecting elements (cell connectors and plate straps). For a fully charged 50 A·h battery for instance, at 20 °C R_i is 5...10 mΩ; whereas at 50% charge and −25 °C it increases to about 25 mΩ. Since R_i together with the remaining resistances in the starter circuit determines the cranking speed when starting, it is one of the decisive parameters for starting behavior.

Terminal voltage

The terminal voltage U_K is the voltage measured between the two terminal posts of a battery. It is a function of the off-load voltage and the voltage drop across the battery's internal resistance Ri (Fig. 6):

Fig. 5

Steady-state voltage of a battery

U_Z Cell voltage,
U_0 Steady-state voltage.

$$U_0 = 12\,\text{V}$$

$+$ $-$

| $U_Z=$ 2V | 2V | 2V | 2V | 2V | 2V |

UME0631Y

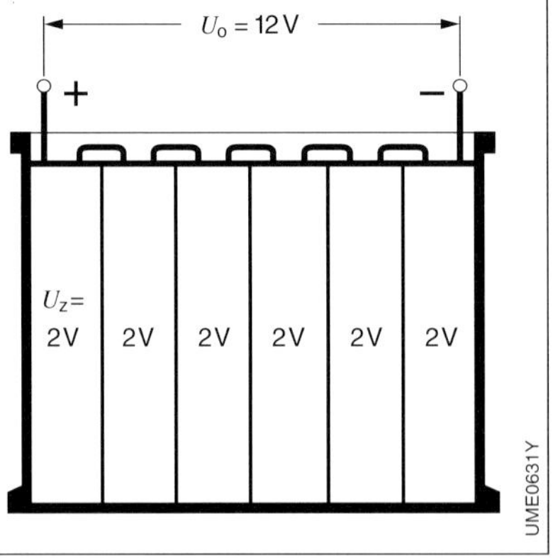

Fig. 6

Battery voltages

I_E Discharge current, R_i Internal resistance,
R_v Load resistance, U_0 Steady-state voltage,
U_K Terminal voltage,
U_i Voltage drop across internal resistance.

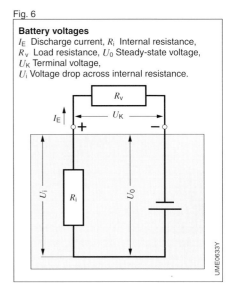

UME0633Y

$U_K = U_0 - U_i$ whereby $U_i = I_E \cdot R_i$

If a current I_E is taken from a battery through a load R_L, a lower terminal voltage is measured than in the no-load state. This is due to the battery's internal resistance. In other words, when a current I_E flows through the cell, a voltage drop U_i, takes place across R_i which increases along with increasing current. Since the internal resistance is a function of among other things temperature and state of charge, this means that the loaded battery's terminal voltage drops at low temperatures and when it is inadequately charged.

Since an unloaded, uncharged battery has a cell voltage of about 2 V and a steady-state voltage of about 12 V between the terminal posts, this means that for practical purposes measuring the terminal voltage of an unloaded battery provides an unreliable indication of its actual state of charge. A general indication of the battery's state of charge can only be obtained by measuring its terminal voltage under load.

Voltage at commencement of gassing
The voltage at commencement of gassing (DIN 40729) is defined as the charge voltage above which the battery clearly starts to gas. Gassing leads to

Fig. 7

Available A·h capacity as a function of temperature and discharge current

Battery; 12 V 100 A·h (referred to 20 h discharge time and 100% state of charge).

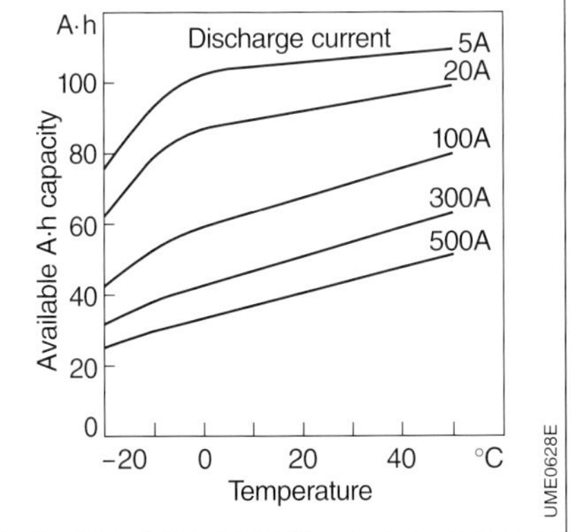

water losses and there is the danger of explosive gas being formed.
According to VDE 0510, a voltage of 2.4...2.45 V per cell applies as a rough limit depending on the particular battery design. For 12 V batteries this voltage limit is about is 14.4...14.7 V.
In order to prevent water loss and the formation of explosive gas, in the case of a 12 V standard battery therefore the maximum voltages of an externally connected battery charger (or of the alternator voltage regulator for 12 V standard batteries) must be limited to 14.4 V (2.4 V/cell). For maintenance-free 12 V batteries, the limit is 13.8 V (2.3 V/cell). When charging maintenance-free closed AS batteries (gel batteries), a 14.1 V (2.35 V/cell) charge voltage is stipulated for a charging time of max. 48 h.

Capacity

Available capacity
The capacity K is the quantity of electricity which the battery can deliver under specified conditions. It is the product of current and time (ampere-hours A·h).
The battery's A·h capacity is determined essentially by the quantity of active material used in its construction. For high powers (such as are needed when starting an IC engine), the active material must have large internal and external surfaces (large number of large-area plates). The large internal surface area is produced by a special form of electrochemical processing known as "forming".
However, the battery's A·h capacity is not a fixed parameter, but depends, among other things, on the following factors (Figs. 7 and 8):
– The level of the discharge current,
– Specific gravity and temperature of the electrolyte,
– Discharging process as a function of time (the remaining A·h capacity is higher when a pause is made during discharge, than when discharge is continuous),
– Battery age (due to the loss of active material from the plates, A·h capacity

decreases as the battery approaches the end of its service life), and
– Is the battery moved around during use, or does it remain stationary? (electrolyte stratification).

The discharge current plays a particularly important role, whereby the higher the discharge current the lower the available A·h capacity. In the example shown in Fig. 8, a discharge current of 2.2 A means that the available 44 A·h capacity can be used for up to 20 h. At a temperature of 20 °C, and with a mean starter current of 150 A, this leads within 8 mins to the available capacity dropping to approx. 20 A·h. The reason for this is that at low discharge currents the electrochemical processes can take place slowly and penetrate deeply into the plate pores, whereas high discharge currents cause these processes to take place mostly on the surface of the plates.

Influence of temperature:
When the battery temperature increases, so do its discharge voltage and A·h capacity, whereas they drop when temperatures decrease. Among other things, the increase is the result of the reduction in the electrolyte's viscosity at higher temperatures and the attendant drop in internal

resistance. The reduction in A·h capacity and discharge voltage described above are due to the fact that the lower the temperature, the less efficient are the electrochemical processes in the battery. When selecting a starter battery therefore, the A·h capacity should be adequate. Otherwise, with a wrongly dimensioned battery, there is the danger that at very low temperatures the engine is not cranked fast enough and long enough to start it.

This fact is demonstrated in Fig. 9:
Curve 1a shows the output power developed by the starting system (battery and starter) as a function of temperature with the battery 20 % discharged. For curve 1b the same applies but for here the battery is heavily discharged. Curve 2 on the other hand, shows the minimum start speed needed by the engine. At low temperatures the starting power needed by the engine is correspondingly higher due to the increase in frictional resistance in the engine (for instance the increase in lube-oil viscosity).

The intersection S_1 of curves 1a and 2 defines the cold-start limit (limit temperature) for a battery which is 20 % discharged. In other words at lower temperatures or with a battery which is even

Fig. 8

Available A·h capacity as a function of discharge current. Battery: 12 V 44 A·h.

Current requirement: **A** 20-hr discharge current; **B** Ignition and lighting; **C** In addition, fan, windshield and rear-window heating, fog lamps, wipers and radio; **D** Mean starter current.

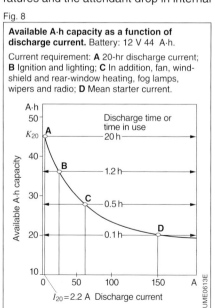

Fig. 9

Influence of temperature on starter speed and the engine's minimum start speed

Example: **1a** Starter speed, battery 20 % discharged; **1b** Starter speed, battery heavily discharged; **2** Minimum engine start speed. **S1, S2** Cold-start limit.

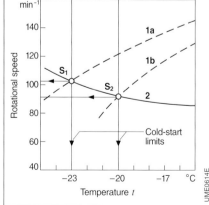

further discharged, it is impossible to start the engine. The power which the battery (or starter) deliver is less than that needed to start the engine. The more the battery is discharged, the more the cold-start limit (intersection S_2) is shifted to higher temperatures.

Nominal capacity

The nominal capacity K_{20} is the battery's rated A·h capacity. According to EN 60095-1, this is the quantity of electricity (in A·h) which, at a fixed discharge current of I_{20}, can be taken from the battery within 20 h until the stipulated final discharge voltage of 10.5 V at $25 \pm 2\,°C$ is reached. The discharge current I_{20} is the discharge current which is allocated to the nominal A·h capacity, and the battery must deliver this current during the total discharge period: $I_{20} = K_{20}/20$ h.

The nominal capacity is a measure for the energy which can be stored by the battery in the as-new condition. It depends upon the quantity of active material and upon the electrolyte's specific gravity.

For instance, a new 44 A·h battery can be discharged with a current of 2.2 A for at least 20 hours (44 A·h/20 h = 2.2 A) until the final discharge voltage of 10.5 V is reached. The nominal capacity K_{20} is therefore a highly important battery specification for the permanent loads in the vehicle's electrical system (Fig. 8).

Cold-discharge test current

The cold-discharge test current I_{KP} is a high-discharge current assigned to the particular battery type. It serves primarily for assessing starting behavior at low temperatures and under defined discharge conditions.

During the cold-start test, the battery (at an initial temperature of $-18 \pm 1\,°C$) is discharged with the stipulated cold-discharge test current until it reaches a final discharge voltage of 6 V (Fig. 10).

According to DIN 43539, the terminal voltage must comply with the following stipulations:
When discharged with I_{KP} and at a temperature of $-18\,°C$,

– 30 s after start of discharge the cell voltage must be at least 1.5 V/cell (9 V for a 12 V battery), and
– 150 s after start of discharge the cell voltage must be at least 1 V/cell (6 V for a 12 V battery).

However, the cold discharge test current is defined using different test conditions from country to country, so that a direct comparison of this specification is impossible.

For an automotive battery which must provide the electrical energy for the starter, starting capabilities at low temperatures are usually more important than the A·h capacity. Since it is referred to low-temperature current delivery, the cold-discharge test current is therefore a measure for starting capability. It is highly dependent upon the total surface area of the active materials (number of plates, surface-area of plates) because the larger the area of contact between the lead material and the electrolyte, the higher is the current that can be withdrawn for a short period. Plate spacing and separator material are two variables which affect the rapidity of the chemical processes in the electrolyte and which also determine the cold-discharge test current.

Fig. 10

Discharge of a 12 V battery with a cold-discharge test current I_{KP} at $-18\,°C$ and at $+27\,°C$

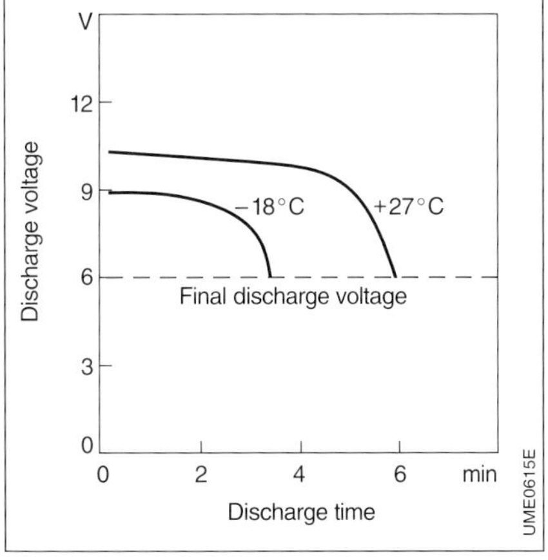

Battery construction

Conventional batteries

The 12 V starter battery contains six series-connected individually partitioned cells in a polypropylene case. Each cell comprises the element (cell pack) which consists of a positive and negative plate set. These, in turn, are composed of the lead plates (lead grid and active mass), together with the microporous insulating material (separators) between the plates of opposite polarity. The electrolyte is diluted sulphuric acid which permeates the pores in the plates and separators and the voids in the cells. Terminals, cell connectors, and plate straps are made of lead. The openings in the partitions for the cell connectors are sealed. A hot-molding process is employed to permanently bond the one-piece cover to the battery case thus providing the battery's upper seal. On conventional batteries, each cell is sealed by its own vent plug which permits initial battery filling and topping-up during service. When screwed in, the vent plugs permit the electrolytic gasses to escape. Maintenance-free batteries often appear to be fully sealed, but they also have escape vents (Fig. 1).

Battery case

The battery case (Fig. 1, item 7) is made of acid-resistant insulating material. Normally, it is provided with bottom rails (8) for mounting purposes. Inside the battery, the plate sets (9, 10) are held by their feet against the element rests running along the floor of the case. The spaces between the element rests, and below the bottom edges of the plates, form sediment chambers which serve to accumulate the solid material which in the course of time is sloughed off from the plates and falls to the bottom of the case. The sediment, which contains lead and is electrically conductive, accumulates in the sediment chambers without contacting the plates. Short-circuits are thus avoided.

Partitions subdivide the battery case into cells, the basic battery assemblies. They contain the elements (9, 10) with positive and negative plates, as well as separators inserted between them. The cells are connected in series using cell connectors (3), which provide the connection through openings in the cell walls.

Fig. 1

Basic design of a lead storage battery (Example: Maintenance-free battery)

1 One-piece cover,
2 Terminal-post cover,
3 Cell connector,
4 Terminal post,
5 Vent plugs underneath the cover plate,
6 Plate strap,
7 Case,
8 Bottom mounting rail,
9 Positive plates inserted into envelope-type separators,
10 Negative plates.

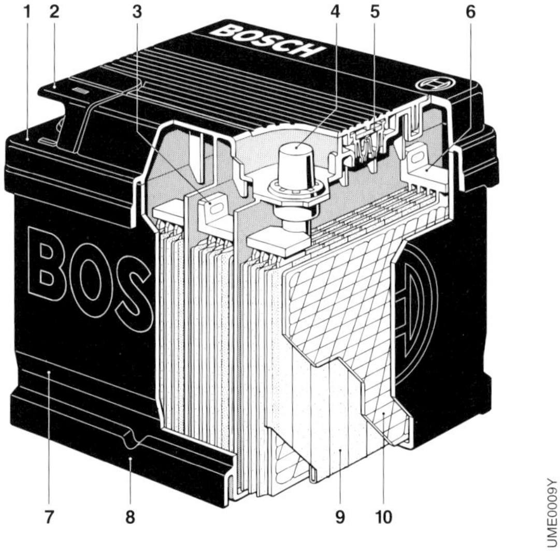

UME0009Y

One-piece cover

The cells are all covered and sealed with a one-piece cover (1), which is provided with an opening above each cell for filling with electrolyte and for servicing. The opening is closed by a screw-in plug with vent opening (5). On 100% maintenance-free batteries, although these vent plugs are no longer accessible they are also provided with vent openings although these are not visible.

Elements

The elements (or cell packs) comprise positive and negative plates and the separators (9) between them. Essentially, the number and surface area of these plates define the cell's ampere-hour capacity.

The plates, so-called grid plates, are constructed of lead grids (carrying "active material"), and the active material itself which is "pasted" onto them. The active material of the positive plate contains lead peroxide (PbO_2, dark brown), and that of the negative plate pure lead in the form of "sponge lead" (Pb, metallic grey). This active material, which is subjected to chemical processes when current flows through it, is porous and therefore provides a large effective surface area.

In an element (or cell pack), all positive plates are welded together with a plate strap as are all the negative plates (Fig. 1). These straps also hold the individual plates in position mechanically. Normally, each element has 1 more negative plate than it has positive plates.

Lead-antimony alloy (PbSb)

Antimony is added to the lead used for the grid to improve the castability of the thin lead grids (very important for high-performance batteries), to speed-up the hardening process, and to provide the lead plates with the strength needed to withstand operation in the vehicle. In other words, the antimony acts as a hardener, this is where the term "hard lead" comes from. However, during the battery's service life the antimony is increasingly separated out due to positive-grid corrosion. It wanders to the negative plate, passing through the electrolyte and the separators on the way, and "poisons" it by forming local voltaic couples. Firstly, these increase the negative plate's self-discharge and reduce the voltage at which gassing starts. Both factors promote the increase of water consumption due to over-charging, and this in turn leads to an increase in the release of antimony. This self-energizing mechanism leads to a continuous drop in power throughout the battery's service life. And in winter, when there is in any case only a relatively low charging current, this is wasted on decomposing the water. The battery then fails to reach an adequate charge level, and the electrolyte level must be checked more often.

The 4...5% antimony content formerly used in the grid lead led to self-discharge of the negative plate, one of the main reasons for starter-battery failure. Depending upon driving conditions, with older batteries the water consumption resulting from increased gassing meant the battery had to be checked every 4 to 6 weeks.

Active material

The active mass is that part of the battery plates which changes chemically when current flows, that is, during the charging and discharging processess (see DIN 40729). During battery manufacture, active materials are produced as follows: Water, dilute sulphuric acid, and maybe other materials together with short plastic fibers are added to lead oxide which usually contains 5...15% of finely distributed metallic lead ("Grauoxid", literally "grey oxide"). A dough-like mass is formed in a mixer or kneader. Basic lead sulphates are produced in this way, whereas lead oxide and metallic lead remain in their original forms. This dough-like mass is then pasted onto the lead grids (the battery grid plates) and left to harden. The active material of the finished plate develops during the subsequent "forming" process. That is, the electrochemical conversion of the active material which usually takes place at the manufacturer.

Separators

Since weight and space-saving considerations predominate in the case of the automotive battery, the positive and negative plates are very close together. They must not contact each other though, neither when they bend nor when particles crumble away from their surface. Otherwise the battery is destroyed immediately by the resulting short-circuit. Partitions (separators) are installed between the individual plates of the elements to ensure that plates of opposite polarity are far enough away, and electrically insulated, from each other. However, these separators must not noticeably oppose the ion migration. And they must be of acid-resistant, porous (micro-porous) material through which electrolyte is free to pass. Such a microporous structure prevents the very fine lead fibers penetrating the separators and causing short-circuits.

Cell connectors

The battery's individual cells (elements) are connected in series by the cell connectors (Fig. 1). To decrease the battery's internal resistance and weight, direct cell connectors are used in modern batteries. Here, the plate straps of individual cells are connected by the shortest path, that is directly through the cell partition. This also reduces the danger of short-circuits due to external influences.

Terminal posts

The plate strap for the positive plates of the first cell is connected to the positive terminal post, and that from the negative plates of the last cell to the negative terminal post. The terminal voltage is then available between these two terminal posts (in other words approx. 12 V).

The battery cables are attached to the terminal posts by special cable terminals. A number of measures prevent mix-up of cables and terminals: The positive post is thicker than the negative post, the posts are marked with their polarity, and positive and negative cable terminals have different opening diameters to fit the respective terminal posts.

Maintenance-free batteries

Battery plates

The alloys used for the hard-lead grids of conventional and maintenance-free batteries differ. After reducing the lead-grid antimony content over a period of many years, a point had finally been reached where no further real improvements in battery characteristics could be realised. This meant that the search had to be intensified for some other form of hardener to replace antimony.

Lead-calcium alloy (PbCa)

In maintenance-free starter batteries, antimony is superseded by calcium, which is electrochemically inactive under the prevailing potential conditions in the lead storage battery. This means that negative-plate poisoning is avoided and self-discharge is stopped. Of even more importance is the fact that the gassing voltage remains stable at a high level throughout the whole of the battery's service life. Polyethylene foil which is resistant to the effects of oxidation and acid is used as the separator material. This is formed into envelopes which surround (and separate) the negative and positive battery elements (Fig. 2).

Water consumption

When new, starter batteries without antimony as well as those with a reduced antimony content, feature a far lower water consumption than the maximum $6 \, g/A \cdot h$ as stipulated by DIN. As a rule the lead-calcium battery has a long-term figure of less than $1 \, g/A \cdot h$.

The fact that the gassing voltage remains at its high initial level throughout the battery's complete service life is responsible for the water-consumption figures for the maintenance-free batteries being so favorable. High gassing voltage leads to minimum water decomposition.

This has the following advantages:

– In the case of the maintenance-free batteries, the charging voltage only exceeds the gassing voltage at high temperatures. This means that only rarely does gassing (water decomposition) occur at all, so that topping up with distilled water remains unnecessary throughout the battery's service life.

This is the reason why the vent plugs are covered by a plate, and can only be opened when necessary by an authorized workshop. Or there are no vent plugs fitted at all.

– It is no longer possible to forget to top-up with distilled water when needed, or to use contaminated water, or even to use so-called "battery improvers".

– Injuries and the danger of damage due to contact with sulphuric acid are a thing of the past.

– Maintenance and servicing costs are reduced.

– It is no longer necessary to install the battery in the vehicle in an easily accessible position.

Characteristics

A vast amount of development work and manufacturing experience are behind the batteries from Bosch. Although this has its price, it certainly provides the user with a high level of reliability and a long service life. These are advantages which are by no means a matter of course for every battery presently being sold on the market.

Bosch starter batteries are completely maintenance-free, and in addition to complying with the performance values stipulated by DIN, they also fulfill the following requirements:

Fig. 2

Battery plate with envelope-type separator

1 Separator cut open.

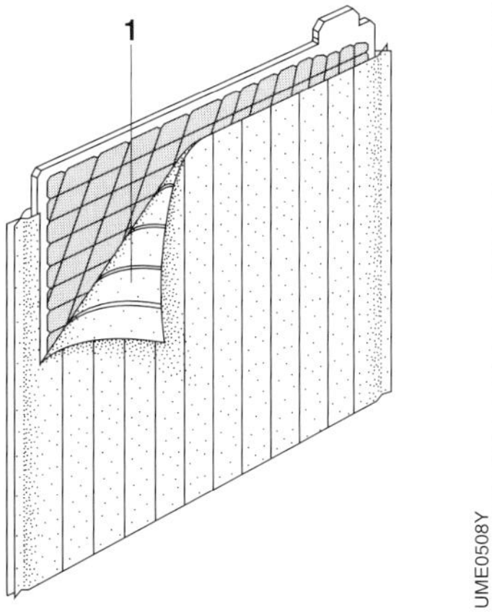

UME0508Y

– Freedom from maintenance for the battery's complete usable life.

– No impairment of performance data and charging characteristic as a result of water consumption.

– Performance data and charging characteristic remain as constant as possible throughout the battery's complete usable life.

– Following exhaustive discharge and subsequent extended period of non-use while connected into the vehicle's electrical system, it must still be possible to recharge the battery under normal operating conditions.

– In case of seasonal operation without intermediate charging (but with disconnected ground cable), there is to be no reduction in usable life compared to all-year-round operation.

– It must be possible to store the filled battery for long periods.

External features

The maintenance-free starter battery from Bosch has the following external features:

– Terminal posts protected by caps against inadvertent short-circuit.

– Cover plate over the vent-plug trough prevents the accumulation of dirt and moisture, and serves to cover the vent plugs.

– Grips facilitate easy transportation.

– Labyrinth cover with central gas vent prevents the escape of electrolyte in case the battery is tipped-up inadvertently for a brief period.

Internal features

The internal features of the Bosch maintenance-free batteries are:

– Plate grids of lead-calcium alloy. In some versions, the pasted-on active material is alloyed with silver.

– Battery case without sludge ribs. The plates reach down to the floor of the case (= increased plate surface) with which they are in contact along their complete length (increased stability).

– Microporous envelope-type separators prevent the active material crumbling off of the plates, as well as preventing the formation of short circuits at the bottoms

and at the side edges of the plates. The mean pore diameter is smaller than that of conventional separators by the factor of 10, and in addition to the reduced electrical resistance, this is also an effective measure in preventing short-circuits.
– The very low self-discharge level means that all lead-calcium batteries can be filled with sulphuric acid at the factory. This procedure takes place under optimal conditions, and rules out the danger of acid being splashed during the filling and mixing process.

Self-discharge

The self-discharge of the negative and positive plates is inherent in the principle of the lead storage battery. Even without external loads connected, the effects of temperature and other factors lead to the battery becoming electrically "empty" after a given period of time. Considering conventional starter batteries, antimony poisoning leads to an increase of the self-discharge reaction at the negative plates, and the rate of discharge climbs considerably along with increasing battery age. In the field, this means that after 6-month storage at room temperature (20 °C), new, conventional-type starter batteries only have an electrolyte specific gravity of 1.20 kg/l. This corresponds to a state of charge of approx. 65 %. Under certain circumstances, old batteries drop to this value within a few weeks.

On the other hand, with the maintenance-free starter batteries from Bosch, the electrolyte specific gravity is still 1.26 kg/l after the same period of time (corresponding to a 90 % state of charge), and it takes 18 months for it to drop to 1.20 kg/l (Fig. 3). The lead-calcium alloy system used in maintenance-free batteries for the plate grids means that the accelerated self-discharge due to antimony poisoning is a thing of the past. The level of negative and positive plate self-discharge stays at a low level throughout the battery's useful life. The self-discharge phenomenon is of particular significance for vehicles used only during specific seasons (agriculture, forestry, and construction sites). It is also important though for second cars and mobile homes which are not driven in winter, or only rarely. It also applies to new vehicles which, since there is no stop in production, are manufactured but have to be parked for long periods due to seasonal fluctuations in sales trends or the transportation time to the salesroom. In all these cases, the battery must have as high a state of charge as possible before the vehicle is taken out of operation, and it must disconnected from quiescent-current loads the whole time it is not in use. This is done, for instance, by disconnecting the ground cable.

Fig. 3

Electrolyte specific gravity as a function of storage time (20 °C room temperature)

1 Conventional starter battery (PbSb),
2 Maintenance-free starter battery (PbCa).

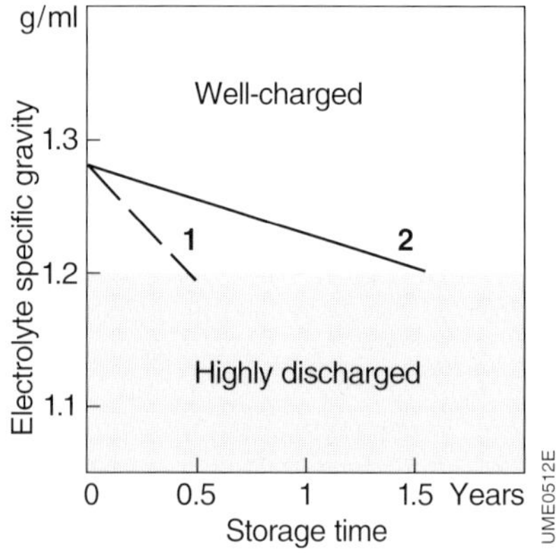

Fig. 4

Starting power as a function of useful life

1 Conventional starter battery (PbSb),
2 Maintenance-free starter battery (PbCa).

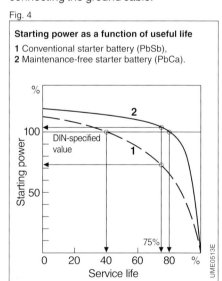

Starting power

The Bosch maintenance-free starter battery features a higher starting power than a conventional battery. Basically speaking, this is due to the envelope separators with their low specific resistance, and to the increase in plate surface due to the omission of the sediment chamber.

Furthermore, thanks to the lead-calcium alloy used for the plate grids, the maintenance-free battery's starting power remains practically unchanged for years. It doesn't drop to below the DIN figure for new batteries until towards the end of its useful life. Whereas after 75% of its useful life the maintenance-free battery is still above the DIN figure, the conventional battery drops far below this DIN figure considerably sooner (at approx. 40% of useful life). In practice it has already lost about a third of its original starting power after only 75% of its useful life (Fig. 4).

Power input

Low-antimony and antimony-free batteries behave almost the same during the DIN power-input test. In the field, differences are negligible when conventional regulator curves are used, and if the tendency for better power consumption on the part of the maintenance-free lead-calcium-alloy battery at a state of charge below 50% is ignored. When charging with regulators which take battery temperature into account and which can increase the voltage to above 14.5 V, the maintenance-free battery has clear advantages. This is due to the phenomenon applying to every battery, namely the fact that the lower the temperature, the higher the charging voltage required for a given state of charge. Since its gassing curve is higher during charging, the maintenance-free battery stores the increased charging current without losses due to gassing. In other words it achieves a higher level of charge and therefore provides more starting power.

Useful life

Considering the extremely varied electrical and mechanical loadings placed upon the starter battery in the field, a brief laboratory test on a battery which is immovably clamped down on a test bench is totally inadequate to ascertain its useful life. Most durability tests are aligned to antimony-content batteries and are therefore only of limited use when maintenance-free batteries are concerned. This is why Bosch tests its batteries under field conditions. During such a test, the starting capability of a conventional battery drops to 50% after 60 months, whereas with lead-calcium batteries it takes 80 months for this level to be reached. Under the cyclical loading conditions encountered on taxis in densely populated areas, on town buses, and on delivery vans, maintenance-free starter batteries have come to the forefront due to the advantages inherent in their separator concept and the reliable protection it affords against premature failure.

Resistance to overcharge

Overcharge is a decisive factor with regard to the battery's usable life. It applies for instance to vehicles which have very high annual mileages, as well as to courier and agricultural vehicles, construction-site machinery, and long-distance haulage trucks. In such cases, the battery is fully charged, the engine turns at high speed, and the alternator only supplies a few loads. Under these circumstances, the charge current leads to overcharging, corrosion, and loosening of the active material. In laboratory tests to simulate these conditions at 40 °C electrolyte temperature and 14 V charge voltage, the maintenance-free starter battery has a considerably longer useful life than the antimony-content battery.

Resistance to exhaustive discharge

To check the battery's resistance to exhaustive discharge, lamps are used to discharge it completely, after which it is left 4 weeks in the short-circuit condition. The battery must then recharge under the conditions of normal vehicle electrical-system connection, it must still be fully operational, and it must only display certain specified reductions in performance.

Substitute batteries

Battery types

Type designation

In order to ensure interchangeability of the products from different manufacturers (compatibility), the versions and designations of the various starter batteries are defined in standards.

In addition to an in-company code and such standard data as the nominal voltage, the Ah capacity, and the cold-discharge test current, e.g. 12 V, 66 A·h, 300 A, the battery usually also carries a code in accordance with national standards. This code can be supplemented by company data.

Many of the West-European manufacturers use a DIN designation in addition to their own in-company code.

The DIN designation comprises 5 digits:
- The 1st digit defines the nominal voltage and gives an approximate indication of the battery's Ah capacity.
  5 ⇒ 12 V starter battery < 100 A·h
  6 ⇒ 12 V starter battery > 100 A·h
  7 ⇒ 12 V starter battery > 200 A·h
- The 2nd and 3rd digits define the starter battery's K_{20} Ah capacity, for instance
  66 ⇒ 66 A·h

- Digits 4 and 5 are defined numbers.

In the USA, an SAE (Society of Automotive Engineers) code is used, and in Japan a JIS (Japanese Industrial Standard) code.

The European battery type-number system (ETN) provides for the following 9-digit type-number system:

Group	A	B	C
Code	536	946	033

Group A
5 ⇒ Nominal voltage 12 V
36 ⇒ Nominal K_{20} Ah capacity 36 A·h

Group B
9 ⇒ Version (here: closed battery)
46 ⇒ Defined numbers

Group C
033 ⇒ The cold-discharge test current (I_{cc} = 330 A) divided by 10 and preceded by 0 results in these three digits.

Designs

All batteries are contained in standard lists which, apart from the electrical specifications, also contain stipulations for the geometrical dimensions of the battery case and the terminal posts. In addition, in order to ensure full interchangeability irrespective of manufactur-

Fig. 1

Example of a starter-battery designation

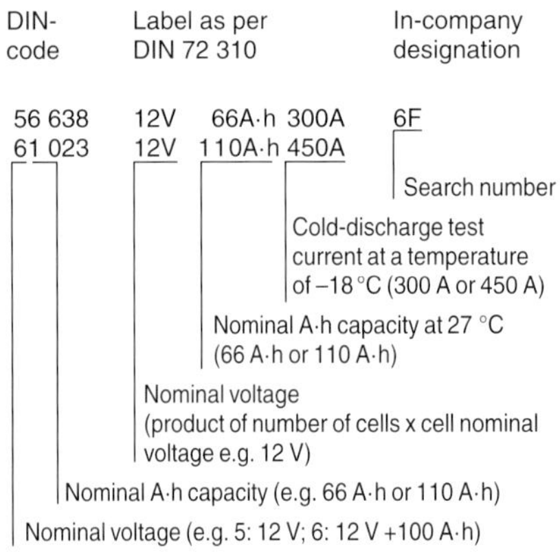

DIN-code	Label as per DIN 72 310			In-company designation
56 638	12V	66A·h	300A	6F
61 023	12V	110A·h	450A	

Search number

Cold-discharge test current at a temperature of −18 °C (300 A or 450 A)

Nominal A·h capacity at 27 °C (66 A·h or 110 A·h)

Nominal voltage (product of number of cells x cell nominal voltage e.g. 12 V)

Nominal A·h capacity (e.g. 66 A·h or 110 A·h)

Nominal voltage (e.g. 5: 12 V; 6: 12 V +100 A·h)

Fig. 2

Electrical connection of the battery cells

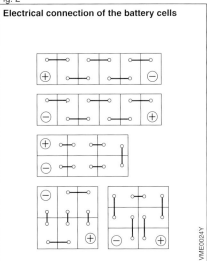

VME0024Y

er, details of mounting variants are given, as well as of the cell configuration and connections. When ordering batteries from the aftermarket program, these are listed with these design variants as product features. The design variants will be discussed in more detail in the following chapters.

Electrical connections
Depending upon the available space and the layout of the equipment in the vehicle, batteries with the most varied dimensions and terminal configurations are needed. These requirements can be fulfilled within very wide limits by appropriate arrangement of the cells (lengthways or transverse installation) and their interconnections. An overview of the most common connection designs is given in Fig. 2.

Terminal posts and battery-cable terminals
With the exception of the terminal posts on the batteries for some motorcycles and scooters to which the cables are connected directly, the battery terminal posts are normally designed for the attachment of battery-cable terminals. These represent the link between the battery and the vehicle electrical system (Figs. 3 and 4).
The terminal posts are conical in shape in order to ensure good electrical contact and tightly fitting cable terminals. The positive terminal post is thicker than the negative one, a measure which is intended to prevent false-polarity connection.
Since the terminal posts contain not only copper but also lead or lead compounds, it is recommended that persons who frequently handle batteries wash their hands thoroughly with special soap to protect against the effects of lead.

There are two main designs of battery-cable terminal (Fig. 4):
– Screw type and
– Solder type.
Cable connections must remain firm even in the case of sustained vibration or collisions. Otherwise, contact resistances occur which lead to excessive power

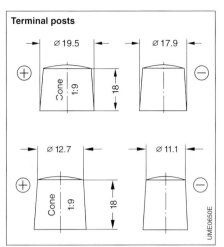

Terminal posts

Fig. 3

Fig. 4

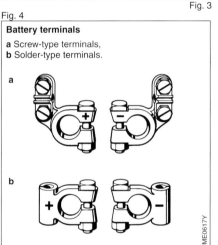

Battery terminals
a Screw-type terminals,
b Solder-type terminals.

a

b

Fig. 5

One-piece cover

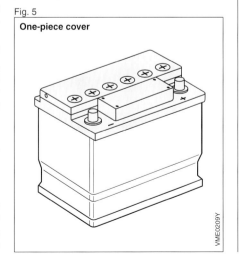

losses due to the high currents that they cause.

Battery covers
Depending upon battery type there are two battery-cover versions (Figs. 5 and 6):
– One-piece cover and
– Mono cover.

Mounting
The battery must be mounted in the vehicle so that it cannot move of its own accord. The battery is therefore clamped onto the support structure by means of
– A hold-down frame,
– A bracket with clamping screw,
– A clamping claw with clamping screw (base mounting), or similar means (Fig. 7).
The various forms of recess in the base of the battery case for the above forms of mounting are common, and are therefore also regulated by standardization.
Correct fastening is important for safety reasons. After all, in extreme situations or even in only a minor tail-end collision, an insecure battery could work loose and cause a fire due to short-circuit. The form of attachment provided complies with all safety regulations and should not be modified.

Battery versions

Maintenance-free battery (as per DIN)
Due to its low level of gas generation, the DIN maintenance-free battery "WFD" is characterized by very low water losses. This prolongs the electrolyte checks to
– Every 15 months or 25,000 km for all low-maintenance batteries, and
– Every 25 months or 40,000 km for maintenance-free batteries (as per DIN).
Maintenance-free batteries are characterized by the following features:
– Cold-start reliability,
– Dry-charged. When delivered, the already mixed electrolyte is added before installation,
– Long service life, and
– Suitability for all vehicles which are subject to normal loading.

100 % maintenance-free battery
There is never any need (and generally no means provided) to check the electrolyte level on the 100 % maintenance-free (lead-calcium) battery "SR". With the exception of two small vent orifices, this type of battery is completely sealed. As long as the vehicle electrical system is operating normally (that is, constant voltage limited to a maximum level), water decomposition is reduced so far that the electrolyte reserves above the plates will last for the life of the battery. This type of (lead-calcium) battery also has the advantage of extremely limited self-dis-

Fig. 6

Mono battery cover

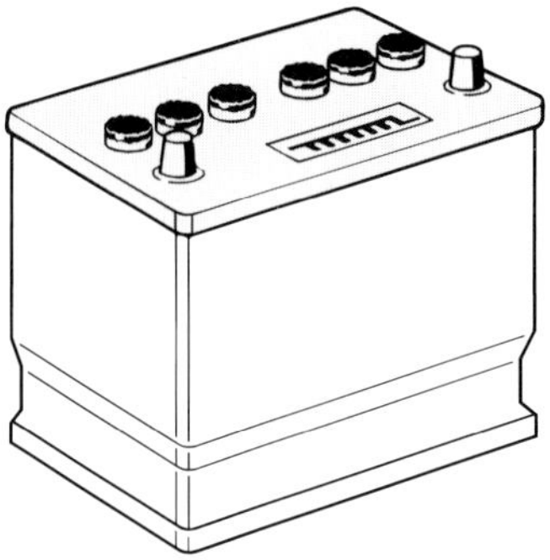

VME0210Y

Fig. 7

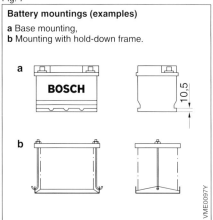

Battery mountings (examples)

a Base mounting,
b Mounting with hold-down frame.

a

BOSCH

10.5

b

VME0097Y

charge, making it possible to store a fully charged battery for months.

When an absolutely maintenance-free battery is charged remote from the vehicle's electrical system, the charge voltage is not to exceed 2.3...2.4 V per cell since overcharging at constant current, or with battery chargers using the W charge curve, decomposes the water so that electrolytic gas is formed.

The outstanding technical features of these batteries are:

- Multiple safety reserves (designation: SR, or SRII with integrated folding handles),
- Even more reliable starting,
- Delivered ready-to-install and start,
- Increased service life for extreme long-distance traffic, and
- Resistant to prolonged overcharge.

The 100% maintenance-free batteries at present on the market incorporate further safety features: For instance, a labyrinth battery cover with vent orifices prevents the escape of acid should the battery be tilted by up to 70°, and provides protection against return current.

Deep-cycle-resistant battery

Due to their design characteristics, standard automotive batteries are poorly suited for applications in which frequent, extreme discharges occur (so-called cyclical loading). Standard batteries respond to these operating conditions with high rates of wear at the positive plates due to the separation and "shedding" (sedimentation) of the active material. In taxis, buses, and delivery vans etc., which are characterized by the short distances travelled between engine starts, and the associated high current consumption, the battery is subjected to very high loading. This can lead to the battery being discharged almost completely so that it is impossible to adequately recharge it with the vehicle's alternator. Added to this are the high levels of current consumption when the vehicle is at a standstill due to fans, air-conditioner, auxiliary heating, lighting, car radio, and two-way radio etc. The so-called deep-cycle-resistant starter battery (identified by "Z") is particularly suited to this range of applications. It can be "deep" discharged more often than a normal battery without impairing its service life. In this starter-battery version, separators with fiber-glass mats are used to give the positive mass the extra support needed to prevent premature "shedding", which otherwise would lead to sludge formation. The service life of such a battery, measured in charge/discharge cycles, is approximately double that of a standard battery. Deep-cycle starter batteries featuring pocket separators and felt layers have an even longer useful life.

Vibration-proof battery

In the "RF" vibration-proof battery, an anchor of cast resin and/or plastic prevents the plate stacks (battery elements)

Fig. 8

Battery versions

1 100% maintenance-free SRII starter battery,
2 Standard starter battery (maintenance-free as per DIN),
3 Drive/traction battery for long-time current output,
4 HD starter battery Extra,
5 HD starter battery,
6 RF starter battery.

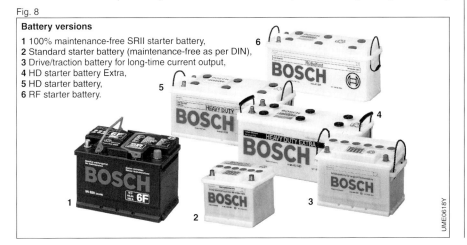

from moving relative to the battery case. According to DIN standards, this type of battery must survive 20 hours of sinusoidal vibration at a frequency of 22 Hz and a maximum acceleration of 6 g. These requirements are approximately 10 times stricter than those for standard batteries. Vibration-proof batteries are installed primarily on trucks, construction machinery and tractors used on building sites, on ski slopes, and for off-road applications in agriculture, and forestry. This battery's identification designation is "RF".

Heavy-Duty battery
The dry-charged "HD" (HD = Heavy Duty) battery is maintenence-free as per DIN, and features a combination of the measures applied in the deep-cycle-resistant and vibration-proof batteries. It ensures a reliable power supply even under conditions of high permanent loading caused by a large number of electrical loads. This type of battery is installed in commercial vehicles which are subject to high levels of vibration and cyclical loading.

The HD-Extra battery features even more far-ranging characteristics. These are needed in order to withstand unusually high levels of loading:
– Extreme cold-start reliability (up to 20 % higher starting reserves),
– Extremely long service life,
– Extra vibration-proof,
– Extra deep-cycle resistance.

Battery for extended current output
This battery shares the basic design of the deep-cycle version. It has less plates though, and these are thicker. Although no cold-discharge test current is specified for this battery, its starting power lies well below (35 ... 40 %) that of comparably sized standard starter batteries.

It is used in applications characterized by extreme cyclic variations, and sometimes even for traction purposes (drive battery). For instance for forklift trucks whose batteries need no starting power although they must often be recharged.

These batteries are also used to provide the drive energy for small drive units (e.g. in sweeping machines and wheelchairs),

Table 1: Battery types (12V).

K_{20} A·h	Features Codes[1]	K_{20} A·h	Features Codes[1]
35	SR, SRII	75	WFD, HD
36	SRII	85	WFD
40	SRII	88	WFD, SRII, RF
43	SRII	90	WFD, RF
44	WFD, SRII	92	WFD
45	WFD, SR, SRII	95	WFD, HD
48	WFD	100	WFD, RF, SRII
50	WFD, SRII	110	WFD, RF, HD
54	SRII	115	WFD, HD
55	WFD, SRII	120	WFD, HD, RF
58	WFD	125	WFD, RF
60	WFD, SR, SRII	135	WFD, RF, HD, HDE
61	WFD	140	WFD, HD, RF, HDE
62	WFD, SRII	143	WFD, RF, HD
63	WFD, SRII	155	WFD, HD
64	WFD, SRII	165	WFD, HD
65	WFD	170	WFD, RF, HD
66	WFD, RF, SRII	180	WFD, RF, HDE
68	WFD	200	WFD, HD
70	WFD, SR, SRII	205	WFD, HD
71	WFD	210	EFD, RF
72	WFD, SRII	215	WFD, HD
74	WFD, SRII	225	WFD, HDE

[1] **WFD**: Maintenance-free as per DIN; **SR**: 100 % maintenance-free with safety reserve; **SRII**: As SR, but with handles; **RF**: Vibration-proof; **HD**: Heavy Duty; **HDE**: HD Extra Heavy Duty.

and the energy for signal installations, construction-site lighting, boats, and auxiliary equipment, and for leisure time and hobby applications.

Battery selection

Same replacement battery
Usually, under normal operating conditions, and in a balanced climate, the batteries and alternators fitted in the vehicle as standard are easily able to provide the currrent needed by the starter and by the other vehicle loads. As long as these factors do not change, the same battery type can be fitted again when replace-

ment becomes necessary. A check can be made for possible alternatives by referring to the applications section in the Aftermarket Program

More powerful replacement battery
If additional loads are installed, or unusual operating conditions are encountered, a check must be made as to whether the presently fitted battery should be superseded by a more powerful version. In case of such changes, it is always advisable to consult the application recommendations from the vehicle-manufacturer or from the Bosch Aftermarket Catalogs, or to contact the expert from a Bosch Service Agent.
A higher capacity battery, e.g. 55 A·h instead of 36 A·h, has a lower internal resistance. This means that when starting, the voltage drop caused by the high current draw is lower and hence there is more voltage available across the starter. The current consumption and the torque at the instant of cranking (short circuit at zero speed) are thus considerably higher. In the most unfavorable case, the battery starting power can be above the value that is permissible for the starter in question. One result can be that excessive current flows during cranking which either leads to the burn-out of the starter winding or to mechanical damage at pinion or flywheel ring gear.
For special operating conditions, there are also batteries available which are particularly suitable for cyclical loading, conditions of vibration, or extreme cold-starting conditions. Bosch has the appropriate battery available for every vehicle.

Fig. 9

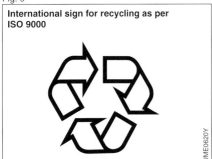

International sign for recycling as per ISO 9000

UME0620Y

Substitute battery: Removal and installation

Removal
Before removing the battery, the engine must be stopped and all electrical loads switched off. The negative cable is disconnected first (ground cable), and then the positive cable. The following basic points must be observed: Do not tip the battery so far that electrolyte can escape through the vent orifices.

Installation
Fasten the battery with the mounting hardware provided. Its terminal posts and the contact surfaces of the cable terminals must be clean and free from grease. Note the positions and polarities of the terminal posts. Connect the positive cable (red) first and then the negative cable (ground). Firmly tighten the cable terminals.
The danger of corrosion necessitates the terminal posts and cable terminals being coated with acid-free grease once the battery has been installed.

Transport
During transport, batteries must be secured against damage, overturning, and sliding back and forth. Bosch batteries have a red cap over the positive terminal post as a protection against short-circuit. A battery can be irreparably damaged by blows against its case or by being dropped. Particular care is required in the case of filled batteries since the caustic electrolyte which then escapes can result in further damage. Transport of batteries weighing more than 250 kg is regulated by law in Germany.

Battery disposal
Used or faulty batteries are special waste, and must be disposed of accordingly. It is best to hand them over to the Bosch Service Agent when the replacement battery is purchased.
Batteries marked with the ISO 7000 sign (Fig. 9) can be reintroduced to the valuable-substance cycle again ("recycled").

Maintenance

Maintenance on conventional batteries

Mixing and filling the battery electrolyte

As a rule, maintenance-free batteries according to DIN are stored in the dry-charged state, and are not filled with electrolyte until just before installation. The quantity of electrolyte poured into the battery depends upon the battery size, and "acid packs" are available comprising a number of bottles each with precisely the correct amount of already mixed acid for a single cell. Filling the battery with the electrolyte is a job which should be left to the Bosch Service Agent or another specialized workshop.

When mixing in the workshop, always pour concentrated sulphuric acid into the water or battery electrolyte, and never vice versa. When mixing, stir with an acid-resistant rod (glass or plastic). Pouring the water into the acid is dangerous and can lead to explosions. Work rooms must be well ventilated. Table 1 shows how the required acid concentrations can be obtained by mixing.

The battery is to be filled with diluted sulphuric acid having a specific gravity of approx. 1.28 kg/l (VDE 0510). In the tropics, this figure is reduced to about 1.23 kg/l (see Table 2). The battery is filled so that the electrolyte is 15 mm above the top of the plates. Electrolyte temperature should be approx. 15 °C. Wait 20 mins before connecting the battery. This gives the battery enough time to reach a state of equilibrium.

The sulphuric acid must have a specific level of chemical purity and must not contain impurities such as metals or chlorine (DIN 43 530). If it is above the stipulated specific gravity, it can be brought to the required level by diluting with distilled water.

In order to prevent over-filling, HD and drive batteries with glass-wool separators are filled in two stages: First fill to the top edge of the plates, wait briefly, and then fill to the required electrolyte level.

Electrolyte specific gravity and state of charge

The electrolyte specific gravity is the major indicator for the battery's state of charge. In the field therefore specific-gravity measurements are used to define the battery's charge level.

Table 2 shows a number of specific-gravity figures and the associated freezing thresholds at different states of charge.

Table 1: **Mixing the electrolyte.**

Desired specific gravity in kg/l	1.23	1.24	1.25	1.26	1.27	1.28	1.30	1.33
Volume ratio of concentrated sulphuric acid (96 %) to distilled water	1:3.8	1:3.6	1:3.4	1:3.2	1:3.0	1:2.8	1:2.6	1:2.4

Table 2: **Acid values of the diluted sulphuric acid.**

State of charge	Battery version	Electrolyte specific gravity kg/l[1]	Freezing threshold °C
Charged	Normal	1.28	−68
	For tropics	1.23	−40
Half-charged	Normal	1.16/1,20[2]	−17 ... −27
	For tropics	1.13/1,16[2]	−13 ... −17
Discharged	Normal	1.04/1,12[2]	−3 ... −11
	For tropics	1.03/1,08[2]	−2 ... −8

[1] At 20 °C: When the temperature rises, the specific gravity sinks (and vice versa) by approx. 0.01 kg/l for each 14 °C change in temperature.

[2] Lower figure: High acid utilization. Higher figure: Low acid utilization.

Electrolyte specific gravity and operating temperature

High temperatures accelerate chemical processes in the battery. Not only the Ah capacity is increased but also starting-system power. On the other hand though, the increase in temperature has a negative effect upon the plates (active material falls out, and grid corrosion increases), and self-discharge rises. Since these are less aggressive, electrolytes with a lower specific gravity (ρ = 1.23 kg/l instead of 1.28 kg/l) are chosen for batteries used in the tropics or for batteries which need not generate high starting powers.

Electrolyte specific gravity and freezing point

The more the battery is discharged, the more the electrolyte is diluted. This leads to the freezing point rising to more unfavorable levels. Presuming a fully charged battery with electrolyte specific gravity of 1.28 kg/l, the freezing point is $-60...-68$ °C. If the battery is empty (discharged) though its freezing point rises to only $-3...-11$ °C. In other words it can

Fig. 1

Acid tester

1 Acid lifter, **2** Areometer,
3 Scale with specific-density markings.

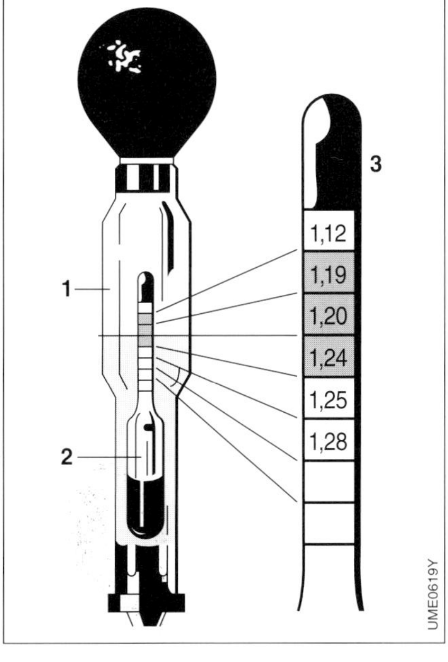

1,12
1,19
1,20
1,24
1,25
1,28

UME0619Y

freeze-up (Table 2). A battery with frozen electrolyte can only deliver a very low current and cannot be used for starting. A polypropylene battery case remains stable even with frozen electrolyte. Since the fluid does not fully crystallize out, it is unlikely that the case will fracture. A frozen battery is not to be charged because the frozen electrolyte then starts to swell. The battery must first of all thaw-out.

Measuring the specific gravity

On conventional batteries, an acid tester (hydrometer) is used for measuring the specific gravity (Fig. 1). This consists of an acid lifter (glass tube with a rubber suction ball), containing an areometer – a float marked with a scale which shows electrolyte concentration. In the case of 100 % maintenance-free batteries, there is no provision made for measuring the specific gravity since this is no longer necessary.

Battery care

Handling in general

On conventional batteries, the electrolyte level should be regularly checked, and if necessary the battery topped-up with distilled water to the "Max" mark. The battery should be kept clean and dry. Before the cold season starts, the state of charge should be checked by measuring the specific gravity. If this is below 1.20 kg/l, the battery should be recharged.

Removal and recharging

- Work only in well-ventilated rooms. Sparks are to be avoided,
- Disconnect the negative terminal (ground) before the positive terminal. Remove the battery,
- Do not tip the battery or shake it,
- Frozen batteries are to be thawed-out before attempting to charge them,
- If vent plugs are fitted, these are to be removed before starting to charge,
- Connect the battery to the battery charger (positive to positive, negative to negative),
- Set the charging parameters as described in the Chapter "Battery chargers",

– Do NOT switch on charger before battery is connected (sparks),
– Ensure that there Is adequate ventilation in the rooms concerned during the charging process (danger of electrolytic-gas explosion).

Charging is to be stopped at the latest
– When electrolyte temperature exceeds 55 °C (case more than hand warm),
– When the battery starts to gas, or
– When the electrolyte's specific gravity or the charge voltage has not changed over a period of 2 hours.

When charging is finished, first switch off the charger, then remove the cables individually from the battery and the charger (that is, unclip one cable from battery and charger, and then unclip the other).

Reinstallation

– Install the battery in the vehicle mechanically and tighten the mounting hardware,
– Then connect the positive cable to the battery, and then the ground cable,
– Check that the battery-cable terminals are firmly attached to the terminal posts (minimum contact resistance),
– Apply acid-free, acid-resistant grease to the terminals and the posts (protection against corrosion also applies to modern maintenance-free batteries).

Battery storage

The following storage times are stipulated for new batteries on the after-market:
– Unfilled: Unlimited,
– Filled, conventional: 3 (max. 6) months
– 100% maint.-free: 18 months.

If the battery is to be stored for longer periods, it must be regularly recharged as per the normal charge characteristic. Batteries must be stored in a cool and dry state and in a good state of charge.

The older a battery gets, the less time it can be stored. As far as possible, the battery is to be trickle-charged with a very low current. If the battery is to remain in the vehicle during the storage period, its ground cable is to be disconnected.

Starter-battery checks

Characteristics and test methods for starter batteries are given in DIN 43 539. Although suitable for determining (monitoring) the quality of new starter batteries, these tests do not claim to comply with all the demands encountered in the field.

Defects

Defects in the battery

Malfunctions due to internal battery faults (such as short circuits due to separator wear or loss of active material, broken cell connectors or plate straps) are irreparable and the battery must be replaced. Major variations in specific gravity from cell to cell (differences of at least 0.03 kg/l) indicate internal short circuits. With cell-connector fracture, low currents can often be taken from the battery and it can be charged, but even when fully charged, when an attempt is made to start the engine the voltage collapses.

Defects in the vehicle electrical system

If a battery defect cannot be defined but the battery is over-charged (excessive water consumption), or exhausted (no starting power, low specific gravity in all cells), this indicates a faulty electrical system. The following can be the cause:
– Alternator (permanent deep discharge of battery, impossible to start the vehicle),
– V-belt (alternator drive),
– Voltage regulator (fluctuation in lamp brightness when engine is accelerated, loss of battery water),
– Faulty relays (loads remain on when the engine has been switched off),
– Accessories (e.g. radio, clock, car alarm) consume excessive current.

Sulphation

If a storage battery is left for any length of time in a discharged state, the fine crystalline lead sulphate formed during discharge may, under unfavorable circumstances, change into coarse lead sulphate crystals which can only be transformed back to the fine crystalline state with difficulty, or not at all. Such a battery is then "sulphated". Sulphation is a result of careless maintenance. It causes an increase in the battery's internal resistance which impedes the chemical processes and makes charging more difficult.

A sulphated battery gets very hot when charged with a "W" charging characteristic. The charge voltage rises sharply as soon as charging starts. If sulphation is only slight, the lead sulphate is broken down slowly and the charge voltage drops steadily. As soon as the lead sulphate has been broken down completely (regenerated), the voltage climbs again the same as it does when charging a serviceable, non-sulphated, battery (Fig. 2).

Regeneration of lead batteries
If the sulphation is not severe, the battery can be regenerated by applying a minimal charge current of approx. 25 mA per A·h battery nominal capacity ($\approx 0.5...2$ A) for about 50 hours.

On the other hand, if the acid has converted completely it will be impossible to regenerate the battery. It is unusable.

Trouble-shooting
If the battery fails to start the engine, this can be due to insufficient charge or to a battery defect. The state of charge and starting power can be measured using the

Fig. 2

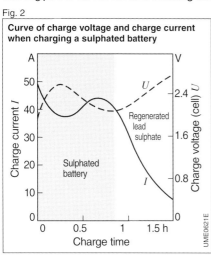

Curve of charge voltage and charge current when charging a sulphated battery

Regenerated lead sulphate

Sulphated battery

U

I

Charge current I

Charge voltage (cell) U

Charge time

UME0621E

Bosch battery tester which only needs a few moments to provide clear information on the battery's condition. Such details are needed for defining the steps to be taken for remedying the problems.

The battery's exact state of charge can be determined by measuring the electrolyte specific density and also by precise measurement of the open-circuit voltage. Excessive water consumption, and/or charging immediately before the test, will falsify the results in the direction of a more favorable verdict. The charge on the plate surfaces which falsifies the results can be compensated for by pausing briefly in the test sequence so that the plate charge can reduce.

Assessment of starting capability is only possible by loading the battery with a standard current for a stipulated time, e.g. 30 seconds, and measuring the voltage drop. Taking into account the battery's initial state of charge, the measured value is compared with a standard value stored in the tester. The results of this comparison are displayed as the percentage starting capability of the battery compared to that of a fully functional battery.

The complex test procedure means that its sequence must follow a program in the tester. Test results are only possible on batteries which have a residual charge. Due to the generation of a mean value for a number of cells, and external influences, this test only provides reliable results in approx. 90 % of the cases. In borderline cases, in order to stabilise the chemical reactions an intermediate charge is required or the test must be repeated.

Charging with the battery charger

Safety requirements
In order to avoid accidents, the battery charger must feature efficient electrical isolation between the mains supply and its charge clips or terminals. It must also be possible to switch off the battery charger before disconnecting the clips from the battery. The battery charger also features protection against incorrect

polarity so that no sparks are generated should it be operated incorrectly.

Battery chargers

If it is impossible for the alternator to charge the battery adequately, a battery charger must be used. This applies when the battery has been out of use for a long time, or directly before it is removed from the vehicle and put into storage. Battery chargers with voltage limitation must be used on maintenance-free batteries. Otherwise overpressure is generated during charging and the battery dries out. A fully charged modern battery can be stored for 6 months without difficulty. A battery can only be charged completely with a relatively low current (max. 1 A).

Electronic battery charger

The LW electronic battery charger enables the battery to be charged in the vehicle without disconnecting it from the electrical system. The charge voltage is free from voltage peaks, and controlled electronically so that the battery cannot be overcharged. During charging, the electronic components such as engine ECU's and airbag triggering units, etc. are protected against damage. These battery chargers are designed for trickle charging and floating operation, particularly with maintenance-free batteries.

Rapid-start battery charger

The rapid-start charger with start-assist stages provides starting help for trucks and passenger cars. Charging times are very short even for large batteries. These chargers are provided with a charge-monitoring facility which ensures that charging is efficient and does not overload the battery.

Workshop battery charger

Some of today's workshop battery chargers feature electronic controls for trickle charging and floating operation. They are also provided with start-assist and rapid-charge functions, and are particularly suited for the strenuous operation encountered in a workshop.

Home battery chargers

So-called home battery chargers are particularly suitable for charging the batteries in small-power drives. They are also useful for home and hobby requirements.

Charging methods

Normal charging

Generally speaking, normal charging takes place using the I_{10} charge current. This corresponds to 10 % of the battery's nominal Ah capacity:

Fig. 3

Battery chargers

1 Rapid-start battery charger,
2 Home battery charger,
3 Electronic battery charger,
4 Battery tester.

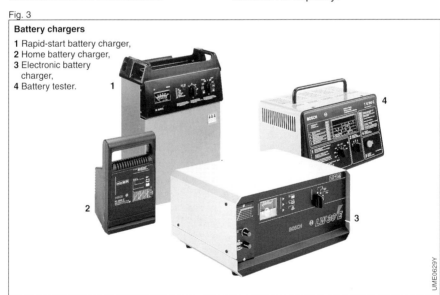

UME0629Y

$I_{10} = 0.1 \cdot K_{20} \cdot$ A/A·h. Depending upon the process used, the charging time can be up to 14 hours.

Boost charging
Boost charging can bring an empty battery back up to about 80 % of its rated capacity so that it can withstand the loading which is typical for automotive applications. Boost charging has no negative effects. Below gassing voltage, high charge currents present no problems. Such currents can be in the vicinity of the number for the nominal Ah capacity (relative charge current $I_{100} = K_{20} \cdot$ A/A·h). Once the gassing voltage is reached, the charging current must either be switched off or a change made to normal charging.

Gassing voltage:
At 20 °C, the gassing voltage is approx. 14.4 V. If this voltage is exceeded during charging, the battery starts to gas very distinctly. This leads to water losses, and there is the danger of electrolytic gas being generated. In order to avoid this, the battery charger's voltage limits must not exceed 14.4 V (2.4 V/cell) for 12 V standard batteries and 13.8 V (2.3 V/cell) for maintenence-free batteries.

Trickle-charging
To compensate for the self-discharge losses in stored batteries (for instance when caravan or mobile-home batteries are stored during the winter), the battery is left connected to a battery charger for an extended period of time, whereby charging current is limited to 1 mA/A·h.

Floating operation
With this type of operation, battery charger and load are permanently connected to the battery. Depending upon the load, a change-over continuously takes place between charging and discharging, whereby the battery charger's electronic circuitry prevents battery overcharge.

Charging curves
There are various methods for charging the battery, each of which is characterized by its own charging curve:

W	Constant resistance (charge current drops when the voltage increases)
U	Constant charge voltage
I	Constant charge current
a	Automatic switch-off
e	Battery charger switches on again automatically
o	Automatic switch-over to other charging curves

Whereby, it is possible to combine the various charging curves. For instance:

WU	As W curve but charge voltage remains constant above a given value (e.g. gassing voltage)
IU	Constant charge current up to a value above which voltage remains constant and the charge current falls
WoW	Switch-over from one charging curve to the other. For instance from a high to a low charge current

The commonly used W charging curve (on home and workshop battery chargers for instance) is defined by the charger's internal resistance. These chargers respond to increasing battery voltage by steadily lowering the charge current. Using the normal charge method as per Fig. 4, with a relatively low initial 20 A charge current, charging takes about 12 hours. Since reduced current continues to flow above the gassing voltage, the battery charger must be switched off once full charge has been reached.

Since the battery-charging process must be monitored, this technique is only applicable to a limited extent on maintenance-free batteries.

On IU-curve battery chargers, charge current and charge voltage are automatically controlled to a constant level. This means that the current remains constant, irrespective of mains-voltage fluctuations, until the gassing voltage is reached, after which it drops sharply due to constant voltage limiting (Fig. 5). This means that the charge current can be considerably higher (in the example given: 3 h charging time with 50 A initial charge current). This method permits a high level of charge in a short time whereby over-charge is avoided. Similar results are obtained from battery chargers whose charge voltage is limited (WU curve), or which automatically switch to a lower current upon reaching a given limit voltage (WoW curve), or which terminate the charging process completely upon reaching this limit voltage (Wa curve).

Charge-voltage setting

The nominal voltages of the battery and the charger must be identical. Since the charge current results solely from the difference in voltage between the battery and battery charger, it is highly affected by the magnitude of the voltage differential. Undervoltage can lead to inadequate charging, and overvoltage on battery chargers without charge-voltage limitation can lead to over-charging. Maintenance-free batteries therefore should only be charged using voltage limiting, and if they are to be charged over a longer period, charging is to take place with a reduced voltage limit (e.g. 2.3 V/cell instead of 2.4 V/cell). On the other hand, if voltage limiting is applied, very cold batteries cannot be charged. Here, a higher charge voltage is necessary. This usually presents no problems,

and can be implemented for instance by a Summer/Winter change-over facility.

Direct charging in the vehicle

In modern vehicles, more and more electrical and electronic "aids" contribute to a high level of safety and comfort. Prime examples are the airbag, car telephone, car radio, and ECU, to name but a few. These highly sensitive devices though must be protected against voltage peaks when charging the battery. Previously, for charging, the battery had to be disconnected from the vehicle electrical system. Electronic battery chargers make this step a thing of the past. Since charging is faster and there is no danger to the vehicle's electronic equipment, this is of course a considerable help in the vehicle repair shop because:
- The time-consuming disconnection/ reconnection and removal/installation of the battery is no longer necessary,
- Data memories in the car radio, ECU's, car telephone, trip computer etc. are no longer erased due to battery disconnection,
- Electrical consumers (airbag, ECU's etc.) are protected,
- No danger of damage due to incorrect handling,

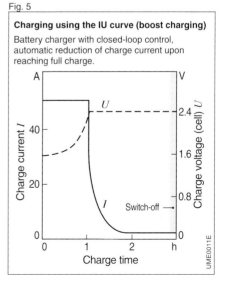

Fig. 4

Normal charging using the W curve

Battery charger without closed-loop control. Charge-current switch-off is therefore necessary upon reaching full charge.

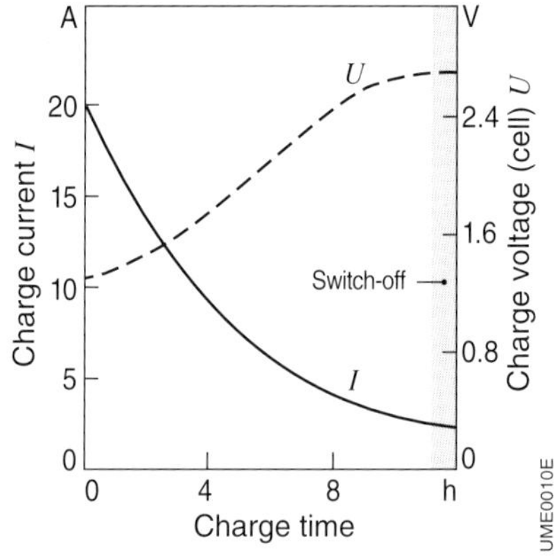

Fig. 5

Charging using the IU curve (boost charging)

Battery charger with closed-loop control, automatic reduction of charge current upon reaching full charge.

- No dangerous battery gassing during trickle charging,
- Batteries can be recharged without disconnecting the current consumers/loads (floating operation), and
- Brief charge times due to the high power reserves inherent in the IU/IWU charging curves.

Back-up operation
When replacing the battery, the LW 30 E battery charger is operated in the back-up mode to retain the data in the data storages of such equipment as car radio, car telephone etc. In this mode, current output is limited to approx. 2 A.

Incorrect-polarity protection
In case of falsely connected battery terminals, this protection facility prevents battery short-circuit and destruction of the battery charger.

Start-assist using the battery charger
The battery charger's start-assist function supports the battery when starting the engine. Its increased short-time power output makes it possible to supply the required high current.
Caution! Start-assist is only permitted on vehicles for which the manufacturer has not imposed limitations or bans on start assist in the operating instructions.

Start-assist with jumper cables
The battery from another vehicle can be used to provide start assistance. In such cases, the battery of the vehicle needing assistance must remain installed and connected, and manufacturer instructions must be complied with. For efficient starting assistance, only standardized jumper cables are to be used (DIN 72 553) with a conductor cross section of at least 16 mm² for SI engines and 25 mm² for diesel engines. Both of the batteries (or battery chargers) which are to be connected must have the same nominal voltage.
Working steps:
- Determine the cause of battery discharge. If the fault is in the electrical system, do not use start assistance since this can damage the battery

charger, or the battery and alternator of the vehicle providing the assistance.
- Connect the positive terminal of the empty battery to the positive terminal of the outside source of power.
- Connect the negative terminal of the outside source of power to a bare metallic surface (remote from the battery) on the vehicle receiving assistance, for instance to the engine ground braid.
- Check that the battery jumper cables are firmly attached (good electrical contact).
- Start the vehicle which is providing assistance. After a brief pause also start the disabled vehicle.
- Once starting has been successful, disconnect the cables in the opposite order to that given above.

Safety instructions

Before starting work on the vehicle's electrical system, or near the battery, the ground cable is to be disconnected provided all consumers/loads have been switched off. This is necessary in order to prevent short circuits (for instance with tools) which can cause sparks and burns. For the same reasons, particular care must be taken when connecting/disconnecting battery-charger cables or battery jumper cables. The following basic rules concerning safety must be followed when working with batteries:
- When handling sulphuric acid or when topping-up with distilled water, always wear protective goggles and rubber gloves.
- Do not fill up with electrolyte to above the MAX mark.
- Do not tip batteries to extreme angles from the vertical and do not keep them tilted for long periods.
- Due to the danger of electrolytic-gas detonation, smoking and naked flames are forbidden and sparks must be avoided when batteries are being charged (connect and disconnect cables in the prescribed order with battery charger switched off).
- Battery-charging rooms are to be well ventilated.

Special cases

Commercial-vehicle circuits

Parallel and series circuits

If a given vehicle requires a voltage higher than 12 V, a number of 12 V batteries can be connected in series. Two series-connected 12 V batteries provide 24 V, although their Ah capacity does not change.

On the other hand, if the A·h capacity is to be increased, the batteries must be connected in parallel, whereby in this case the voltage does not change. The A·h capacity is the sum of the individual capacities of the batteries. For parallel connection, positive pole is connected to positive pole and negative to negative. In order to achieve uniform current distri-bution when charging and discharging the batteries, it is advisable to ensure that the battery Ah capacities differ as little as possible from each other. From the wiring viewpoint, connections should also be as symmetrical as possible, with identical lengths of connection cable and conductor cross sections.

Battery changeover 12/24 V

A number of heavy trucks have mixed 12/24 V installations. In such cases, all electrical components including the alternator are designed for nominal 12 V operation. The starting motor on the other hand is designed for nominal 24 V so that it can generate the cranking power necessary for starting large diesel engines. The installation is equipped with a battery changeover relay and two 12 V batteries. Under normal driving conditions, or when the engine is not turning, the two batteries are connected in parallel for sup-

Fig. 1

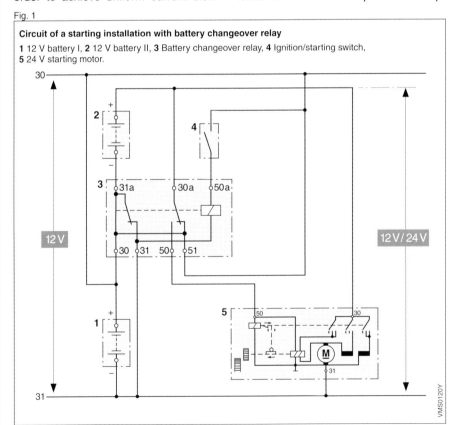

Circuit of a starting installation with battery changeover relay
1 12 V battery I, **2** 12 V battery II, **3** Battery changeover relay, **4** Ignition/starting switch,
5 24 V starting motor.

plying the electrical loads. The vehicle system voltage is 12 V.

When the starting switch is turned, the battery changeover relay automatically switches the two batteries in series so that 24 V is applied across the starting-motor terminals during the cranking process (Fig. 1). All other loads/consumers are still supplied with 12 V.

As soon as the engine has started, that is when the starting switch has been released and the starting motor switched off, the battery changeover relay automatically connects the batteries in parallel again. With the engine turning, the 12 V alternator recharges both batteries.

Components

Battery master switch

Generally, the vehicle's electrical installation is wired such that when the key is pulled from the starting switch the electrical lines leading from the switch are "dead". That is, the lines to the ignition system, the ECU (Motronic, ABS), and to the wipers etc. On the other hand, the lines leading to the ignition/starting switch, to the starting motor, and to the light switch remain "live". In other words there is still voltage on these lines, and if they have frayed or worn-through points these can lead to low resistances which

can cause leakage currents or short circuits which result in a discharged battery. And there is even the possibility of fire breaking out. These dangers can be alleviated by fitting a battery master switch (Fig. 2).

The single-pole battery master switch is installed in the battery's ground line (negative terminal) as near to the battery as possible. It should be within convenient reach of the driver. On installations equipped with alternators, due to the danger of voltage peaks (with the attendant destruction of electronic components), it is forbidden to operate the vehicle without the battery connected. On such installations therefore, the battery master switch may only be actuated with the engine at standstill.

Battery relay

Legislation stipulates that in buses, road tankers etc., a battery relay (Fig. 3) must be installed as the master switch to separate the vehicle electrical system from the battery. Not only short circuits are avoided (during repair for instance), but also the decomposition effects due to creepage currents on current-carrying components.

For this type of installation with alternator, in order to prevent excessive voltage peaks it is necessary to fit a 2-pole electromagnetic battery master switch. This prevents the alternator being sepa-

Fig. 2

Battery master switch

Fig. 3

Battery relay

VKE0583Y

VKE0159Y

rated from the battery when the engine is running.

Battery changeover relay

The battery changeover relay is used for connecting two 12 V batteries in series or in parallel. It is used for instance in commercial vehicles which have a 12 V electrical system but a 24 V starting motor (see "Commercial-vehicle circuits").

Battery cutoff relay

The battery cutoff relay separates the starter battery from a second battery used for ancillary equipment. It protects the starter battery against discharge when the alternator is not delivering charge current. This relay is provided with a diode for protection against false polarity, and a decay diode to suppress the inductive voltage peaks caused by switching (Fig. 4).

Battery charging relay

The battery charging relay is needed when an additional 12 V battery is to be charged in a 24 V vehicle system. It is provided with resistors across which at 10 A a voltage drop takes place which reduces the charging voltage to 12 V. This of course necessitates the 24 V alternator being able to generate the additional 10 A.

Fig. 4

Installation with battery cutoff relay

G1 Battery for ancillary equipment, **G2** Starter battery, **G3** Alternator,
H Charge-indicator lamp, **K** Battery cutoff relay, **M** Starting motor,
N Alternator regulator, **S1** Ignition/starting switch, **S2** Driving switch.

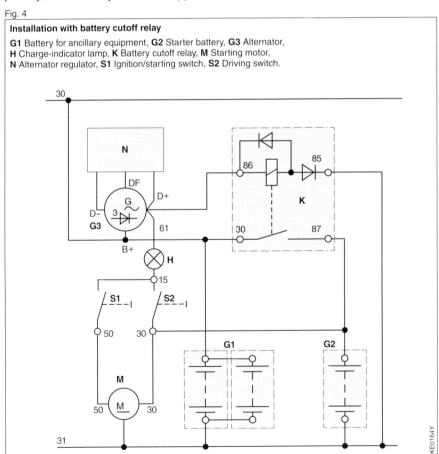

Battery history

When considering the history of the battery, a number of renowned scientists and inventors deserve particular credit. Above all, such personalities as Luigi Galvani (1789), Alessandro Graf Volta (around 1800), Johan Ritter (around 1800), Gaston Planté (1859) and Camille Faure were decisive in forcing the development of the battery (accumulator). At the end of the 19th century grid plates were already being manufactured, the basic principles of which are still retained in today's lead batteries.

In other words, basically speaking the lead battery has hardly changed up to the present day: Still cells, still grid plates, still sulphuric acid.

But that's way off the mark! If we examine the battery more closely, we find that not only has the energy density increased immensely, but also that the previously used materials (separators and case were of wood) have been superseded to a great extent by plastic, and 100% freedom from maintenance is today standard practice for starter batteries.

And in exceptional cases, these can achieve a service life comparable to that of the automobile itself.

1905: The first batteries were fitted in a motor vehicle (at first only for lighting purposes).
1914: The first starter battery was fitted in a motor vehicle.
1922: The first Bosch motorcycle batteries were installed.
1926: The first Bosch battery charger was introduced.
1927: Starting in 1927, Bosch also developed automotive starter batteries, and in 1936 started to mass-produce them.
After the Second World War, the development of Bosch automotive batteries was marked by
– The introduction of plastics to battery manufacture (e.g. polystyrene in 1955; polypropylene in 1971),
– The improvement of individual battery components (e.g. the folded-rib separator in 1956; the one-piece battery cover for 6 V batteries in 1964 and for 12 V batteries in 1966; direct cell connectors in 1971; expanded-metal technology for the negative grid in 1985), and
– The production of special-type batteries (e.g. "deep-cycle resistant" in 1969; "low maintenance" in 1979; "vibration-proof" in 1980; "maintenance-free" in 1982, and "100% maintenance-free" in 1988).

Starter battery: 1951 version

1 Joining bar,
2 Vent plug,
3 Terminal post,
4 Cell cover,
5 Sealing compound,
6 Plate strap,
7 Negative plate,
8 Wooden separator,
9 Hard-rubber separator,
10 Positive plate,
11 Cell connector,
13 Battery case.

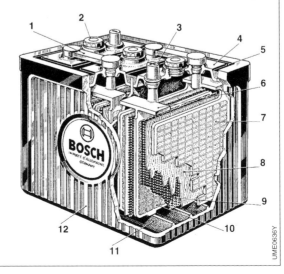

Drive/traction batteries

Electric drive

Applications

The electric drive is an alternative drive system for all vehicles for which low levels of noise and exhaust gas are stipulated, but for which relatively restricted ranges of operation suffice. Depending upon requirements, a difference is made between electric drive (and therefore the drive battery) for industrial trucks and similar vehicles and that for road vehicles:

– Industrial trucks for transport inside business premises (e.g. fork-lift trucks), railway-station luggage trolleys, vehicles for handicapped persons, passenger and materials transport vehicles for sporting and similar events, and sweeping machines etc. are examples of electrically driven vehicles (electric vehicles: EV) which only travel short distances, and which are not licensed for use on public roads. Up to now, electric drive using conventional drive batteries was adequate for such applications.

– There is still no suitable battery available as an alternative source of energy to achieve ranges which can compare with those of vehicles powered by an IC engine. This means therefore that when considering medium and long distances, there are still limitations on the operation of electrically powered road vehicles such as passenger cars and light trucks. The source of energy for such a road-vehicle drive must feature a higher energy density than that of the starter battery.

This drive battery (also known as a traction battery) not only provides the vehicle electrical system with energy while the vehicle is being driven, but also supplies the energy which is actually needed for traction purposes.

Components

Generally, an electric-vehicle drive comprises the following major components (Fig. 1):
– Drive battery, drive control system,
– Motor(s), and
– Gearbox.
The drive control system converts the

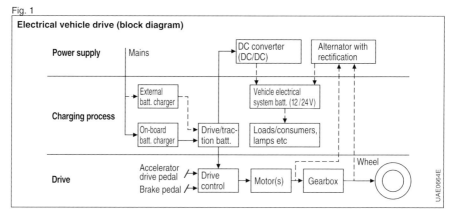

Fig. 1

Electrical vehicle drive (block diagram)

Power supply	Mains → DC converter (DC/DC), Alternator with rectification
Charging process	External batt. charger, On-board batt. charger, Drive/traction batt., Vehicle electrical system batt. (12/24V), Loads/consumers, lamps etc
Drive	Accelerator drive pedal, Brake pedal → Drive control → Motor(s) → Gearbox → Wheel

UAE0664E

accelerator-pedal position to correspond-ing voltage and current values at the motor. Usually, the drive torque is select-ed with the accelerator pedal the same as with an IC engine.

In contrast to the IC engine though, when considering electric drives a difference must be made between the power which is available for a short time only, and that which is available for an extended period of time. For electric-vehicle (EV) drives, the so-called "half-hour power" is used to define the maximum power available for an extended period. It is limited by the maximum permissible motor tempera-ture. For the majority of battery systems, it is necessary to distinguish between short-time and continuous output (factor 1.5 ... 3 depending upon type of drive).

DC series-motor drive

This electric drive demands only the simplest form of drive control system. The motor voltage is set to a value cor-responding to the required motor current. Additional components are needed if braking-energy recovery is to take place, and in the case of road-going EV's fitted with this system it would in fact be necessary to utlilize a multi-stage gear-box. Notwithstanding this disadvantage, and even though its efficiency level is relatively low, this form of drive is still very often used in industrial trucks. This has a number of decisive reasons:
- Simple design,
- Low costs, and
- Due to the low maximum speeds in-volved, the possibility of using a single-speed step-down gearbox.

Externally-excited DC drive

With this form of drive, the motor's mag-netic excitation is set by an integral ac-tuator (field rheostat). The commutating pole required by this motor makes it more complicated than the series motor. Due to the mechanical commutator, motor speeds are limited to approx. 7000 min^{-1}. This form of drive is often equipped with a multi-speed gearbox in order to keep down the motor costs and weight. Effi-cient braking-energy recovery is pos-

sible, and no extra components are re-quired.

Asynchronous drive

The asynchronous drive features the simplest and lowest-cost motor design. In principle, the costs for the drive-control system for 3-phase drives are higher than those for DC drives. Since a mecha-nical commutator is unnecessary, appro-priate motor designs permit speeds up to 20000 min^{-1}. Drive concepts with a single step-down ratio for road-going ve-hicles and high-efficiency braking-energy recovery are possible.

Permanent-magnet synchronous drive

Due to the use of permanent magnets for generating the excitation field, this drive-system variant achieves very high effi-ciency levels even in the part-load range. The fact that speed-adjustment ranges comparable to those of asynchronous drives are not possible, means that this drive must be combined with a two-speed or multi-speed gearbox.

Power supply

Even if the EV population were to in-crease rapidly, there would be no prob-lems with the infrastructure for the supply of battery-charging power.

The batteries of an electrically powered passenger car or small delivery van can be charged easily at all domestic wall sockets which, with the fuses normally fitted, permit a maximum charging power of 3.7 kVA. Theoretically speaking, an on-board battery charger with the above rating would only need about 4 hours to charge a 10 kWh battery. On the other hand though, longer charging times are needed for a full charge when the various battery-charging characteristics are taken into account. For the above example, the time taken would be about 6...8 hours. Three-phase AC connections permit the use of high-rating battery chargers, so that for certain battery types the charging time can be considerably reduced.

Battery systems

Overview

Dictated by costs considerations, the lead-acid storage battery at present still predominates on all types of EV. In addition to this type of battery, the cadmium-nickel battery is also used for traction purposes, for instance in driverless transport systems and is at present undergoing testing in the passenger car (Table 1). A number of other battery systems are on the threshold of volume production.

Lead storage battery

Basically speaking, the design of the lead-acid battery (with liquid electrolyte) is identical to that of the conventional starter battery, although material composition and actual design of the battery cell are aligned to the specific traction requirements.

The electrolyte as used in the lead-gel battery comprises diluted sulphuric acid bound in a multi-component gel. This type of battery is characterized by very high resistance to deep cycling. Self-discharge is very low and is less than 2% per month at 20 °C. In other words, the battery can be stored as long as 18 months following production. It is 100% maintenance-free and unspillable (Table 2). Safety valves in the battery cover release excess pressure which may build up in the battery, and at the same time prevent it drying-out. Under normal conditions, road-going EV's powered by lead-acid batteries have ranges of approx. 50 ... 70 km for each

Table 2: **12 V traction batteries from Bosch.**

Lead-acid battery		Lead-gel battery	
Capacity A·h	Model[1]	Capacity A·h	Model[1]
60	WA	48	WF
70	WA	60	WF
80	WA	75	WF
90	WA	85	WF
125	WA	200	WF
130	WA		
180	WA		
230	WA		

[1] WA Low-maintenance, WF 100% maintenance-free.

battery charge (see Table 1). Recharging when the vehicle is not in use permits the range to be increased considerably.

With lead storage batteries, the amount of energy which can be taken from the battery drops along with falling temperature. Depending upon the climatic conditions, some form of heating becomes necessary in winter in order to prevent a severe drop in vehicle range. Due to the battery's high thermal capacity, the battery warm-up which takes place during charging usually suffices as the "heating". The fact that the electrolyte takes part in the chemical reactions involved in the lead battery means that the amount of current that can be drawn from these is a function of the discharge time. If, for instance, the discharge time is reduced from 2 hours to 1 hour the available capacity is reduced by approx. 20%. In other words, the battery's Ah capacity can only be utilized 100% if the discharge time is chosen accordingly. For road-going EV's, the mean discharge time is 2 hours or less.

Fleet tests with EV passenger cars have proved that the batteries used can have

Table 1: **Automotive applications: Practical examples.**

Vehicle type	Battery type	Acceleration 0...50 km/h	Max. speed	Typical range per charge	Unladen weight	Pay-load	Typical mains/line power consumption
Van	Lead-acid	12 s	80 km/h	70 km	2,400 kg	800 kg	40 kW·h/100 km
Pass. car	Lead-gel	12 s	100 km/h	60 km	1,500 kg	350 kg	25 kW·h/100 km
Pass. car	Ni-Cd	9 s	90 km/h	80 km	1,050 kg	300 kg	18 kW·h/100 km

a service life of approx. 5 years or 700 cycles.

Future battery systems

Nickel-battery systems

In the case of nickel batteries, closed systems are employed for powering equipment and appliances, whereas open cadmium-nickel cells are often used for traction applications. The low cell voltage of only 1.2 V necessitates a high proportion of inactive components. A battery service life of up to 10 years, or approx. 2,000 cycles appears possible, but testing under automotive conditions has not yet been completed.

Ni-Cd batteries, no matter whether of the open or closed type, must be cooled in road-going EV's. Heating is unnecessary unless the operating temperature drops below −20°C. The battery's Ah capacity is practically independent of the discharge time.

Typical ranges for passenger-car EV's powered by nickel-cadmium batteries are approx. 80...100 km.

The nickel-metal-hybrid system is a promising new battery development. Here, the cadmium of the Ni-Cd battery is replaced by hydrogen which requires a storage medium composed of a number of different metals. Compared to the Ni-Cd battery, the nickel-metal-hybrid battery features a slightly higher energy density together with a longer service life.

Generally speaking, nickel batteries have a higher energy density coupled with a longer service life, and this makes them particularly interesting for hybrid vehicles.

Natrium-battery systems

Both the natrium-nickel-chloride and the natrium-sulphur battery have a solid electrolyte comprised of an ion-conducting aluminum ceramic.

To be able to participate in the chemical reaction, the electrodes (which are solid at ambient temperatures) must be liquified by means of high operating temperatures. For both natrium-battery systems the usual operating temperature is approx. 300 °C, whereby the natrium-nickel-chloride battery can be operated at lower temperatures than the natrium-sulphur system. Some form of "super insulation" must be provided in order to keep the heat losses of these battery systems within justifiable limits. Both systems have been subjected to intensive safety tests and have proved their suitability for use in road-going EV's. With natrium-battery systems, road-going EV's become feasible for ranges well in excess of 100 km.

Lithium-battery systems

Whereas the lithium-battery systems permit aproximately the same energy density as the natrium-battery systems, they can be operated at ambient temperatures. They also feature higher cell voltages of 3.5 V.

Table 3: **New battery systems** (discharge time 2 hrs, charge time 8 hrs).

	Lead-gel battery systems	Nickel-battery systems	Natrium-battery systems	Lithium-battery systems
Cell voltage	2 V	1.2 V	2...2.5 V	3.5 V
Energy density	25...30 W·h/kg	50...80 W·h/kg	90...100 W·h/kg	ca. 100 W·h/kg
Energy efficiency without heating	70...85 %	60...85 %	80...90 %	85...90 %
Service life in cycles	600...900	Projected 1,000...2,000	Projected 1,000	Projected > 1,000
Maintenance-free	Yes	Partially	Yes	Yes
Operating temp.	0...55 °C	−20...55 °C	300...380 °C	−20...60 °C

Alternators

Generation of electrical power in the motor vehicle

On-board electrical power

In order to supply the power required for the starter, for ignition and fuel-injection systems, for the ECU's to control the electronic equipment, for lighting, and for safety and convenience electronics, motor vehicles need their own efficient and highly reliable source of energy which must always be available at any time of day or night.

Whereas, with the engine stopped, the battery is the vehicle's energy store, the alternator becomes the on-board "electricity generating plant" when the engine is running. The alternator's task is to supply power to all the vehicle's current-consuming loads and systems (Fig. 1).

Fig. 1

Alternator principle

The alternator rectifies the 3-phase alternating current to provide DC for supplying the electrical devices and for charging the battery.

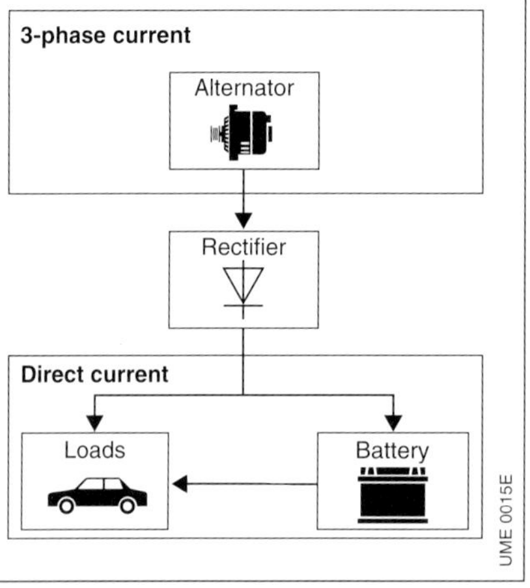

In order that the entire system is reliable and trouble-free in operation, it is necessary that the alternator output, battery capacity, and starter power requirements, together with the remaining electrical loads, are matched to each other as optimally as possible. For instance, following a normal driving cycle (e.g., town driving in winter) the battery must always be sufficiently charged so that the vehicle can be started again without any trouble no matter what the temperature. And the ECU's, sensors and actuators for the vehicle's electronic systems (e.g., for fuel management, ignition, Motronic, electronic engine-power control, antilock braking system (ABS), traction control (TCS) etc. must always be ready for operation.

Apart from this, the vehicle's safety and security systems as well as its signalling systems must function immediately, the same as the lighting system at night or in fog. Furthermore, the driver-information and convenience systems must always function, and with the vehicle parked, a number of electrical loads should continue to operate for a reasonable period without discharging the battery so far that the vehicle cannot be started again.

As a matter of course, millions of motorists expect their vehicle to always be fully functional, and demand a high level of operational reliability from the electrical system. For many thousands of miles – in both summer and winter.

Electrical loads

The electrical loads have differing duty cycles (Fig. 2). A differentiation is made between permanent loads (ignition, fuel injection etc.), long-time loads (lighting, car radio, vehicle heater etc.), and short-time loads (turn signals, stop lamps etc.). Some electrical loads are only switched

Fig. 2

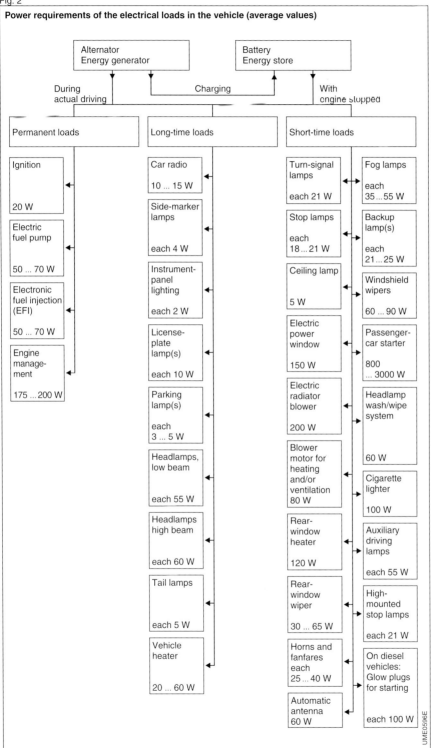

on according to season (air-conditioner in summer, seat heater in winter). And the operation of electrical radiator fans depends upon temperaure and driving conditions.

Charge-balance calculation

A computer program is used to determine the state of battery charge at the end of a typical driving cycle. Here, such influences as battery size, alternator size, and load input powers must be taken into account.

Rush-hour driving (low engine speeds) combined with winter operation (low charging-current input to the battery) is regarded as a normal passenger-car driving cycle. In the case of vehicles equipped with an air conditioner, summer operation can be even more unfavorable than winter.

Vehicle electrical system

The nature of the wiring between alternator, battery, and electrical equipment also influences the voltage level and, as a result, the state of battery charge.

If all electrical loads are connected at the battery, the total current (sum of battery charging current and load current) flows through the charging line, and the resulting high voltage drop causes a reduction in the charging voltage.

Conversely, if all electrical devices are connected at the alternator side, the voltage drop is less and the charging voltage is higher. This may have a negative effect upon devices which are sensitive to voltage peaks or high voltage ripple (electronic circuitry).

For this reason, it is advisable to connect voltage-insensitive equipment with high power inputs to the alternator, and voltage-sensitive equipment with low power inputs to the battery. Appropriate line cross-sections, and good connections whose contact resistances do not increase even after long periods of operation, contribute to keeping the voltage drop to a minimum.

Electrical power generation using alternators

The availability of reasonably priced power diodes as from around 1963, paved the way for Bosch to start with the series production of alternators. The alternator's design principle results in it having a far higher electromagnetic efficiency than the DC generator. This fact, together with the alternator's much wider rotational-speed range, enables it to deliver power, and cover the vehicle's increased power requirements, even at engine idle. Since the alternator speed can be matched to that of the engine by means of a suitable transmission, this means that the battery remains at a high charge level even in winter during frequent town driving.

The increased power requirements mentioned above, result from the following factors: The increase in the amount of electrical equipment fitted in the vehicle, the number of ECU's required for the electronic systems (e.g., for engine management and for chassis control), and the safety, security and convenience electronics. The expected power requirements up to the year 2000 are shown in Fig. 3.

Fig. 3

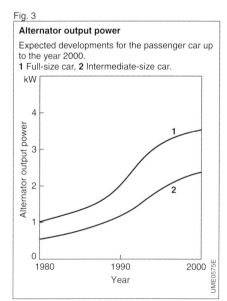

Alternator output power

Expected developments for the passenger car up to the year 2000.
1 Full-size car, **2** Intermediate-size car.

Apart from these factors, typical driving cycles have also changed, whereby the proportion of town driving with extended stops at idle has increased (Fig. 4).

The rise in traffic density has led to frequent traffic jams, and together with long stops at traffic lights, this means that the alternator also operates for much of the time at low speeds corresponding to engine idle. Together with the fact that longer journeys at higher speeds have become less common, this has a negative effect on the battery's charge balance.

It is imperative that the battery continues to be charged even when the engine is idling.

At engine idle, an alternator already delivers at least a third of its rated power (Fig. 5).

Alternators are designed to generate charging voltages of 14V (28V for commercial verhicles). The three-phase winding is incorporated in the stator, and the excitation winding in the rotor.

The three-phase AC generated by the alternator must be rectified, the rectifiers also preventing battery discharge when the vehicle is stationary. The additional relay as required for the DC generator can be dispensed with.

Design factors

Rotational speed
A generator's efficiency (energy generated per kg mass) increases with rotational speed. This factor dictates as high a conversion ratio as possible between engine crankshaft and alternator. For passenger-cars, typical values are between 1:2 and 1:3, and for commercial vehicles up to 1:5.

Temperature
The losses in the alternator lead to increased alternator-component temperatures. The input of fresh air to the alternator is a suitable means of reducing component temperature and increasing alternator service life and efficiency.

Vibration
Depending upon installation conditions and the engine's vibration patterns, vibration accelerations of between 500...800 m/s^2 can occur at the alternator. Critical resonances must be avoided.

Further influences
The alternator is also subjected to such detrimental influences as spray water, dirt, oil, fuel mist, and road salt.

Fig. 4

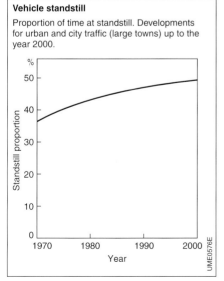

Vehicle standstill

Proportion of time at standstill. Developments for urban and city traffic (large towns) up to the year 2000.

Standstill proportion (%) vs Year

UME0576E

Fig. 5

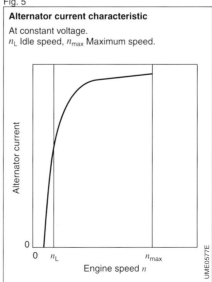

Alternator current characteristic

At constant voltage.
n_L Idle speed, n_{max} Maximum speed.

Alternator current vs Engine speed n

UME0577E

Electrical power generation using DC generators

Originally, the conventional lead-acid battery customarily fitted in motor vehicles led to the development of the DC generator, and for a long time this generator system was able to meet the majority of the demands made upon it. Consequently, until the middle of the seventies, most vehicles were equipped with DC generators. Today though, these have become virtually insignificant in the automotive sector and will not be dealt with in detail here.

With the DC generator, it proved to be more practical to rotate the magnetic lines of force, while locating the electrically excited magnet system in the stationary housing. The alternating current generated by the machine can be rectified relatively simply by mechanical means using a commutator, and the resulting direct current supplied to the vehicle electrical system or the battery.

Requirements to be met by automotive generators

The type and construction of an automotive electrical generator are determined by the necessity of providing electrical energy for powering the vehicle's electrical equipment, and for charging its battery.

Initially, the alternator generates alternating current (AC). The vehicle's electrical equipment though requires direct current (DC) for keeping the battery charged and for powering the electronic subassemblies. The electrical system must therefore be supplied with DC. The demands made upon an automotive generator are highly complex and varied:

- Supplying all connected loads with DC,
- Providing power reserves for rapidly charging the battery and keeping it charged, even when permanent loads are switched on,
- Maintaining the voltage output as constant as possible across the complete engine speed range independent of the generator's loading,
- Rugged construction to withstand all the under-hood stresses (e.g., vibration, high ambient temperatures, temperature changes, dirt, dampness etc.),
- Low weight,
- Compact dimensions for ease of installation,
- Long service life,
- Low noise level,
- A high level of efficiency.

Characteristics (summary)

The alternator's most important characteristics are:

- It generates power even at engine idle.
- Rectification of the AC uses power diodes in a three-phase bridge circuit.
- The diodes separate alternator and battery from the vehicle electrical system when the alternator voltage drops below the battery voltage.
- The alternator's far higher level of mechanical efficiency means that they are designed to be far lighter than DC generators.
- Alternators feature a long service life. The passenger-car alternator's service life corresponds roughly to that of the engine (up to 150,000 km), which means that no servicing is necessary during this period.
- On vehicles designed for high mileages (trucks and commercial vehicles in general), brushless alternator versions are used which permit regreasing. Or bearings with grease-reserve chambers are fitted.
- Alternators are able to withstand such external influences as vibration, high temperatures, dirt, dampness.
- Normally, operation is possible in either direction of rotation without special measures being necessary, when the fan shape is adapted to the direction of rotation.

Basic principles

Electrodynamic principle

The basis for the generation of electricity is electromagnetic "induction". The principle is as follows: When an electric conductor (wire or wire loop) cuts through the lines of force of a DC magnetic field, a voltage is induced in the conductor. It is immaterial whether the magnetic field remains stationary and the conductor rotates, or vice versa.

A wire loop is rotated between the North and South poles of a permanent magnet. The ends of this wire loop are connected through collector rings and carbon brushes to a voltmeter. The continuously varying relationship of the wire loop to the poles is reflected in the varying voltage shown by the voltmeter. If the wire loop rotates uniformly, a sinusoidal voltage curve is generated whose maximum values occur at intervals of 180°. Alternating current (AC) flows as soon as the circuit is closed (Fig. 1).

How is the magnetic field generated?

The magnetic field can be generated by permanent magnets. Due to their simplicity, these have the advantage of requiring a minimum of technical outlay, and are used for small generators (e.g., bicycle dynamos).

On the other hand, electromagnets through which DC current flows permit considerably higher voltages and are controllable. This is why they are applied for generation of the (exciter) magnetic field.

Electromagnetism is based on the fact that when an electric current flows through wires or windings these are surrounded by a magnetic field.

The number of turns in the winding and the magnitude of the current flowing through it determine the magnetic field's strength, which can be further increased by using a magnetisable iron core, which when it rotates induces an alternating voltage in the armature coil. In practical generator applications, in order to increase the effects of induction, a number of wire loops are used to form the "winding".

When this principle is applied to the generator or alternator, a decisive advantage lies in the fact that the magnetic field, and with it the induced

Fig. 1

Induced single-phase alternating voltage

Voltage curve generated during one revolution of a winding rotating in a magnetic field.
The position of the rotor as shown on the left corresponds to Position 3.

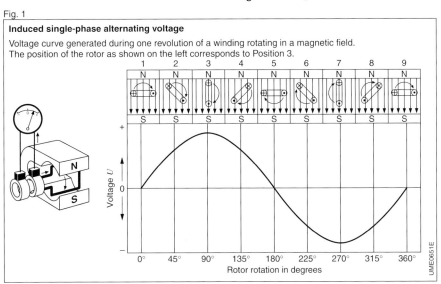

UME0651E

voltage, can be strengthened or weakened by increasing or decreasing the (excitation) current flowing in the (excitation) winding.

Except for a slight residual or remanence magnetism, the electromagnet in the form of the excitation winding loses its magnetism when the excitation current is switched off. If an external source of energy (e.g., battery) provides the excitation current, this is termed "external excitation". If the excitation current is taken from the machine's own electric circuit this is termed "self-excitation".

In electric machines, the complete rotating system comprising winding and iron core is referred to as the rotor.

Principle of the alternator

3-phase current (3-phase AC, Fig. 2) is also generated by rotating the rotor in a magnetic field, the same as with single-phase AC as described above. One of the advantages of 3-phase AC lies in the fact that it makes more efficient use of the electrical generator's potential. The generator for 3-phase AC is designated an "alternator" and its armature comprises three identical windings (u, v, w) which are offset from each other by 120°.

The start points of the three windings are usually designated u, v, w, and the end points x, y, z. In accordance with the laws of induction, when the rotor rotates in the magnetic field, sinusoidal voltages are generated in each of its three windings. These voltages are of identical magnitude and frequency, the only difference being that their 120° offset results in the induced voltages also being 120° out-of-phase with each other, as well as being out-of-phase by 120° as regards time.

Therefore, with the rotor turning, the alternator generates a constantly recurring 3-phase alternating voltage.

Normally, with the windings not connected, an alternator would require 6 wires to output the electrical energy that it has generated (Fig. 3a). However, by interconnecting the 3 circuits the number of wires can be reduced from 6 to 3. This joint use of the conductors is achieved by the "star" connection (Fig. 3b) or "delta" connection (Fig. 3c).

In the case of the "star" connection the ends of the 3 winding phases are joined to form a "star" point. Without a neutral conductor, the sum of the 3 currents towards the "star" is always 0 at any instant in time.

Discussions up to this point have centered on the alternator version with

Fig. 2

Induced three-phase alternating voltage

Voltage curve generated during one revolution of three windings (phases) rotating in a magnetic field.
The windings are offset from each other by 120°.
The connection of the individual phase voltages results in a 3-phase alternating voltage.

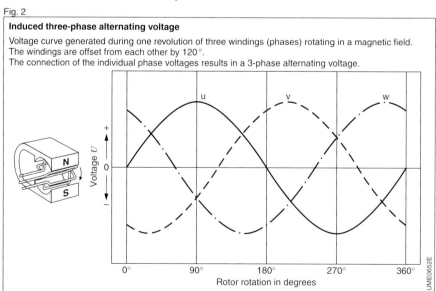

stationary excitation field and rotating armature winding in which the load current is induced.

For automotive alternators though, the 3-phase (star or delta connected) winding system is in the stator (the stationary part of the alternator housing) so that the winding is often also referred to as the stator winding.

The poles of the magnet together with the excitation winding are situated on the rotor.

The rotor's magnetic field builds up as soon as excitation current flows through the excitation winding.

When the rotor rotates, its magnetic field induces a 3-phase alternating voltage in the stator windings which provides the 3-phase current to power the connected loads.

Rectification of the AC voltage

The 3-phase AC generated by the alternator cannot be stored in the vehicle's battery nor can it be used to power the electronic components and ECU's. To do so, the three-phase AC must first of all be rectified. One of the essential prerequisites for this rectification is the availability of high-performance power diodes which can operate efficiently throughout a wide temperature range.

Rectifier diodes have a reverse and a forward direction, the latter being indicated by the arrow in the symbol. A diode can be compared to a non-return valve which permits passage of a fluid or a gas in only one direction and stops it in the other.

The rectifier diode suppresses the negative half waves and allows only positive half waves to pass. The result is a pulsating direct current. So-called full-wave rectification is applied in order to also make full use of all the half waves including those that have been suppressed.

Fig. 3

Connection of the three alternator windings

a Windings not connected.
b Star connection. Alternator voltage U and phase voltage U_p differ by the factor $\sqrt{3} = 1.73$.
The alternator current I equals the phase current I_p
$U = U_p \cdot \sqrt{3} \cdot I = I_p$.
c Delta connection. Alternator voltage U equals the phase voltage U_p.
The alternator current I and the phase current I_p differ by the factor $\sqrt{3} = 1,73$.
$U = U_p. \ I = I_p \cdot \sqrt{3}$.

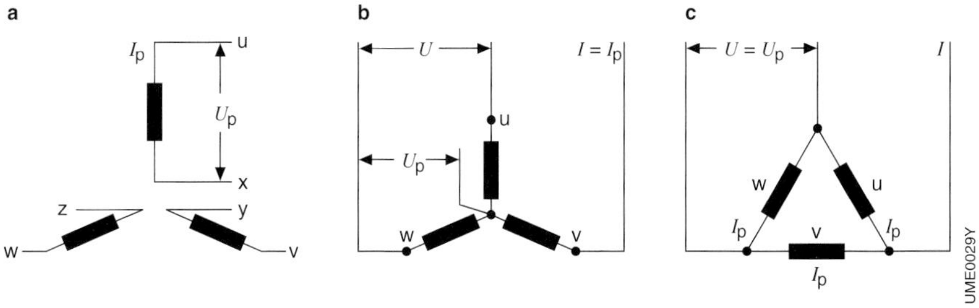

UME0029Y

Bridge circuit for the rectification of the 3-phase AC

The operating principle of the diode in the rectification of an alternating current is shown in Fig. 4.

Half-wave rectification is shown in Fig. 4a and full-wave rectification in Fig. 4b.

The AC generated in the 3 windings of the alternator is rectified in an AC bridge circuit using 6 diodes (Fig. 5).

Two power diodes are connected into each phase, one diode to the positive side (Term. B+) and one to the negative side (Term. B–).

The positive half-waves pass through the positive-side diodes and the negative half waves through the negative-side diodes. With full-wave rectification using a bridge circuit, the positive and negative half-wave envelopes are added to form a rectified alternator voltage with a slight ripple (Fig. 5c).

This means that the direct current (DC) which is taken from the alternator at Terminals B+ and B– to supply the vehicle electrical system is not ideally "smooth" but has a slight ripple. This ripple is further smoothed by the battery, which is connected in parallel to the alternator, and by any capacitors in the vehicle electrical system.

The excitation current which magnetizes the poles of the excitation field is tapped off from the stator winding and rectified by a full-wave bridge rectifier comprising the 3 "exciter diodes" at Term.

Fig. 4

Rectifier circuits

a Half-wave rectification, **b** Full-wave rectification.
$U_{G\sim}$ AC voltage before the diodes,
U_{G-} Pulsating DC voltage after the diodes.
1 Battery, **2** Excitation winding (G),
3 Stator winding, **4** Rectifier diodes.

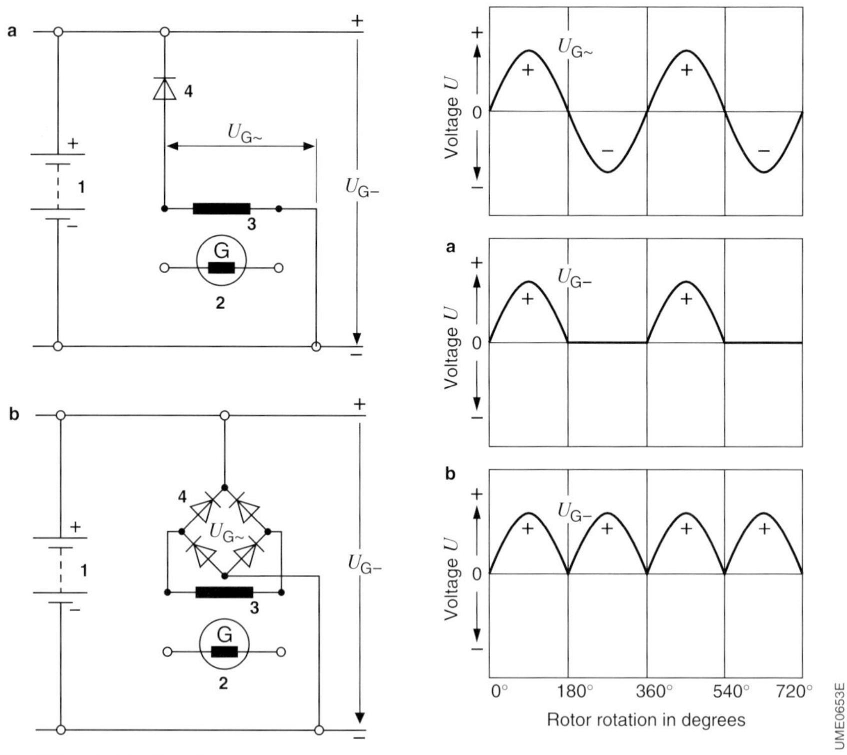

0°   180°   360°   540°   720°
Rotor rotation in degrees

UME0653E

D+, and the 3 power diodes at Term. B-(negative side).

With the aim of increasing power output at high speeds (above 3,000 min⁻¹), with star-connected versions so-called "auxiliary diodes" can be used to make full use of the alternator voltage's harmonic component.

Reverse-current block

The rectifier diodes in the alternator not only rectify the alternator and excitation voltage, but also prevent the battery discharging through the 3-phase winding in the stator. With the engine stopped, or with it turning too slowly for self-excitation to take place (e.g., during cranking), without the diodes battery current would flow through the stator winding.

With respect to the battery current, the diodes are polarized in the reverse direction so that it is impossible for battery-discharge current to flow. Current flow can only take place from the alternator to the battery.

Fig. 5

3-phase bridge circuit

a Three-phase AC voltage,
b Formation of the alternator voltage by the envelope curves of the positive and negative half-waves,
c Rectified alternator voltage.
U_P Phase voltage,
U_G Voltage at the rectifier (negative not to ground),
U_{G-} Alternator DC voltage output (negative to ground),
U_{Gms} r.m.s. value of the alternator DC voltage output.
1 Battery,
2 Excitation winding,
3 Stator winding,
4 Positive diodes,
5 Negative diodes.

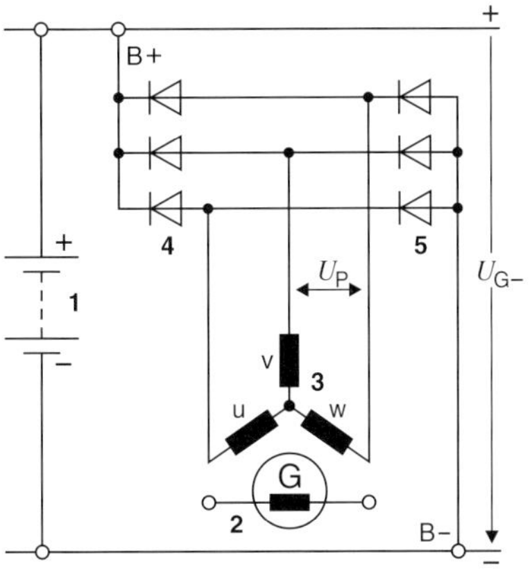

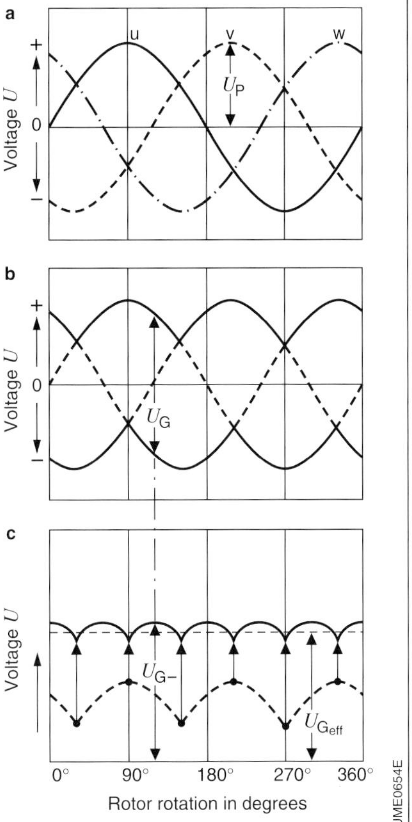

Rectifier diodes

Regarding their operation, the power diodes on the plus and negative sides are identical with each other. The difference between them lies merely in their special design for use as rectifiers in the alternator. They are termed positive and negative diodes, and in one case the diode's knurled metal casing acts as a cathode and in the other as an anode. The metal casing of the positive diode is pressed into the positive plate and functions as a cathode. It is connected to the battery's positive pole and conducts towards B+ (battery positive). The metal casing of the negative diode is pressed into the negative plate and functions as an anode. It is connected to ground (B–). The diode wire terminations are connected to the ends of the stator winding (Fig. 6). The positive and negative plates also function as heat sinks for cooling the diodes.

The power diodes can be in the form of Zener diodes which also serve to limit the voltage peaks which occur in the alternator due to extreme load changes (load-dump protection).

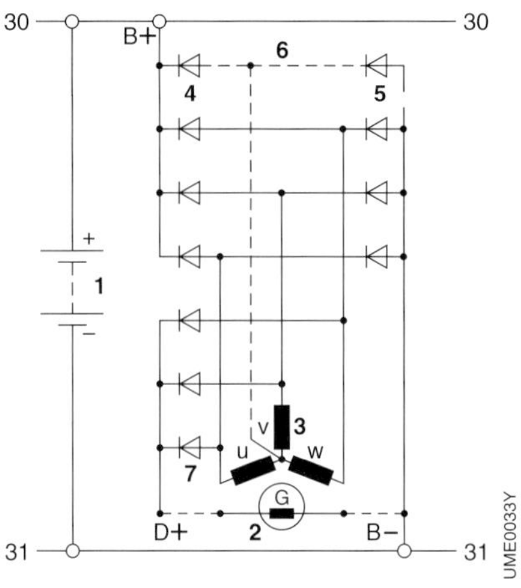

The circuits of the alternator

The following three circuits are standard for the alternator:

– Pre-excitation circuit (separate excitation using battery current)
– Excitation circuit (self-excitation)
– Generator or main circuit.

Pre-excitation circuit

When the ignition or driving switch (4) is operated, the battery current I_B first of all flows through the charge indicator lamp (3), through the excitation winding (1d) in the stator, and through the voltage regulator (2) to ground. In the rotor, this battery current serves to pre-excite the alternator.

Why is pre-excitation necessary ?

On most alternators, the remanence in excitation winding's iron core is very weak at the instant of starting and at low speeds, and does not suffice to provide the self-excitation needed for building up the magnetic field.

Fig. 6

Rectification of excitation current

1 Battery, **2** Excitation winding (G),
3 Stator winding, **4** Positive-plate diodes,
5 Negative-plate diodes, **6** Auxiliary diodes,
7 Exciter diodes.

Fig. 7

Pre-excitation circuit

1 Alternator, **1a** Exciter diodes, **1b** Positive-plate diodes, **1c** Negative-plate diodes,
1d Excitation winding, **2** Voltage regulator,
3 Charge indicator lamp, **4** Ignition switch,
5 Battery.

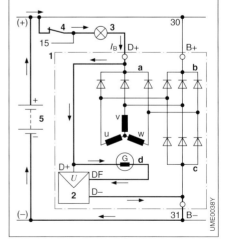

Self-excitation can only take place when the alternator voltage exceeds the voltage drop across the two diodes (2 x 0.7 = 1.4 V).

This serves to support the pre-excitation current which flows through the charge-indicator lamp from the battery. It generates a field in the rotor which in turn induces a voltage in the stator which is proportional to the rotor speed. When the engine is started, in order that alternator self-excitation can "get going", the engine must turn at a speed which enables the induced voltage to exceed the voltage drop across the diodes in the excitation circuit. Since the charge-indicator lamp increases the pre-excitation circuit resistance compared to that of the excitation circuit, this speed is above the engine idle speed. It is therefore affected by the charge-indicator lamp's wattage rating.

Charge indicator lamp

When the ignition or driving switch (4) is operated, the charge indicator lamp (3) in the pre-excitation circuit functions as a resistor and determines the magnitude of the pre-excitation current. A suitably dimensioned lamp provides a current which suffices to generate a sufficiently strong magnetic field to initiate self-excitation. If the lamp is too weak, as is the case for instance with electronic displays, a resistor must be connected in parallel to guarantee adequate alternator self-excitation. The lamp remains on as long as the alternator voltage is below battery voltage.

The lamp goes out the first time the speed is reached at which maximum alternator voltage is generated and the alternator starts to feed power into the vehicle electrical system.

Typical ratings for charge indicator lamps are:
2 W for 12 V systems,
3 W for 24 V systems.

Excitation circuit

During alternator operation, it is the task of the excitation current I_{exc} to generate a magnetic field in the rotor so that the required alternator voltage can be induced in the stator windings. Since alternators are "self-excited", the excitation current must be tapped off from the current flowing in the 3-phase winding.

Referring to Fig. 8, the excitation current I_{exc} takes the following route: Exciter diodes (1a), carbon brushes, collector rings, excitation winding, Term. DF of the voltage regulator (2), Term. D– of the voltage regulator, and back to the stator winding through the power diodes (1c). With the alternator operating, no external power source is required for self-excitation.

Generator circuit

The alternating voltage induced in the three phases of the alternator must be rectified by the power diodes in the bridge circuit before it is passed on to the battery and to the loads. The alternator current I_G flows from the three windings

Fig. 8

Excitation circuit

1 Alternator, **1a** Exciter diodes,
1b Positive-plate diodes,
1c Negative-plate diodes,
1d Excitation winding,
2 Voltage regulator, **3** Charge indicator lamp,
4 Ignition switch, **5** Battery.

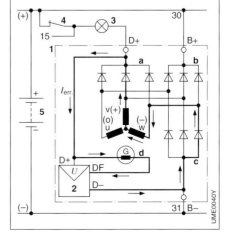

and through the respective power diodes to the battery and to the loads in the vehicle electrical system.

In other words the alternator current is divided into battery current and load current. In Fig. 10, the curves of the stator-winding voltages are shown as a function of the angle of rotation of the rotor. Taking a rotor with six pole pairs for instance, and an angle of rotation of 30°, the voltage referred to the star point at the end of winding v is positive, for winding w it is negative, and for winding u it is zero.

The resulting current path is shown in Fig. 9.

Current flows from the end of winding v and through the positive diodes (b) to alternator terminal B+ from where it flows through the battery, or the load, to ground (battery terminal B–) and via the negative diodes (c) to winding end w. Taking a 45° angle of rotation, current from the v and w winding ends takes the same path to winding end u. In this case, there is voltage present across all of the phases. Both examples though are momentary values. In reality, the phase voltages and currents continually change their magnitude and direction, whereas the DC supplied for battery charging and for the electrical loads always maintains the same direction.

This is due to the fact that irrespective of the rotor's position, all the diodes are always involved in the rectification process. For current to flow from the alternator to the battery, the alternator voltage must be slightly higher than that of the battery.

Voltage regulation

Why is it necessary to regulate the alternator voltage?

The regulator has the job of maintaining the alternator voltage, and thus the vehicle system voltage, at a constant level across the engine's complete speed range, independent of load and rotational speed.

The alternator voltage is highly dependent upon the alternator's speed and loading. Notwithstanding these continually changing operating conditions, steps must be taken to ensure that

Fig. 9

Alternator circuit

1 Alternator, **1a** Exciter diodes,
1b Positive-plate diodes,
1c Negative-plate diodes,
1d Excitation winding,
2 Voltage regulator, **3** Charge indicator lamp,
4 Ignition switch, **5** Battery.

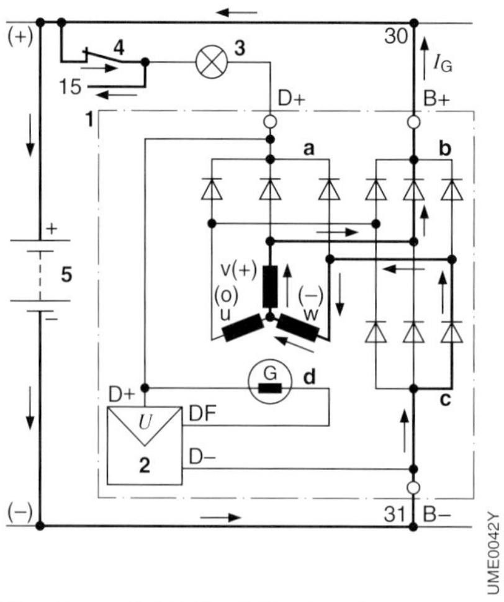

Fig. 10

Voltages in the stator windings

Voltage curves as a function of the angle of rotation of a rotor with 6 pole pairs.

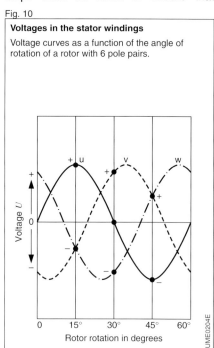

Rotor rotation in degrees

alternator voltage is limited to the specified level. This voltage limitation protects the electrical equipment against over-voltage, and prevents battery over-charge.

In addition, the battery's electro-chemical properties must be taken into account during charging. This means that normally the charging voltage must be slightly higher in cold weather in order to compensate for the fact that the battery is slightly more difficult to charge at low temperatures.

Principle of voltage regulation

The voltage generated by the alternator increases along with alternator speed and the excitation current. Considering a fully excited alternator which is not connected to the battery, and which is being driven without load, the voltage without regulation would increase linearly with alternator speed until it reaches about 140 V at a speed of 10,000 min^{-1}. The voltage regulator controls the level of the alternator's excitation current, and along with it the strength of the rotor's magnetic field as a function of the voltage generated by the alternator (Fig. 11).

This enables the alternator terminal voltage U_{G-} (between terminals B+ and B–) to be maintained constant up to the maximum current.

The voltage regulation tolerance zone for vehicle electrical systems with 12 V battery voltage is around 14 V, and for systems with 24 V battery voltage around 28 V. The regulator remains out of action as long as the alternator voltage remains below the regulator response voltage.

Within the tolerance range, if the voltage exceeds the specified upper value, the regulator interrupts the excitation current. Excitation becomes weaker and the alternator voltage drops as a result. As soon as the voltage then drops below the specified lower value, the regulator switches in the excitation current again, the excitation increases and along with it the alternator voltage. When the voltage exceeds the specified upper value again, the control cycle is repeated. Since these control cycles all take place within the milliseconds range, the alternator

Fig. 11

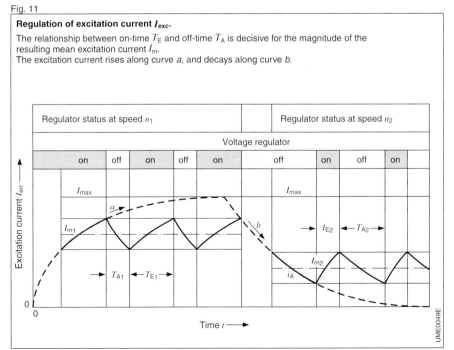

Regulation of excitation current I_{exc}.
The relationship between on-time T_E and off-time T_A is decisive for the magnitude of the resulting mean excitation current I_m.
The excitation current rises along curve a, and decays along curve b.

Regulator status at speed n_1					Regulator status at speed n_2			
Voltage regulator								
on	off	on	off	on	off	on	off	on

Excitation current I_{err}

Time t

UME0049E

mean voltage is regulated in accordance with the stipulated characteristic.

The infinitely variable adaptation to the various rotational speeds is automatic, and the relationship between the excitation current "On" and "Off" times is decisive for the level of the mean exciting current (Fig. 11). At low rotational speeds, the "On" time is relatively long and the "Off" time short, the exciting current is interrupted only very briefly and has a high average value. On the other hand, at high rotational speeds the "On" time is short and the "Off" time long. Only a low excitation current flows.

Influence of ambient temperature

The regulator characteristic curves, that is the alternator voltage as a function of temperature, are matched to the battery's chemical characteristics. This means that at low temperatures, the alternator voltage is increased slightly in order to improve battery charging in the winter, whereby the input voltages to the electronic equipment and the voltage-dependent service life of the light bulbs is taken into account. At higher temperatures, alternator voltage is reduced slightly in order to prevent battery overcharge in summer.

Fig. 12

Voltage-regulator characteristic

Permissible tolerance band for the alternator voltage (14 V) as a function of the intake-air temperature.

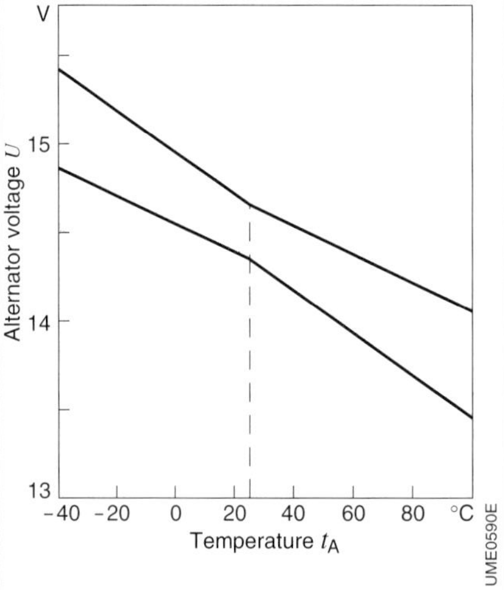

Temperature compensation is achieved through suitable choice of regulator components, e.g., of the Z-diode. Fig. 12 shows the characteristic curves for 14 V alternator voltage. The voltage level is 14.5 V and has an incline of −10 mV/K.

Alternator design

The theoretical principles and inter-relationships discussed so far are reflected in the technical construction of modern alternators. Individual versions can differ from each other in certain details according to their particular application. At present, the claw-pole alternator with compact diode assembly is still being installed in the majority of vehicles, but the compact alternator is coming more and more to the forefront. The major design differences between these two alternator types are the compact alternator's two internally-mounted fans, its smaller collector rings, and the location of the rectifier outside the collector-ring end shield.

The basic construction of a compact alternator is shown in Fig. 13:

− Stator (2) with 3-phase stator winding. The stator consists of mutually insulated, grooved laminations which are pressed together to form a solid laminated core. The turns of the stator winding are embedded in the grooves.

− Rotor (3), on the shaft of which are mounted the pole-wheel halves with claw-shaped magnet poles, the excitation winding, both fans, the ball bearings, and the two collector rings. The excitation winding consists of a single toroidal coil which is enclosed by the claw-pole halves. The relatively small excitation current is supplied via the carbon brushes which are pressed against the collector rings by springs.

− The pulley for the belt drive is also mounted on the rotor shaft. Alternator rotors can be rotated in either direction. The fan design must be changed in accordance with clockwise or counterclockwise rotation.

– The stator is clamped between the collector-ring end shield and the drive end shield. The rotor shaft runs in bearings in each end shield.

– Rectifiers with heat sinks (6). At least six power diodes for rectification of the 3-phase AC are pressed into the heat sinks.

– Collector rings (5). The excitation current flows to the rotating excitation winding through the carbon brushes and collector rings.

– Electronic regulator (4) forms a unit with the brush holder for alternator mounting.

– Electronic regulator for mounting on the vehicle body (not shown). Used in rare cases on commercial vehicles as an alternative to the alternator-mounted version. Mounted at a protected location on the vehicle body, this regulator is electrically connected to the brush holder by plug-in connection.

Fig. 13

Design of the compact alternator

1 Housing,
2 Stator,
3 Rotor,
4 Transistor voltage regulator with brush holder,
5 Collector rings,
6 Rectifier,
7 Fan.

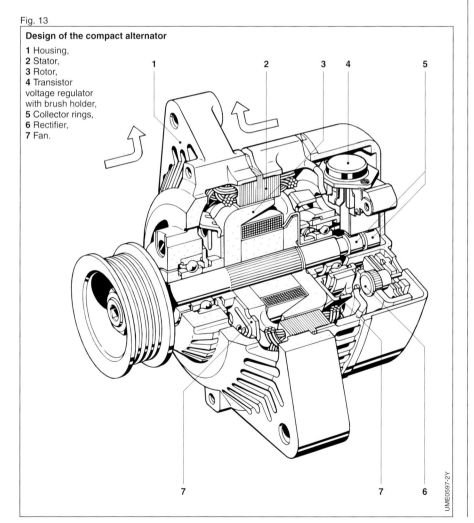

UME0597-2Y

127

Alternator versions

Design criteria

The following data are decisive for alternator design:
- Vehicle type and the associated operating conditions,
- Speed range of the engine with which the alternator is to be used,
- Battery voltage of the vehicle electrical system,
- Power requirements of the loads which can be connected,
- Environmental loading imposed on the alternator (heat, dirt, dampness, etc.),
- Specified service life,
- Available installation space, dimensions.

The requirements to be met by an automotive alternator differ very considerably depending upon application and the criteria as listed above. Regarding economic efficiency the criteria also vary along with the areas of application. It is therefore impossible to design an all-purpose alternator which meets all requirements. The different areas of application, and the power ranges of the vehicle types and engines concerned, led to the development of a number of basic models which will be described in the following.

Electrical data and sizes

The vehicle size is not decisive for determining the required alternator output power. This is solely a function of the loads installed in the vehicle.

The selection of the correct alternator is governed primarily by:
- The alternator voltage (14 V/28 V),
- The power output as a product of voltage and current throughout the rotational-speed range,
- The maximum current.

With these electrical data, it is possible to define the electrical layout, and therefore the required alternator size.

The different alternator sizes are identified by a letter of the alphabet, and increase in alphabetical order. A further important feature is the alternator or rotor system (e.g., claw-pole alternator as a compact alternator or alternator with compact diode assembly, or with salient-pole rotor or windingless rotor). This characteristic is identified by numbers or letters. In addition, the various alternators are identified by an alpha-numeric code e.g., GC, KC, NC, G1, K1, N1 for passenger cars, and K1, N1, T1 for commercial vehicles and buses.

Further variations are possible with regard to the type of mounting, the fan shape, the pulley, and the electrical connections.

Claw-pole alternators with collector rings

Claw-pole alternators with collector rings feature compact construction with favorable power characteristics and low weight. This leads to a correspondingly wide range of applications. These alternators are particularly suited for use in passenger cars, commercial vehicles, and tractors etc.

Table 1: **Alternator types.**

Design	Application	Type	No. of poles
Compact	Passenger cars, motorcycles	GC, KC, NC	12
Compact diode assembly	Pass. cars, commercial vehs., tractors, motorcycles	G1	
	Pass. cars, commercial vehs., tractors,	K1, N1	
	Buses	T1	16
	Long-haul trucks, construction machinery	N3	12
Standard	Special vehicles	T3	14
	Special vehicles, ships/boats	U2	4, 6

The T1 is a high-power version and is intended for vehicles with high power requirements (e.g., buses). The basic construction is shown in Fig. 1.

Features

The ratio of length to diameter is carefully selected to guarantee a maximum of power together with a low outlay on materials. This results in the compact shape with its large diameter and short length which is typical for this type of alternator. Furthermore, this shape also permits excellent heat dissipation.

The designation claw-pole alternator derives from the shape of the alternator's magnetic poles. The two oppositely-poled pole halves are attached to the rotor shaft, and the claw-shaped pole half fingers mesh whith each other in the form of alternating north and south poles which envelop the toroidal excitation winding on the pole body (Fig. 2). The number of poles which can be realised in practice is limited. On the one hand, a low number of poles leads to a low machine efficiency, whereas on the other the more poles there are, the higher is the magnetic leakage. For this reason, such alternators are designed as 12-pole or 16-pole machines depending upon the power range.

Compact alternators
Types GC, KC, and NC

Application

This series comprises the alternators size GC, KC, and NC. Compact-design alternators are intended for use in passenger cars with high current consumption, and are particularly suitable for modern vehicle engines with their low idle speeds. The increase in the alternator's maximum permissible speed (briefly up to 20,000 min^{-1}) permits a higher transmission ratio to be used so that these alternators generate as much as 25% more power for a given engine speed than the conventional compact-diode-assembly models.

Operating principle

Fig. 3 shows a 12-pole compact alternator. The magnetic flux flows through the pole body and the left-hand pole half and its pole fingers, across the air gap to the stationary laminated stator core with stator winding, from where it flows back to the pole body through the right-hand pole half and completes the magnetic circuit.

When the rotor turns, this field of force cuts through the three phases of the stationary stator winding and during a complete 360° rotation induces six complete sinusoidal waves in each phase.

The generated current is divided into primary current and excitation current. After rectification, the primary current flows as

Fig. 1

Basic construction of the claw-pole alternator with collector rings

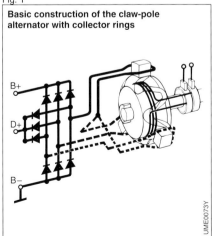

Fig. 2

The components of a 12-pole claw-pole rotor

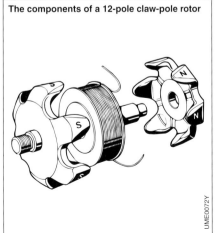

operating current via terminal B+ to the battery and to the loads.

Design and construction

Compact alternators are dual-flow-ventilated, self-excited alternators with synchronous claw-pole rotor, small-dimensioned collector rings, and Zener-type power diodes. The 3-phase 12-pole AC winding is located in the 12-pole stator and the excitation system in the rotor (Fig. 3).

The stator core is fastened at its center laminations in the casing and centered using the end shields. These measures lead to a high degree of precision in the assembly of the alternator, and a low level of "magnetic" noise.

Considerable reduction in noise was achieved by chamfering the claw poles, and by clamping the stator core at its center laminations. Two interior-mounted fans ventilate the alternator by drawing air through it from each end.

This ventilation system results in a lower fan noise due to the reduction of the noise-emission level. It also means that the designer has a higher degree of freedom regarding the alternator installation point on the engine.

The considerably reduced collector-ring diameter leads to a drop in their peripheral speed. This in turn means that there is less wear at the collector-ring surface as well as at the carbon brushes so that carbon-brush wear is no longer a limiting factor for the alternator's service life. The transistor voltage regulator is integrated in the carbon-brush holder. The Z-diode rectifiers are designed in "sandwich" form and are protected against corrosion by a plastic coating. Z-diodes provide additional protection against overvoltages and voltage peaks.

Compact alternators Series B

Application

The Series B compact alternator for passenger-cars and commercial-vehicles is a further development of the first-generation compact alternator. Notwithstanding its higher output power, the Series B alternator features a longer service life, smaller

Fig. 3

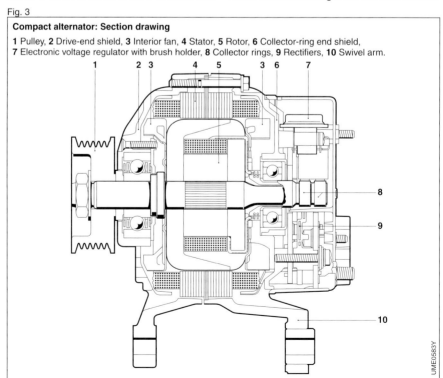

Compact alternator: Section drawing

1 Pulley, **2** Drive-end shield, **3** Interior fan, **4** Stator, **5** Rotor, **6** Collector-ring end shield, **7** Electronic voltage regulator with brush holder, **8** Collector rings, **9** Rectifiers, **10** Swivel arm.

UME0583Y

Table 2: Series B compact alternators

Size	Nominal voltage	Rated current at	
	V	1,800 min^{-1} A	6,000 min^{-1} A
GCB1	14	22	55
GCB2	14	37	70
KCB1	14	50	90
KCB2	14	60	105
NCB1	14	70	120
NCB2	14	90	150
KCB1	28	25	55
NCB1	28	35	80
NCB2	28	40	100

The introduction of a new-model rectifier permits higher air throughflow and therefore improved cooling. The three center laminations of the stator lamination pack are clamped and centered between the two end shields around their complete circumference. Compared to the 1st-generation compact alternator, these measures not only lead to improved vibration strength, but also to an improvement in the heat transfer from the stator core to the end shields.

The Series B compact alternators are equipped with multifunction voltage regulators (Chapter "Voltage-regulator versions").

dimensions, and lower weight. The B Series is comprised of six different sizes for 14 V nominal voltage and three for 28 V. The close spacing between the different sizes enables optimum adaptation to the particular power demand and to the cramped engine-compartment conditions encountered in modern-day vehicles.

Design and construction
The basic design of the Series B compact alternator (Fig. 4) is the same as that of the conventional model.

Alternators with compact diode assembly Type G1, K1, and N1

Application
The compact alternator is increasingly superseding the compact-diode-assembly alternator as the preferred alternator for passenger cars and commercial vehicles. For special applications (in particular for special-purpose commercial

Fig. 4

Series B compact alternator (part sectional drawing)

1 Housing with double-pass ventilation, **2** Inboard fan, **3** Stator, **4** Rotor, **5** Voltage regulator, **6** Outboard collector rings, **7** Outboard rectifier.

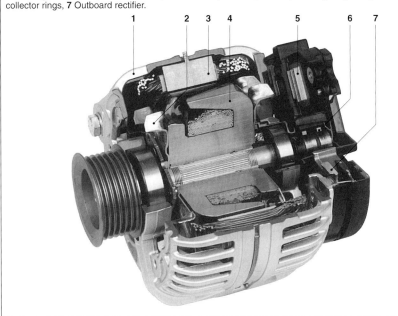

UME0648Y

vehicles) which for instance necessitate a specific form of corrosion protection, or a version with hose-connection adapter for fresh-air input, a compact-diode-assembly alternator is still installed.

Operating principle

Compact-diode-assembly alternators and compact alternators have the same operating principles. Fig. 5 shows a K1 alternator.

Design and construction

Compact-diode-assembly alternators have 12 poles, they are single-flow internally ventilated and feature self-excitation. The stator core is clamped between the drive end shield and the collector-ring end shield. The rotor runs in bearings in each of these end shields. The fan and the pulley are attached to the drive-end-shield side of the rotor shaft. The excitation winding is supplied with excitation current through the carbon brushes. These are mounted in the collector-ring end shield and pressed against the collector rings by springs.

The six power diodes for rectification of the generator voltage are press-fitted in the heat sinks of the collector-ring end shield. In the majority of versions, the electronic voltage regulator forms a unit together with the brush holder and is attached to the outside face of the collector-ring end shield.

The K1 and N1 alternators are equipped as follows for special applications:

– In case of very high outside temperatures, cool air is drawn in through a hose and connection adapter.

– The maximum rotational speed can be increased to 18,000 min^{-1}.

– Special anti-corrosion treatments are available for extreme high-exposure applications.

– Zener power diodes are used to protect sensitive components against the voltage peaks that accompany sudden drops in current demand (load dump) and when the battery is out of circuit.

Fig. 5

Claw-pole alternator with compact diode assembly: Section drawing

1 Pulley, **2** Fan, **3** Drive-end shield, **4** Stator core, **5** Excitation winding, **6** Collector-ring end shield, **7** Collector rings, **8** Swivel arm, **9** Transistor voltage regulator with brush holder.

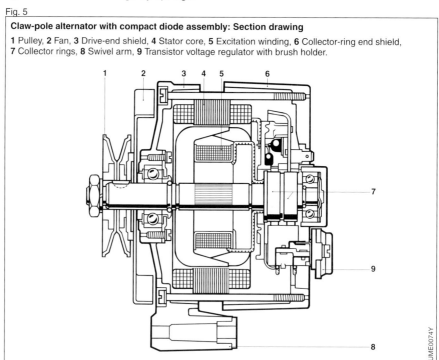

UME0074Y

Compact-diode-assembly alternators Series T1

Applications
Series T1 alternators featuring compact-diode assemblies are designed for vehicles with high current consumption, with buses representing the chief application. The alternators in urban buses must maintain high output levels throughout a wide engine-speed range extending down to idle. T1 alternators are ideal for these kinds of applications.

Operating principle
The basic concept is identical to that employed in the G1, K1 and N1 compact-diode-assembly alternators.

Design and construction
The T1 is a self-exciting alternator with single-flow internal ventilation. This 16-pole unit features integral rectifier diodes and encapsulated collector rings. The 3-phase alternating current winding is incorporated within the stator, while the excitation assembly is located within the

rotor. T1 alternators are available as swivel-mounted units with left and right-mount brackets for both flexible and rigid installation. Extra-wide roller bearings are installed with a large-volume grease packing to provide extended maintenance-free service. Ventilation is furnished by fans designed to operate in either direction.

Special anti-corrosion measures are employed to protect the alternator against the salt and spray encountered in winter operation.

An adapter and air-induction duct assembly, designed to maintain a supply of dry, dust-free cooling air to the alternator, is available for severe-duty (heat and dust) applications.

Special version Series DT1
Increasing demands for comfort and convenience in buses have been accompanied by consistent increases in the performance required from their alternators; the double T1 alternator is designed to satisfy these demands. The unit basically consists of two electrically and mechani-

Fig. 6

Cutaway view of a double T1 alternator with two stators and two excitation circuits

1 Fan, 2 Drive end shield, 3 Pulley, 4 Drive-end ball-bearing assembly, 5 Swivel arm, 6 Stator winding 1, 7 Excitation winding 1 (rotor), 8 Stator winding 2, 9 Excitation winding 2 (rotor), 10 External wiring socket, 11 Brush holder, 12 Collector-ring roller-bearing assembly, 13 Collector ring, 14 Collector-ring bearing end cover, 15 Rectifier assembly.

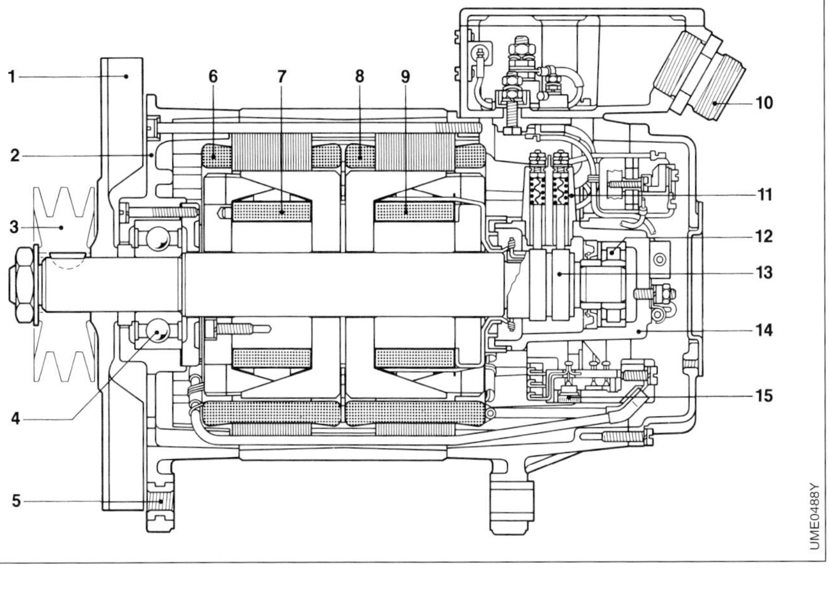

UME0488Y

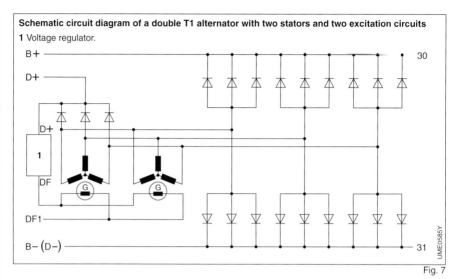

Schematic circuit diagram of a double T1 alternator with two stators and two excitation circuits

1 Voltage regulator.

Fig. 7

cally linked T1 alternators encased in a common housing (Fig. 6). The electronic voltage regulator is installed in the alternator, while carbon brushes and collector rings are housed within a dust-protected collector-ring chamber. A 100-ohm resistor is installed between D+ and D– to trigger the charge indicator lamp when the field current is interrupted. Figure 7 is a schematic diagram illustrating the circuits within a double T1 alternator featuring two stators and two excitation systems.

Salient-pole alternators with collector rings

Series U2
The salient-pole alternator combines a wide performance range with a high level of specific output.

Application
Salient-pole alternators with collector rings are usually installed in 24V systems in large vehicles with high current requirements (> 100 A). These units are thus ideal for use in buses, rail vehicles, large special-purpose vehicles and in marine applications.

Operating principle
The alternator illustrated in Figure 8 is a self-exciting, four-salient-pole unit. With each rotation the rotor passes four poles, inducing four half waves in each circuit. Under 3-phase operation this is 4 x 3 = 12 half waves for each rotation.

Design and construction
The layout of the 3-phase stator winding and the current flow pattern correspond to those found in the claw-pole alternator.
However, the rotor system on this type of alternator (Fig. 9) differs from that used on the claw-pole unit. The claw-pole rotor includes a central excitation winding for all poles. In contrast, the salient-pole rotor features four or six individual poles with individual local excitation windings.

Fig. 8

Basic structure of a salient-pole alternator with collector rings

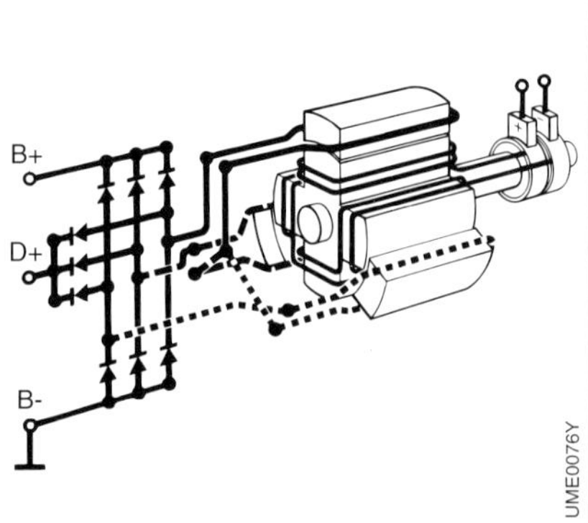

The characteristic shape of the salient-pole alternator – a slim, elongated cylinder – derives from the rotor's form.

The stator with its 3-phase winding is fitted in the alternator housing.

At the end of each housing are the collector ring and drive end shields. The housing also supports the rotor with its integral excitation winding, while the excitation current is delivered via the collector rings and carbon brushes.

Rectifier and regulator are external components. They are installed remote from the alternator at a point protected against engine heat, moisture, and dirt. Alternator and regulator are connected electrically by a 6-line wiring harness.

Encapsulated collector rings and an extra-large grease chamber qualify the unit for high-mileage applications.

Alternators with windingless rotors (without collector rings)

Series N3
On windingless-rotor alternators, the only wear-components are the roller bearings. These units are thus used in applications in which long service life is essential.

Application
Windingless-rotor alternators with high-stability end bearings are installed in construction vehicles, long-haul trucks and heavy-duty special-purpose vehicles. Alternators with windingless rotors withstand extremely severe operating conditions over exceptionally high mileages. Their design concept is based on minimizing the number of wear components in the alternator in order to extend the effective service life. This alternator is practically maintenance-free.

Fig. 9

Cutaway view of a Type U2 salient-pole alternator

1 Drive end shield, **2** Housing, **3** Stator winding, **4** Rotor, **5** 3-phase connection (separate rectifier and regulator), **6** End cap, **7** Fan, **8** Collector-ring end shield, **9** Collector rings, **10** Driveshaft.

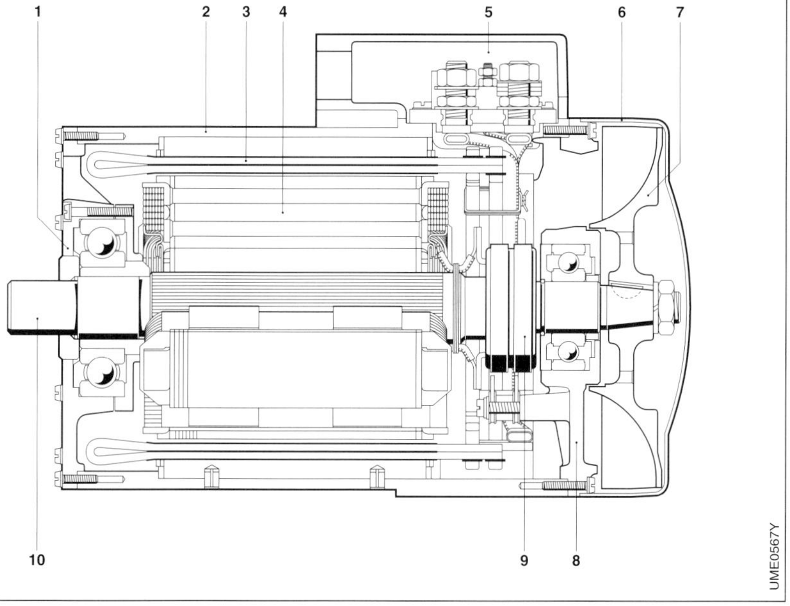

UME0567Y

Operating principle and design

Self-excitation is furnished by a stationary excitation winding on the internal pole. The residual magnetization makes pre-excitation of the alternator field unnecessary. The excitation field magnetizes the alternating pole fingers on the rotating windingless rotor, and the rotating magnetic field from these poles induces 3-phase alternating current in the stator winding. The magnetic flux field extends from the spinning rotor's pole core and through the stationary internal pole to the conductive element before proceeding over its pole finger to the stationary stator core.

The polarity of the claw-pole half section is opposed to that of the first section; it completes the magnetic circuit within the rotor's pole core. Compared to the collector-ring rotor design, here the magnetic flux must flow through two supplementary gaps between the rotating pole wheel and the stationary internal pole.

The most significant characteristics of the N3 alternator are shown in Figure 10. This alternator's special features include the internal pole with excitation winding,

Fig. 10

Section through a windingless rotor

1 Rotor shaft with pole core, **2** Left pole-finger crown, **3** Non-magnetic retainer ring, **4** Right pole-finger crown.

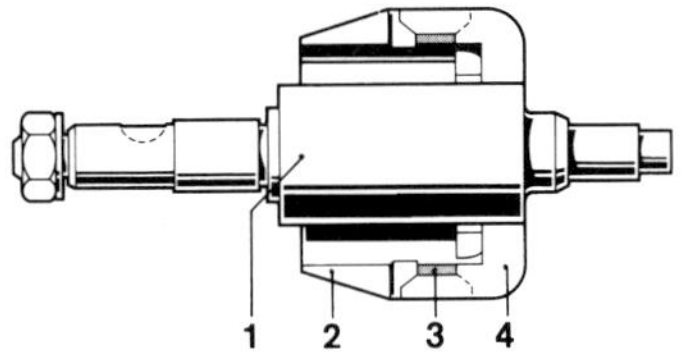

Fig. 11

which, along with the housing and stator core, the heat sinks with power diodes, and the external, electronic voltage regulator, forms a stationary part of the unit. This design does away with the need to transmit an external excitation current to the rotating rotor via collector rings and carbon brushes.

The only rotating component is the rotor with the pole wheel and conductive elements (Fig. 11). Six pole fingers of a single polarity form a single north or south pole-finger crown. The two crowns form claw-pole half sections, and are retained by a non-magnetic ring positioned

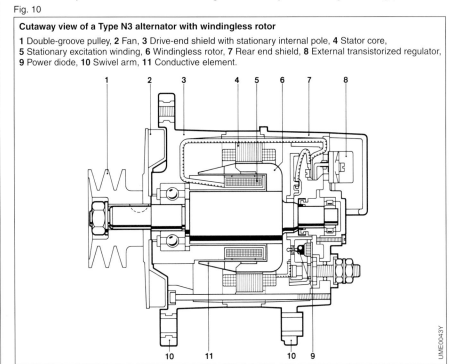

Cutaway view of a Type N3 alternator with windingless rotor

1 Double-groove pulley, **2** Fan, **3** Drive-end shield with stationary internal pole, **4** Stator core, **5** Stationary excitation winding, **6** Windingless rotor, **7** Rear end shield, **8** External transistorized regulator, **9** Power diode, **10** Swivel arm, **11** Conductive element.

below the pole fingers, with their mutual recesses and extensions.

Liquid-cooled compact alternators

In the case of air-cooled alternators, it is the cooling fan which is mainly responsible for the air-flow noise. At high current outputs, a pronounced reduction of noise can only be achieved by using liquid-cooled alternators which utilise the coolant from the engine's cooling circuit.

Application

On modern intermediate and luxury cars, the use of liquid-cooled fully-encapsulated alternators is often the only way to achieve a decisive reduction in vehicle noise. The sound insulation resulting from the coolant surrounding the alternator is particularly effective at high rotational speeds. On air-cooled alternators, it is the air-flow noise which predominates at high speeds.

If the alternator loading is high enough (for instance, due to heating resistors in the inlet air ducts to the passenger compartment), the alternator's heat loss supports coolant heat-up during the engine's warm-up phase. Particularly on modern-day diesel engines with their optimized efficiency levels, this means that not only the engine reaches its operating temperature more quickly, but also that the passenger compartment heats up to the required level in far less time.

Design and construction

Since its high internal temperatures would lead to a conventional brush/collector-ring system wearing out prematurely, this fully-encapsulated liquid-cooled alternator is equipped with a windingless rotor.

The alternator is inserted in a special coolant housing which provides enough space for the coolant between it and the alternator housing. This coolant space is connected to the engine's coolant circuit. The most important sources of heat loss (stator, power semiconductors, voltage regulator, and stationary excitation winding) are coupled to the alternator housing in such a manner that efficient heat transfer (Fig. 12) is ensured. The electrical connections are located at the pulley end of the alternator.

Fig. 12

Liquid-cooled compact alternator with windingless rotor

1 Pulley, **2** Rectifier, **3** Voltage regulator, **4** Drive end shield, **5** Alternator housing, **6** Coolant, **7** Coolant housing for installation on or in the engine, **8** Stationary excitation winding, **9** Stator lamination stack, **10** Stator winding, **11** Windingless rotor alternator with windingless rotor, **12** Non-magnetic intermediate ring, **13** Conductive element.

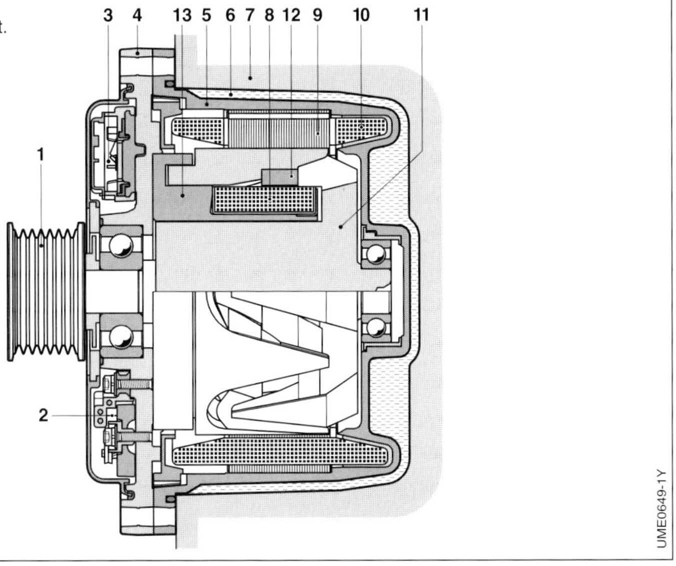

Voltage-regulator versions

The mechanical electromagnetic contact (or vibrating-type) regulators and the electronic (transistor) versions are the two basic versions of voltage regulator. Whereas the electromagnetic regulator is today practically only used for replacement purposes, the (monolithic or hybrid) transistor regulator is standard equipment on all alternator models.

Electromagnetic contact-type regulators

The excitation current is varied by opening and closing a movable contact in the excitation-current circuit which is pressed against a fixed contact by a spring. When the rated voltage is exceeded, the movable contact is lifted off by an electromagnet.

The contact regulators which are suitable for alternator applications are of the single-element type. That is, regulators with a voltage-regulator element comprising an electromagnet, an armature, and a regulating contact.

In the single-element, single-contact regulator (Fig. 1), the contact opens and closes as follows: The magnetic force and the spring force of a suspension and adjusting spring are both applied to the regulating armature.

As soon as the alternator voltage exceeds the set value, the electromagnet pulls in the armature and opens the contact (position "b"). This switches a resistor into the excitation circuit which reduces the excitation current and with it the alternator voltage. When the alternator voltage drops below the set voltage the magnetic force is also reduced, the spring force predominates and closes the contact again (position "a"). This opening and closing cycle is repeated continually.

The single-element double-contact regulator (Fig. 2) operates with a second pair of contacts which permit 3 switching positions.

The regulating resistor is short-circuited in position "a" and a high excitation current flows. In position "b" the resistor and the excitation winding are connected in series and the excitation current is reduced. In position "c", the excitation winding is short-circuited and the excitation current drops to practically zero.

Due to its size and characteristics, this alternator is only suitable for mounting on the vehicle body.

Fig. 1

Diagram of a single-element single-contact regulator

1 Regulator, 2 Alternator, 3 Electromagnet,
4 Regulating contact, 5 Regulating resistor,
6 Excitation winding (G).

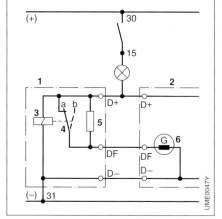

Fig. 2

Diagram of a single-element double-contact regulator

1 Regulator, 2 Alternator, 3 Electromagnet,
4 Regulating resistor, 5 Regulating contact,
6 Excitation winding (G).

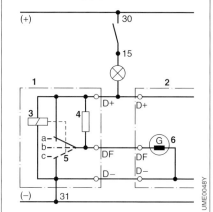

Electronic regulators

Electronic regulators are used solely with alternators. Thanks to its compact dimensions, its low weight, and the fact that it is insensitive to vibration and shock, this regulator can be integrated directly in the alternator.

Whereas the first transistor regulators were built from discrete components, modern-day versions all use hybrid and monolithic circuitry.

The transistor regulator's essential advantages are:
- Shorter switching times which permit closer control tolerances,
- No wear (= no servicing),
- High switching currents permit a reduction in the number of types,
- Spark-free switching prevents radio interference,
- Insensitive to shock and vibration, and climatic effects,
- Electronic temperature compensation also permits closer control tolerances,
- Compact construction allows direct mounting on the alternator, irrespective of alternator size.

Principle of operation

Basically speaking, the operating principle is the same for all electronic regulator types. The Type EE electronic regulator is used here as an example, and Fig. 3 shows its operation between the "On" and "Off" states.

The principle of operation is easier to understand when one considers what happens when the alternator's terminal voltage rises and falls.

The actual value of the alternator voltage between terminals D+ and D− is registered by a voltage divider (R1, R2, and R3). A Zener diode in parallel with R3 functions as the alternator's setpoint generator. A partial voltage proportional to the alternator voltage is permanently applied to this diode.

The regulator remains in the "On" state as long as the actual alternator voltage is below the set value (Fig. 3a).

The Z-diode's breakdown voltage has not yet been reached at this point. That is, no current flows to the base of transistor T1 through the branch with the Z-diode. T1 is in the blocking state.

With T1 blocked, a current flows from the exciter diodes via terminal D+ and resistor R6 to the base of transistor T2 and switches T2 on. Terminal DF is now connected to the base of T3 by the switched transistor T2. This means that T3 always conducts when T2 is conductive. T2 and T3 are connected as a Darlington circuit and form the regulator's driver stage. The excitation current $I_{exc.}$ flows through T3 and the excitation winding and increases during the "On" period, causing a rise in the alternator voltage U_G. At the same time, the voltage at the setpoint generator also rises.

The regulator assumes the "Off" state as soon as the actual alternator voltage exceeds the setpoint value (Fig. 3b).

The Z-diode becomes conductive when the breakdown voltage is reached, and a current flows from D+ through resistors R1, R2 in the branch with the Z-diode, and from there to the base of transistor T1 which also becomes conductive.

As a result, the voltage at the base of T2 is practically 0 referred to the emitter, and transistors T2 and T3 (driver stage) block. The excitation circuit is open-circuited, the excitation decays, and the alternator voltage falls as a result.

As soon as the alternator voltage drops under the set value again, and the Z diode switches to the blocked state, the driver stage switches the excitation current on again.

When the excitation current is open-circuited a voltage peak would be induced due to the excitation winding's self-induction (stored electrical energy) which could destroy transistors T2 and T3. A "free-running diode" D3 is connected parallel to the excitation winding, and at the instant of open-circuiting absorbs the excitation current thereby preventing the formation of a dangerous voltage peak.

The control cycle in which the current is switched on and off by connecting the excitation winding alternately to the alternator voltage or short-circuiting it with the free-wheeling diode is repeated periodically. Essentially, the on/off ratio depends on the alternator speed and the applied load. The ripple on the alternator DC is smoothed by capacitor C. Resistor R7 ensures the rapid, precise switchover of transistors T2 and T3, as well as reducing the switching losses.

Hybrid regulators

The transistor regulator using hybrid technology comprises a hermetically encapsulated case, in which are enclosed a ceramic substrate, protective thick-film resistors, and a bonded integrated circuit (IC) incorporating all the control functions. The power components of the drive stage (Darlington transistors and the free-wheeling diode) are soldered directly onto the metal socket in order to ensure good heat dissipation. The electrical connections are via glass-insulated metal pins.

The regulator is mounted on a special brush holder and directly fastened to the alternator without wiring.

Due to the Darlington circuit in the power

Fig. 3

Circuit diagram of an EE-type electronic regulator

a Excitation current switched on by T3,
b Excitation current switched off by T3.
1 Power stage,
2 Control stage,
3 Voltage divider,
4 Temperature-compensation diodes.
C Voltage-smoothing capacitor,
D3 Free-wheeling diode.

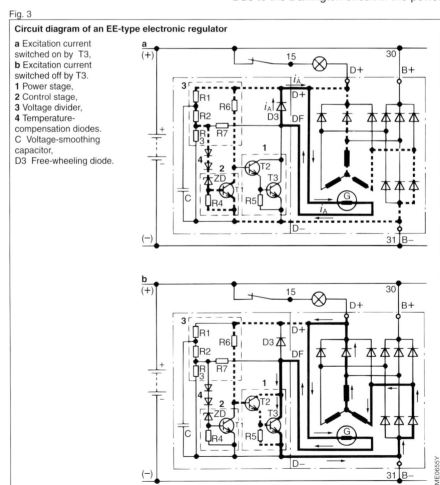

UME0655Y

stage (two transistors), there is a voltage drop of about 1.5 V in the current-flow direction.

The circuit diagram (Fig. 4) shows an alternator fitted with an EL-type hybrid regulator.

The hybrid regulator's advantages can be summed up as follows:

– Compact construction, low weight, few components and connections,
– High reliability in the extreme under-hood operating conditions met in automotive applications.

Normally, hybrid regulators using conventional diodes are used with compact-diode-assembly alternators.

Monolithic regulators

The monolithic regulator has been developed from the hybrid regulator. The functions of the hybrid regulator's IC, power stage, and free-wheeling diode have been incorporated on a single chip. The monolithic regulator uses bipolar techniques.

The compact construction with fewer components and connections enabled reliability to be even further improved.

Since the output stage is in the form of a simple power stage, the voltage drop in the current-flow direction is only 0.5 V. Monolithic regulators in combination with

Z-diode rectifiers are used in compact alternators.

Multifunction voltage regulator

In addition to voltage regulation, the multifunction regulator can also trigger an LED display instead of the charge-indicator lamp, as well as a fault display to indicate under-voltage, V-belt breakage, or excitation interruption.

The alternator no longer needs excitation diodes. The signal for "engine running" can be taken from Terminal L. Terminal W provides a signal which is proportional to engine speed. The actual voltage value is taken from Terminal B+ on the alternator.

The standard version of the Series B compact alternator has further functions available:

In case a load is switched into the vehicle electrical system, the alternator's excitation follows a ramp. This prevents torque jumps in the alternator's belt drive which, for instance, could otherwise interfere with the smooth running of the engine (LRD: Load-Response Drive; LRS: Load-Response Start).

The regulator's on/off ratio can be picked-off via the DFM terminal. This ratio defines the alternator's loading and can be used for selection circuits (e.g. for switch-

Voltage-regulator versions

Fig. 4

Circuit diagram of an alternator equipped with EL-type hybrid electronic voltage regulator

1 Control stage using thick-film techniques with resistors and IC,
2 Power stage (Darlington stage),
3 Free-wheeling diode.

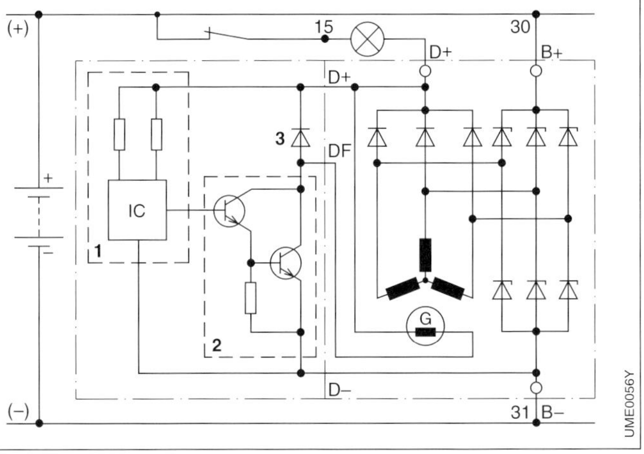

ing off low-priority loads when the alternator must deliver full power). Terminal L is designed for relay triggering up to max. 0.5 A.

The power loss associated with the charge indicator lamp in the instrument cluster is often excessive. It can be reduced by using an LED display instead. Multifunction regulators permit the triggering of lamp bulbs as well as of LED display elements in the instrument cluster.

Overvoltage-protection devices

Usually, with the battery correctly connected and under normal driving conditions, it is unnecessary to provide additional protection for the vehicle's electronic components. The battery's low internal resistance suppresses all the voltage peaks occurring in the vehicle electrical system.

Nevertheless, it is often advisable to install overvoltage protection as a precautionary measure in case of abnormal operating conditions. For instance, on vehicles for transporting hazardous materials, and in case of faults in the vehicle electrical system.

Reasons for overvoltage

Overvoltage may occur in the vehicle electrical system as a result of the following situations:
– Regulator failure,
– Influences originating from the ignition,
– Switching off of devices with a predominantly inductive load,
– Loose contacts or cable breaks.

Such overvoltages take the form of very brief voltage peaks lasting only a few milliseconds which reach a maximum of 350 V and originate from the coil ignition. Overvoltages are also generated when the line between battery and alternator is open-circuited with the engine running (this happens when an outside

battery is used as a starting aid), or when high-power loads are switched off. For this reason, under normal driving conditions, the alternator is not to be run without the battery connected.

Under certain circumstances though, short-term or emergency operation without battery is permissible.

This applies to the following situations:
– Driving of new vehicles from the final assembly line to the parking lot,
– Loading onto train or ship (the battery is installed shortly before the vehicle is taken over by the customer),
– Service work etc.

With towing vehicles and agricultural tractors it is also not always possible to avoid operation without the battery connected.

The overvoltage protection device guarantees that overvoltages have no adverse effects on operation, although it does require extra circuitry.

Types of protection

There are three alternatives for implementing overvoltage protection:

Z-diode protection

Z-diodes can be used in place of the rectifier power diodes. They limit high-energy voltage peaks to such an extent that they are harmless to the alternator and regulator. Furthermore, Z-diodes function as a central overvoltage protection for the remaining voltage-sensitive loads in the vehicle electrical system.

The limiting voltage of a rectifier equipped with Z-diodes is 25...30 V for an alternator voltage of 14 V, and 50...55 V for an alternator voltage of 28 V.

Compact alternators are always equipped with Z-diodes.

Surge-proof alternators and regulators

The semiconductor components in surge-proof alternators have a higher electric-strength rating. For 14 V alternator voltage the electric strength of the semiconductors is at least 200 V, and for 28 V alternator voltage 350 V.

In addition, a capacitor is fitted between the alternator's B+ terminal and ground. which serves for short-range interference suppression.

The surge-proof characteristics of such alternators and regulators only protect these units, they provide no protection for other electrical equipment in the vehicle.

Overvoltage protection devices

(only for 28 V alternators)

These are semiconductor devices which are connected to the alternator terminals D+ and D– (ground). In the event of voltage peaks, the alternator is short-circuited through its excitation winding.

Primarily, overvoltage protection devices protect the alternator and the regulator, and to a lesser degree the voltage-sensitive components in the vehicle electrical system.

Generally, alternators are not provided with polarity-reversal protection. If battery polarity is reversed (e.g., when starting with an external battery), this will destroy the alternator diodes as well as endangering the semiconductor components in other equipment.

Overvoltage protection devices, non-automatic

This type of overvoltage protection device is connected directly to the D+ and D– terminals on T1 Series alternators, e.g., in buses and heavy trucks (Fig. 1). The unit responds to voltage peaks and consistent overvoltage that exceed its response threshold of approx. 31 V. At this point thyristor Th becomes conductive. The activation voltage is defined by Zener diode ZD, while the necessary response delay is regulated by resistors R1 and R2 along with condenser C. The unit requires only milliseconds to short circuit the regulator and alternator across D+ and D–. The thyristor assumes responsibility for the short-circuit current. Meanwhile, current from the battery triggers the charge indicator lamp to alert the driver. The thyristor remains active, reverting to its off-state only once the ignition is switched off, or the engine and alternator come to rest.

The unit will not provide overvoltage protection if the wires at terminals D+ and

Fig. 1

Circuit diagram of a (non-automatic) overvoltage-protection device for a 24 V vehicle electrical system

1 Battery, 2 Overvoltage-protection device, 3 Driving switch, 4 Voltage regulator, 5 Alternator.

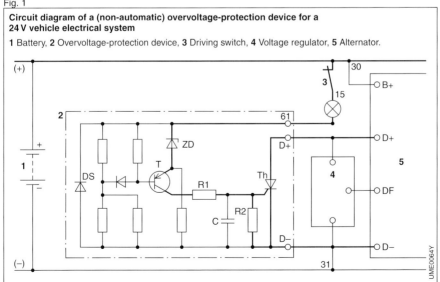

D– are reversed. As the indicator lamp also fails to respond, the problem would remaln unnoticed if a backup diode DS were not installed between terminals D+ and D– to ensure a signal at the lamp. This diode responds to reversed connections by polarizing to allow current flow, and the indicator lamp remains on continuously.

Overvoltage protection devices, automatic

This type of protection device is designed for use with Series T1 alternators (Fig. 2). The unit incorporates two inputs, D+ and B+. These are designed to react to different voltage levels and with varying response times. Input D+ provides rapid overvoltage protection, as on the device described above. The second input, B+, responds only to defects at the voltage regulator, while the alternator voltage continues to climb until it reaches the unit's response voltage of approx. 31 V. The alternator then remains shorted until the engine is switched off. Input B+ is thus designed to prevent consequential damage. This overvoltage protection device makes it possible for the alternator to operate for limited periods without a battery in the circuit. The alternator voltage collapses briefly when the overvoltage device responds; if the load becomes excessive, renewed excitation at the alternator will become impossible.

Voltage peaks which can be generated by the alternator itself when loads are switched off ("load-dump"), cannot damage other devices in the system because the alternator is immediately short-circuited.

Consequential-damage protection device

This protection device is specially designed for use with the double T1 alternator with two stators and two excitation systems (Fig. 3).

While the overvoltage protection device short-circuits the alternator, the consequential-damage protection unit functions as a kind of backup regulator, even with the battery out-of-circuit. Provided that the alternator's speed and the load factor allow, it maintains a mean alternator voltage of approximately 24 V to furnish emergency capacity.

The consequential-damage protection device responds to operation with battery and a faulty, short-circuited regulator by interrupting the alternator's excitation current approx. 2 seconds after the alternator output passes the response threshold of 30 V. The unit's relay contact then assumes a backup voltage-control function by operating as a contact regulator.

When the system is operated with the battery out-of-circuit, the unit reacts to voltage peaks of 60 V or more lasting for more than 1 ms.

The charge indicator lamp flashes to indicate that the system is operating in the backup mode. The system does not charge the battery, as the mean voltages in this mode are very low. Maximum operating times in this backup mode extend to approx. 10 hours, after which the consequential-damage protection device must be replaced.

Free-wheeling diode

The free-wheeling diode (also known as a suppressor diode or anti-surge diode) has already been mentioned in the description of the transistor regulator.

When the regulator switches to the "Off" status, upon interruption of the excitation current a voltage peak is induced in the excitation winding due to self-induction. Sensitive semiconductor components can be destroyed if precautionary measures are not taken. The free-wheeling diode is connected in the regulator parallel to the alternator's excitation winding. Upon the excitation winding being interrupted, the free-wheeling diode "takes over" the excitation current and permits it to decay, thus preventing the generation of dangerous voltage peaks.

A similar effect can occur on vehicles which are equipped with inductive loads remote from the alternator regulator. Thus, when electromagnetic door valves,

solenoid switches, magnetic clutches, motor drives, and relays etc. are switched off, voltage peaks can be generated in the windings of such equipment due to self-induction, and can endanger the diodes and other semiconductor components.

These induced voltages can be rendered harmless by means of a free-wheeling diode.

Fig. 2

Simplified circuit diagram with an automatic overvoltage-protection device for a Type T1 alternator

1 Battery, **2** Overvoltage-protection device, **3** Ignition switch, **4** Regulator, **5** Alternator.

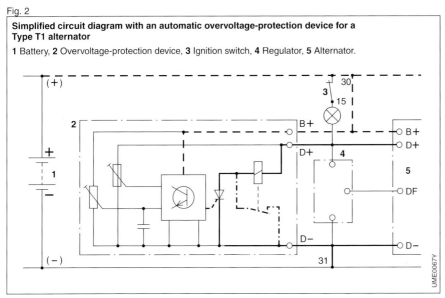

Fig. 4

Circuit diagram with consequential-damage protection device and double T1 alternator

1 Battery, **2** Consequential-damage protection device, **3** Ignition switch,
4 Alternator and voltage regulator.

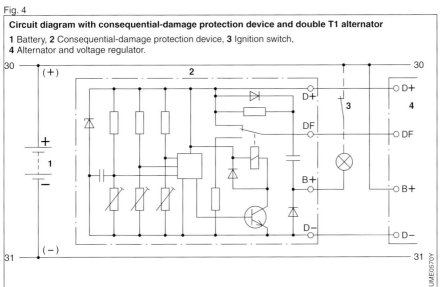

145

Cooling and noise

Due above all to the heat development by the alternator when converting mechanical power into electrical power, and also due to the effects of heat from the engine compartment (engine and exhaust system), considerable increases in component temperature take place. And when the engine compartment is encapsulated for sound-proofing purposes temperatures rise even further. In the interests of functional reliability, service life, and efficiency, it is imperative that this heat is dissipated completely. Depending upon alternator version, maximum permissible ambient temperature is limited to 80...100 °C. Cooling must guarantee that even under the hostile under-hood conditions encountered in everyday operation, component temperatures remain within the specified limits ("worst-case" consideration).

Cooling without fresh-air intake

Through-flow cooling is the most common cooling method applied for automotive alternators. Radial fans for one or both directions of rotation are used.

Since both the fan and the alternator shaft must be driven, the cooling-air throughput increases along with the speed. This ensures adequate cooling irrespective of alternator loading.

In order to avoid the whistling noise (siren effect) which can occur at specific speeds, the fan blades on some alternator types are arranged asymmetrically.

Single-flow cooling

Compact-diode-assembly alternators use single-flow cooling. The external fan is attached to the drive end of the alternator shaft. Fig. 1 shows a series G1 alternator with a clockwise-rotation fan. Air is drawn in by the fan at the collector-ring or rectifier end, passes through the alternator, and leaves through openings in the drive end shield.

Double-flow cooling

Due to their higher specific power, compact alternators are equipped with double-flow cooling. The compact alternator's two fans are mounted inside the alternator on the drive shaft to the left and right of the rotor's active section. The two air streams enter the alternator axially through openings in the drive and collector-ring end shields, and leave through openings around the alternator's circumference (Fig. 2).

One essential advantage lies in the use of smaller fans, with the attendant re-

Fig. 1

Single-flow cooling

Type G1 alternator with compact-diode assembly and fan designed for clockwise rotation.

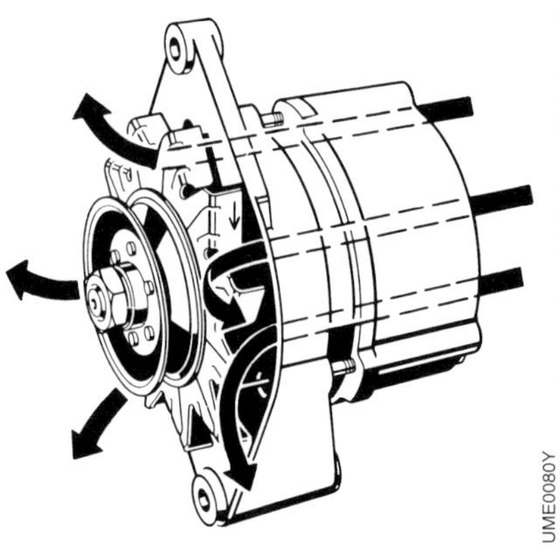

UME0080Y

Fig. 2

Double-flow cooling

Compact-alternator.

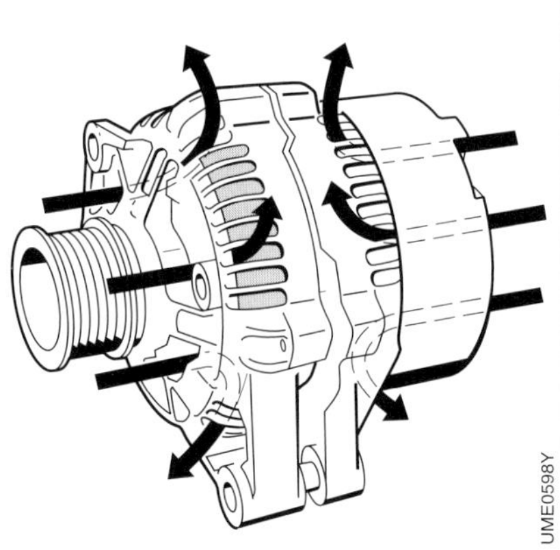

UME0598Y

duction of fan-generated aerodynamic noise.

Cooling with fresh-air intake

When fresh air is used for cooling purposes, a special air-intake fitting is provided on the intake side in place of the air-intake openings. A hose is used to draw in cool, dust-free air from outside the engine compartment.

For instance, with the T1 alternator the cooling air enters through the air-intake fitting, flows through the alternator and leaves again through openings in the drive end shield. With this type of alternator also, the cooling air is drawn through the alternator by the fan.

It is particularly advisable to use the fresh-air intake method when engine-compartment temperatures exceed 80°C and when a high-power alternator is used. With the compact alternator, the fresh-air method can be applied for cooling the rectifiers and the regulator.

Liquid cooling

The liquid-cooling principle utilises the engine's coolant to cool the fully-encapsulated alternator. The space for the coolant between the alternator and the coolant housing is connected to the engine's coolant circuit. The most important sources of heat loss (stator, power semiconductors, voltage regulator, and stationary excitation winding) are coupled to the alternator housing in such a manner that efficient heat transfer is ensured.

Cooling the diodes

The heat levels in semiconductor diodes should not exceed certain limits. Thus arrangements are required to dissipate the warmth that both power diodes and excitation diodes generate as thermal losses.

The diodes are installed by pressing them into heat sinks. These heat sinks combine an extended surface area with high levels of thermal conductivity for efficient heat transfer into the cooling air stream or into the coolant. The alternators usually employ a dual heat sink system for the power diodes. The cathodic ends of three of the diodes are inserted in one heat sink; this heat sink is connected to battery terminal B+. The

Fig. 3

Cooling with air-intake ducts

Type T1 alternator with compact-diode assembly and bi-directional fan.

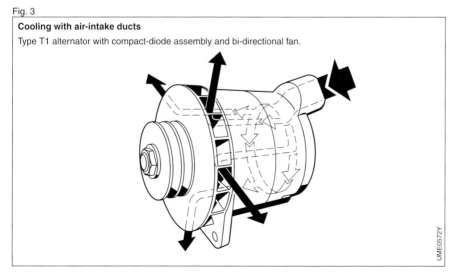

UME0572Y

remaining diodes are installed with the anodic sides in a heat sink connected to B–.

The excitation diodes located between the stator windings and D+ are either separate, without heat sinks, or installed in a third heat sink of their own (e.g., on alternators for heavy commercial vehicles).

The larger 3-phase alternators employ a triple heat-sink system. In each heat sink is inserted the cathodic side of one power diode and the anodic end of one power and excitation diode. With this system, each of the three heat sinks is connected to one phase of the 3-phase stator winding. The remaining legs of the diodes are connected to either B+ or B–, depending upon polarity, while the excitation diodes are connected to D+.

Noise

The more emphasis that is placed on quiet-running vehicles, the more important it is to reduce alternator noise.

Alternator noise is comprised of two main components: Aerodynamic noise and magnetically induced noise.

Aerodynamic noise can be generated by the passage of the cooling air through openings, and at high fan speeds. It can be limited by careful routing of the cooling air and by using smaller fans with asymmetrically arranged blades.

Magnetically induced noises are attributable to strong local magnetic fields and the dynamic effects which result between stator and rotor under load. There are a number of measures which can be taken to limit magnetic noise. These include air-gap increases and tighter manufacturing tolerances etc.

One of the most effective measures for reducing radially radiated noise is the "claw-pole chamfer". Here, the claw-pole's trailing edge is chamfered (Fig. 4). This measure reduces the effects of armature reaction caused by the stator currents. When the alternator is electrically loaded, the armature reaction causes a pronounced field displacement in the air gap which in turn leads to the

generation of noise. Optimization of the claw-pole chamfer method, combined with a reduction of the housing's noise-radiating surface, leads to noise reductions of up to 10 dB(A), this approximates to about 50 % in individual noise perception.

Account must also be taken of the effect of the alternator's position on the engine. Structure-borne noise excites the alternator mounting bracket and affects the alternator's oscillatory characteristics and its noise generation. A resilient alternator mounting can prevent this coupling completely.

In individual cases, mechanical noise can be caused by ball bearings in which hardened grease together with the effects of moisture ingress lead to "stick-slip" oscillations at very low temperatures. The remedy here is to use well-sealed ball bearings together with greases which retain their lubricity even at low temperatures.

Fig. 4

Chamfered claw-pole trailing edge

1 Stator, **2** Rotor, **3** Claw-pole chamfer (exaggerated).

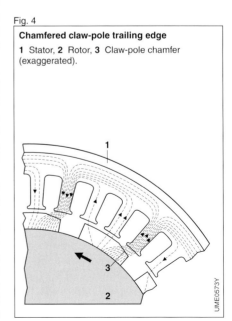

Power losses

Efficiency

Losses are an unavoidable by-product when converting mechanical energy into electrical energy. Efficiency is defined as the ratio between power input to the conversion unit and power taken from it.
The maximum efficiency of an air-cooled alternator is approximately 65%, a figure which drops rapidly when speed is increased.
Under normal driving conditions, an alternator usually operates in the part-load range. Mean efficiency is around 55%.
Presuming the same loading, using a larger, heavier alternator means that operation can take place in a more favorable portion of the efficiency curve (Fig. 1).
On the other hand, the increased efficiency of the larger alternator can be offset by an accompanying increase in fuel consumption. What must be taken into account though is the increased moment of inertia which results in a higher energy input for accelerating the rotor.

The alternator is typical for a permanently operating vehicle assembly, and regarding fuel-consumption must first of all be optimized with respect to efficiency and then with respect to weight.

Sources of power loss

The power losses are shown in Fig. 2 below. The major losses are either "iron losses", "copper losses" or "mechanical losses". Iron losses result from the hysteresis and eddy currents produced by the alternating magnetic fields in the rotor and the stator. They increase superproportionally along with frequency. That is, along with the rotational speed and with the magnetic induction. The copper losses represent the resistive losses in the stator windings. Their extent is proportional to the power-to-weight ratio, i.e., the ratio of generated electrical power to the mass of the effective components. The mechanical losses include friction losses at the rolling bearings and at the sliding contacts as well as the windage losses of the rotor and the fan. At higher speeds, the fan losses increase considerably.

Fig. 1

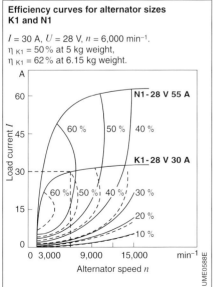

Efficiency curves for alternator sizes K1 and N1

$I = 30$ A, $U = 28$ V, $n = 6,000$ min^{-1}.
$\eta_{K1} = 50\%$ at 5 kg weight,
$\eta_{K1} = 62\%$ at 6.15 kg weight.

Load current I — N1 - 28 V 55 A — K1 - 28 V 30 A

Alternator speed n

UME0588E

Fig. 2

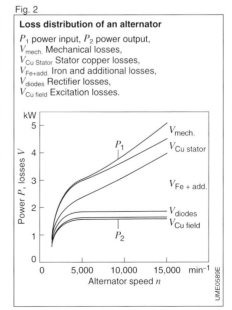

Loss distribution of an alternator

P_1 power input, P_2 power output,
$V_{mech.}$ Mechanical losses,
$V_{Cu\ Stator}$ Stator copper losses,
$V_{Fe+add.}$ Iron and additional losses,
V_{diodes} Rectifier losses,
$V_{Cu\ field}$ Excitation losses.

Power P, losses V — kW — $V_{mech.}$ — $V_{Cu\ stator}$ — $V_{Fe\ +\ add.}$ — V_{diodes} — $V_{Cu\ field}$ — P_1 — P_2

Alternator speed n

UME0589E

Characteristic curves

Alternator performance

The characteristic performance of the alternator at a variety of different speeds is shown by the characteristic curves. Due to the constant transmission ratio between alternator and engine, the alternator must be able to operate at greatly differing speeds.

As the engine takes the alternator from standstill up to maximum speed, the al-ternator passes through certain speeds. Each of these rotational speeds is of particular importance for understanding the alternator's operation and each has therefore been allocated a specific name. Normally, the curves for alternator current and drive power are shown as a function of the rotational speed (Fig. 1). The characteristic curves of an alter-nator are always referred to a constant voltage and precisely defined tempera-ture conditions. For instance, an ambient temperature of 80 °C (or a room tempera-ture of 23 °C) is specified for the limit-temperature test.

Fig. 1

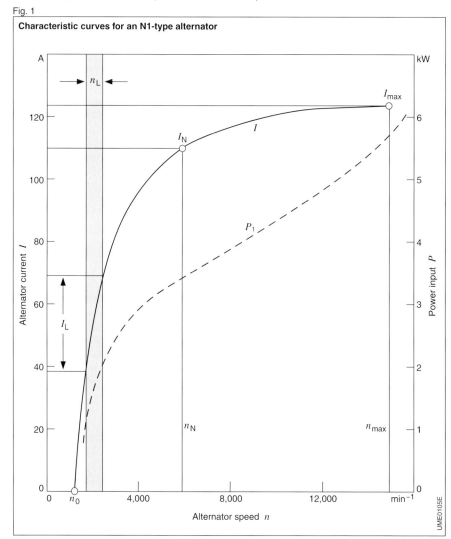

Characteristic curves for an N1-type alternator

150

UME0105E

Current characteristic curve (I)

n_0 0-Ampere speed
The 0-Ampere speed is the speed (approx. 1,000 min^{-1}) at which the alternator reaches its rated voltage without delivering power. This is the speed at which the curve crosses the min^{-1} abscissa. The alternator can only deliver power at higher speeds.

n_L Speed with engine idling
I_L Current with engine idling
With the speed increasing, alternator speed n_L is reached with the engine at idle. This point is shown as an area in Fig. 1 since the precise value depends upon the transmission ratio between engine and alternator. At this speed, the alternator must deliver at least the current required for the long-term consumers. This value is given in the alternator's type designation. In the case of compact-diode-assembly alternators, $n_L = 1,500$ min^{-1}, for compact alternators $n_L = 1,800$ min^{-1} due to the usually higher transmission ratio.

n_N Speed at rated current
I_N Rated current
The speed at which the alternator generates its rated current is stipulated as $n_N = 6,000$ min^{-1} The rated current should always be higher than the total current required by all loads together. It is also given in the type designation.

n_{max} Maximum speed
I_{max} Maximum current
I_{max} is the maximum achievable current at the alternator's maximum speed. Maximum speed is limited by the rolling bearings and the carbon brushes as well as by the fan. With compact alternators it is 18,000...20,000 min^{-1}, and for compact-diode-assembly alternators 15,000...18,000 min^{-1}. In the case of commercial vehicles, it is 8,000... 15,000 min^{-1} depending upon alternator size.

n_A Cutting-in speed
The cutting-in speed is defined as that speed at which the alternator starts to deliver current when the speed is increased for the first time. It is above the idle speed, and depends upon the pre-excitation power, the rotor's remanence, the battery voltage, and the rate of rotational-speed change.

Characteristic curve of power input (P_1)

The characteristic curve of power input is decisive for drive-belt calculations. Information can be taken from this curve concerning the maximum power which must be taken from the engine to drive the alternator at a given speed. In addition, the power input and power output can be used to calculate the alternator's efficiency. The example in Fig. 1 shows that after a gradual rise in the medium-speed range, the characteristic curve for power input rises again sharply at higher speeds.

Explanation of the type designation

Every Bosch alternator carries a rating plate containing type designation and 10-digit Part Number which in the case of alternators always starts with 0 12... .
The type designation gives information on the alternator's most important technical data such as current at engine idle and rated voltage etc.

Example of a type designation
K C ($\rightarrow$) 14 V   40–70 A

K	Alternator size (stator OD),
C	Compact alternator,
($\rightarrow$)	Direction of rotation, clockwise,
14 V	Alternator voltage,
40 A	Current at $n = 1,800$ min^{-1},
70 A	Current at $n = 6,000$ min^{-1}.

Alternator circuitry

Sometimes, the alternator or the vehicle electrical system is confronted with requirements which cannot always be fully complied with by the standard series-production versions.

For such cases, there are special circuitry variants available which can be implemented individually or in combination.

Parallel-connected power diodes

As already dealt with in the section on semiconductor devices, diodes can only be loaded up to a certain current level without damage. At high currents, excessive heat-up would destroy the diodes. This is particularly important when considering the heavily loaded power diodes in the 3-phase AC bridge circuit through which the entire alternator current flows.

However, the alternator's maximum achievable power output is limited by the maximum possible generator current. And high-power alternators feature a generator current which is so high that the six power diodes in the normal 3-phase AC bridge circuit are unable to handle it.

For this reason, such alternators are equipped with two or more parallel-connected power diodes for each phase. As a result, the alternator current is divided between the parallel-connected diodes so that the individual diodes are no longer overloaded.

The circuit diagram of an alternator using this principle is shown in Fig. 1. Two power diodes are connected in parallel for each phase, which means that the 3-phase AC rectifier comprises 12 power diodes instead of six.

Auxiliary diodes at the star (neutral) point

On alternators with star-connected stator windings, the ends of the windings are joined at a single point, the star or neutral point.

Since, at least theoretically, the addition of the three phase currents or phase voltages is always zero at any instant in time, this means that the neutral conductor can be dispensed with.

Due to harmonics, the neutral point assumes a varying potential which changes periodically from positive to negative. This potential is mainly caused by the

Fig. 1

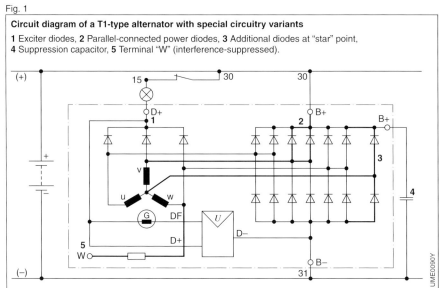

Circuit diagram of a T1-type alternator with special circuitry variants

1 Exciter diodes, **2** Parallel-connected power diodes, **3** Additional diodes at "star" point, **4** Suppression capacitor, **5** Terminal "W" (interference-suppressed).

"third harmonic" which is superimposed on the fundamental wave and which has three times its frequency (Fig. 2). The energy it contains would normally be lost, but instead it is rectified by two diodes connected as power diodes between the neutral point and the positive and negative terminals (Fig. 1). As from around 3,000 min⁻¹, this leads to an alternator power increase of max. 10%. These auxiliary diodes increase the ripple of the alternator voltage.

Operation of alternators in parallel

If demanded by power requirements, alternators with the same power rating can be connected in parallel. Special balancing is not necessary, although the voltage regulators concerned must have the same characteristics, and their characteristic curves must be identical.

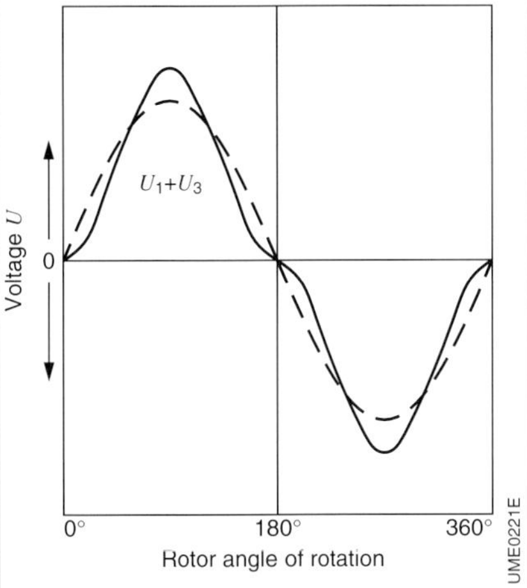

Terminal "W"

For specific applications, Terminal "W" can be connected to one of the three phases as an additional terminal (Fig. 1). It provides a pulsating DC (half-wave-rectified AC) which can be used for measuring engine speed (for instance on diesel engines).

According to the following equation, the frequency (number of pulses per second) depends on the number of pole pairs and upon alternator speed.

$f = p \cdot n/60$

f Frequency (pulses per second),
p Number of pole pairs (6 on Size G, K and N; 8 on Size T),
n Alternator speed (min⁻¹).

Interference-suppression measures

The main source of electrical interference in the SI engine is the ignition system, although some interference is also generated by alternator and regulator, as well as by other electrical loads. If a 2-way radio, car radio, or car telephone etc. is operated in the vehicle itself or in the vicinity, it is necessary to install intensified interference suppression of alternator and regulator. For this purpose, alternators are fitted with a suppression capacitor. In the case of compact-diode-assembly alternators, if not present as standard equipment, the suppression capacitor can be retrofitted on the outside of the collector-ring shield. On compact alternators it is already integrated in the rectifier.

Older versions of the contact regulator are combined with an interference-suppression filter or are replaced by an interference-suppressed version. Transistor regulators do not require additional suppression measures. If Terminal "W" is connected, this can be suppressed with a resistor which is installed in the "W" line (Fig. 1).

Fig. 2

Voltage with third harmonic
U_1 Phase voltage (fundamental wave),
U_3 Third-harmonic voltage.

UME0221E

Operation of alternators in the vehicle

In the vehicle, engine, alternator, battery, and electrical loads, must be considered as an interrelated system.

Energy balance in the vehicle

When specifying or checking alternator size, account must be taken of the battery capacity, the power consumption of the connected loads, and the driving conditions.

Alternator size and battery capacity are specified by the automaker in accordance with the electrical loads installed in the vehicle and the normal driving conditions. Individual circumstances can deviate from the above conditions though. On the one hand because the operator installs extra electrical equipment in the vehicle, and on the other because driving conditions differ considerably from those taken as normal.

These considerations are intended to underline the fact that the total input-power requirements, together with the individual driving conditions, are of decisive importance with regard to the loading of the alternator and battery.

An adequate state of battery charge is the prime consideration. It is decisive for sufficient energy being available to start the engine again after it has been switched off. The battery functions as an energy store which supplies the various loads, and which in turn must continually be charged by the alternator in its function as the energy supplier. On the other hand, if the energy drawn from the battery exeeds that supplied to it, even a high-capacity battery will gradually discharge until it is "empty" (or "flat"). The ideal situation is a balance between input and output of energy to and from the battery (Fig. 1).

Thus, a correctly dimensioned alternator is decisive for an adequate supply of on-board energy. An under-rated (i.e. over-loaded) alternator is not able to keep the battery sufficiently charged, which means that battery capacity cannot be fully utilized.

Consequently, if power demand is increased, for instance as a result of fitting extra equipment, it is advisable to replace the standard fitted alternator by a more powerful version. One of the most important steps to be taken when ascertaining the electrical system's charge balance is the registration of all the installed electrical loads (including retro-fitted equipment), together with their power inputs and the average length of time they are switched on (short-term, long-term, or permanent loads). Similarly, the driving cycles as dictated by the traffic situation must also be considered. These include, for instance, low alternator speeds typical for town traffic coupled with repeated standstills, expressway traffic with congestion, and high alternator speeds on clear first-class roads. The time of day (journeys mainly by daylight or during the dark), and the season (winter or summer driving with the related temperatures and weather), also have an affect.

We can sum up as follows:

Even under the most unfavorable operating conditions, in addition to powering all the electrical loads, the alternator power must suffice to keep the battery sufficiently charged so that the vehicle is always ready for operation.

An expert should be consulted before the final selection of alternator size and its matching to the appropriate battery. The following example illustrates the loading of the vehicle's energy household by the electrical loads under a variety of different conditions:

Fig. 1

Charge balance

Current flow between alternator, battery, and electrical devices, with constant power demands from the loads, and varying alternator operating conditions.

In general, the following applies:

$$I_G = I_W + I_B$$

Where:
I_G Alternator current
I_W Equipment current
I_B Battery current

The battery current may be positive or negative, depending on whether the battery is being charged or is discharging.

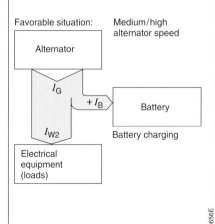

Unfavorable situation: Low alternator speed

Alternator
I_G
$-I_B$ Battery
I_{W1}
Battery discharging
Electrical equipment (loads)

Favorable situation: Medium/high alternator speed

Alternator
I_G
$+I_B$ Battery
I_{W2}
Battery charging
Electrical equipment (loads)

UME0656E

Operation of high-beam headlamps

These are used mainly for overland trips at high engine speeds (i.e., high alternator speeds), and with low traffic density. They are switched to low-beam in the event of oncoming traffic. High-beam headlamps are not required in town traffic with its low engine speeds and short distances. These types of load are no problem for the vehicle's energy household. The alternator is operated at a speed which ensures that all loads are supplied with enough electrical energy and that the battery is charged. Here, the operating conditions are favorable.

Operation of fog lamps

On the other hand, the situation is not so favorable regarding fog lamps. Since fog forces the driver to drive slowly, this means that fog lamps are usually switched on in the lower engine-speed range at which the alternator does not yet generate full power. And even when traffic approaches, the fog lamps are not switched off, so that their overall switched-on time is relatively long. Here, the operating conditions are unfavorable. In many cases, the loading placed on the vehicle's energy household also depends upon the driver's skill and common sense.

Operation of rear-window heating

Although the rear-window heater is a relatively high current consumer, it only remains on briefly until the backlite has cleared. However, if the driver then forgets to switch the heater off, this results in a considerable drain on the vehicle's energy household as provided by the alternator and battery

Operation of other loads

Compared to the examples dealt with above, electronic equipment, turn-signal indicators, horn, and instrument cluster, are insignificant as regards the loading they impose on the system.

Determining the correct alternator

(Example refers to Fig. 2)
The following method enables a check to be made whether the installed alternator version suffices for supplying the vehicle electrical system:

1. Determine the power input for all the loads that are switched on permanently or for prolonged periods at 14 V. The sum results in a power input of $P_{W1} = 350$ W.

2. Determine the power input of all short-term loads at 14 V.

Fig. 2

Checking the alternator size. Alternator type K1-14 V 23//55 A.

1. Power demand (for 14 V) of all loads switched on either continuously or for prolonged periods.

Electrical devices or systems (loads) Factor 1.0	Power W
Ignition system	20
Electric fuel pump	70
Electric gasoline injection	100
Car radio	12
Lower beam	110
Side-marker lamps	8
Tail lamps	10
License-plate lamps	10
Instrument-panel lamps	10
Power 1	$P_{W1} = 350$ W

2. Power demand (for 14 V) of all loads switched on for brief periods.

Electrical devices and systems (loads)	Actual value W	Factor*	Estimated consumed W
Blower for heating and/or ventilation	80	0.5	40
Heated rear screen	120	0.5	60
Wipers	60	0.25	15
Electr. radiator fan		0.1	
Aux. driving lamps		0.1	
Stop lamps	42	0.1	4.2
Turn-signal lamps	42	0.1	4.2
Fog lamps	70	0.1	7
Fog warning lamps	35	0.1	3.5
Power 2			$P_{W2} = 134$ W

Total power
$$P_W = P_{W1} + P_{W2} = 484 \text{ W}$$

3. Generator rated current

P_W (for 14 V) W	350 ... < 450	450 ... < 550	550 ... < 675	675 ... < 800	800 ... < 950
I_N A	45	55	65	75	90

4. Load current at idle

Current of all devices (loads) switched on either continuously or for prolonged periods
$$I_{W1} = P_{W1} / 14 \text{ V} = 25 \text{ A}$$

Alternator characteristic curve

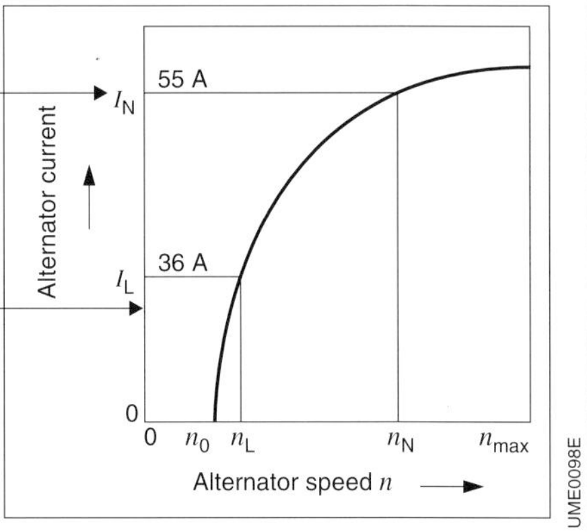

Calculated demand:
$$I_L = 1{,}3 \cdot I_{W1} = 33 \text{ A}$$

Approximation
$$I_L = 36 \text{ A} > 33 \text{ A}$$

* Actual value of load x factor = estimated consumed power

The sum results in a power input of:
$P_{W2} = 134$ W (rounded off).
The system's total power input P_W results from the addition of P_{W1} und P_{W2}:
$P_W = 484$ W.
3. Using the reference table, it is now possible to determine the minimum rated current necessary: $I_N = 55$ A. Provided the correct size of alternator has been fitted, this rated current, or a higher figure, appears in the type designation – in our example 55 A.
4. A further check can be made using the alternator current I_L at engine idle.
I_L can be taken from the alternator's characteristic curve, provided that the alternator speed n_L at engine idle is known. In our example, the alternator speed is:
$n_L = 2,000$ min^{-1}.
Practical experience has shown that for passenger cars, at engine idle I_L should exceed the current I_{W1} by a factor of 1.3. I_{W1} results from the input power P_{W1} for all permanent and long-term loads. This ensures efficient battery charging even at engine idle and when only short distances are travelled.
In the example:
At idle, the alternator delivers a current of $I_L = 36$ A. The current I_{W1} is calculated from the power P_{W1} ($I_{W1} = P_{W1}/14$ V). This results in $I_{W1} = 25$ A from which a required current of 33 A is calculated. Since $I_L = 36$ A, this means that the power demand is safely covered.

Alternator installation and drive

Installation
The motor-vehicle operator usually has little say concerning the alternator or regulator fitted in his vehicle. And in every vehicle, the alternator's installation position is dependent upon the conditions prevailing in the engine compartment due to construction and design.
However, certain basic factors must always be borne in mind concerning installation:
– Good accessibility for readjusting the

V-belt tension and for any maintenance work which may be required,
– Adequate cooling for alternator waste heat as well as for heat conducted and radiated from the engine.
– Protection against dirt, moisture, shock, impact, fuel and lubricants (ingress of gasoline leads to the danger of fire and explosion, and diesel fuel damages the carbon brushes and collector rings).
Almost without exception, alternators which are driven by the engine through normal V-belts are attached by means of a swivel-arm mounting. In addition to the mounting using a swivel bearing, an adjustment facility (to pivot the alternator around a swivel arm) is provided for adjusting the V-belt tension.
If the alternator is driven through a ribbed V-belt (poly-V belt), the alternator is usually rigidly mounted. The belt is adjusted using a belt tensioner (Fig. 3).

Fig. 3: V-belt and ribbed V-belt alternator drive

Fig.: VW

UME0574Y

In special cases, large alternators are cradle-mounted in a recess directly on the engine.

Irrespective of the type of mounting, all alternators must have good electrical connection to the engine block. Furthermore, since current return from the electrical system is in the most cases via ground, there must be a highly conductive ground connection of adequate cross-section between engine and chassis.

Buses and special-purpose vehicles are often equipped with extra return lines in order to reduce voltage losses and thereby increase safety.

Electric cables and lines only provide efficient connection if they are provided with properly attached terminals or plug connectors.

Alternator drive

Alternators are driven directly from the vehicle engine. As a rule, drive is via V-belts. Less frequently, flexible couplings are used.

The belt drive (using V-belts, ribbed-V belts etc.) is the most important element in the transmission of power, and as such it is subject to exacting requirements:

– The belt material must have very high flexural strength,
– Belt slip leads to heat-up and wear, and in order to prevent it, longitudinal stretch should remain at a minimum as the belt gets older.

Investigations conducted by ADAC (Germany's largest automobile club) have revealed that V-belt damage is a frequent cause of breakdown. It is therefore important to use V-belts which comply with the above requirements and which are capable of a long service life. Typical for automotive applications are the "open-flank" belt and the ribbed-V belt.

The "open-flank" design (Fig. 4) features high flexibility, coupled with extreme lateral rigidity and resistance to wear. Particularly with small-diameter pulleys, this leads to improved power transmission and a longer service life compared to conventional rubber-jacketed V-belts.

The high flexibility of the ribbed-V belt (Poly-V belt, Fig. 5) permits very small bending radiuses. This in turn means that small-diameter pulleys (minimum dia.: 45 mm) can be fitted to the alternators thus permitting higher transmission ratios. The back of the belt may also be

Fig. 4

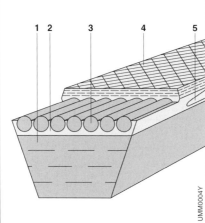

Construction of an "open-flank" V-belt

1 Short-cut fiber mixture,
2 Embedding compound,
3 Specially prepared cord,
4 Cover fabric,
5 Cut flanks.

UMM0004Y

Fig. 5

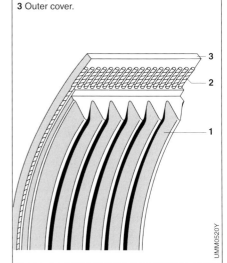

Construction of a ribbed V-belt (Poly-V-belt)

1 Carcasse,
2 Cord,
3 Outer cover.

UMM0520Y

used to transmit power, thus enabling a number of aggregates (alternator, radiator fan, water pump, power-steering pump etc.) to all be driven from a single belt with an adequate wrap angle around each pulley wheel.

Usually, a single V-belt suffices to drive small-power alternators. With large-power alternators on the other hand, two V-belts are more common in order to overcome the alternator's resistance to turning which is inherent in its higher power.

Depending upon application, pulley wheels and fan wheels are available which have either been turned, or stamped from sheet metal, and which can be combined with each other as required (Fig. 6).

The pulley wheel's correct diameter depends upon the required transmission ratio between engine and alternator.

Being as the speed ranges covered by the multitude of engines concerned differ considerably from each other, there is a wide variety of different pulley-wheel diameters available.

The transmission ratio must take into account the fact that the alternator's permitted maximum speed must not be exceeded at the engine's maximum speed.

Notes on operation

Battery and regulator must be connected when the alternator is operated. This is the normal operating setup and the installed electronic equipment and semiconductor devices perform efficiently and safely.

Emergency operation without the battery connected results in high voltage peaks which can damage equipment and components. Here, efficient emergency operation is only possible if precautionary measures are taken.

There are three alternatives:
– Zener diodes in the rectifier
– Surge-proof alternator and regulator
– Overvoltage protection devices.

Fig. 6

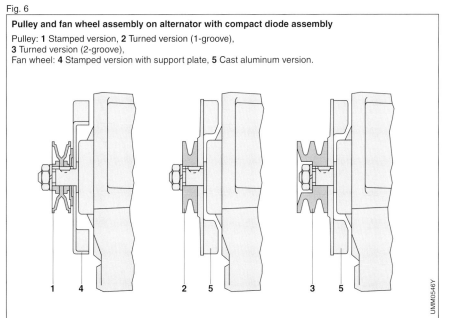

Pulley and fan wheel assembly on alternator with compact diode assembly

Pulley: **1** Stamped version, **2** Turned version (1-groove),
3 Turned version (2-groove),
Fan wheel: **4** Stamped version with support plate, **5** Cast aluminum version.

Connecting the battery into the vehicle's electrical system with the wrong polarity, immediately destroys the alternator diodes, and can damage the regulator, no matter whether the engine is switched off or running. The same damage can occur if an external voltage source is used as a starting aid and the terminals are reversed.

Special circuitry is available to safeguard against reverse-polarity damage. When the battery is falsely connected, engine start is blocked in order to protect alternator and regulator.

The charge indicator lamp acts as a resistor in the alternator circuit. If a correctly rated lamp is fitted which draws enough current, the resulting pre-excitation current provides a magnetic field which is strong enough to initiate alternator self-excitation.

When the charge indicator lamp lights up, this merely indicates to the driver that the ignition or driving switch is switched on and that the alternator is not yet feeding power into the electrical system. The lamp goes out as soon as the alternator's self-excitation speed is reached and the alternator supplies energy to the electrical system. The lamp therefore provides an indication that alternator and regulator are functioning correctly, that they are correctly connected, and that the alternator is supplying current.

The charge indicator lamp gives no indication as to whether, and as of what speed, the battery is being charged. When the alternator is heavily loaded, it can happen that even though the lamp has gone out, the battery is not being charged but discharged. The lamp gives no information concerning the state of battery charge even though it is erroneously referred to as the "charge indicator lamp".

If the lamp is defective (broken filament), this means that pre-excitation current cannot flow and self-excitation first sets in at very high speeds. This error is noticeable when the lamp fails to light up with the engine at standstill and the ignition switched on.

If there is an open-circuit in the excitation circuit, in the pre-excitation line, or in the alternator ground line, and the alternator breaks down completely as a result, the driver is not warned of this fact even though the charge indicator lamp is intact.

Here, it is necessary to connect in an additional resistor (Fig. 7) so that the charge indicator lamp lights up to inform the driver of the open-circuited excitation circuits.

If the charge indicator lamp fails to go out even at high speeds, this indicates a fault in the alternator itself, at the regulator, in the wiring, or at the V-belt.

Service life, mileage, maintenance intervals

Using a variety of statistical methods, and taking typical operating conditions into account, it is possible to calculate specific average service lives, mileages, and driving cycles for different categories of vehicle (passenger car, commercial vehicle, long-haul truck, town and long-distance buses, and construction machinery).

Fig. 7

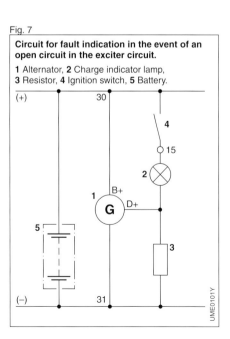

Circuit for fault indication in the event of an open circuit in the exciter circuit.

1 Alternator, **2** Charge indicator lamp, **3** Resistor, **4** Ignition switch, **5** Battery.

Considering the different fields of application of these vehicle categories, the requirements and criteria for the economic efficiency of their alternators also differ. This leads to there being a range of alternators available for different service lives and maintenance intervals.

Depending upon version and application, passenger-car alternators with encapsulated ball bearings have service lives of 150,000 ... 250,000 km.

Presuming that the engine's service life until it is replaced or has a major overhaul corresponds to that of its alternator, this makes specific maintenance work on the alternator unnecessary. The grease in the bearings suffices for this period.

Due to the use of particularly wear-resistant components, the alternators installed in trucks and buses for instance achieve mileages of 200,000 ... 300,000 km. One prerequisite is that they are equipped with suitable ball bearings, featuring enlarged grease chambers for instance.

It is recommended that engines which are designed to cover mileages exceeding 300,000 km before their first major overhaul are equipped with alternators featuring windingless rotors and high-rating rolling bearings.

Provided the alternator is installed in a location which is relatively free from dirt, oils, and grease, the carbon-brush wear is negligible due to the low excitation currents involved.

Workshop testing techniques

Bosch customer service

Customer service quality is also a measure for product quality. The car driver has more than 10,000 Bosch Service Agents at his deposal in 125 countries all over the world. These workshops are neutral and not tied to any particular make of vehicle. Even in sparsely populated and remote areas of Africa and South America the driver can rely on getting help very quickly. Help which is based upon the same quality standards as in Germany, and which is backed of course by the identical guarantees which apply to customer-service work all over the world. The data and performance specs for the Bosch systems and assemblies of equipment and precisely matched to the engine and the vehicle. In order that these can be checked in the workshop, Bosch developed the appropriate measurement techniques, test equipment, and special tools and equipped all its Service Agents accordingly.

Testing technology for alternators

Average driving cycles and operating times can be determined for all types of different vehicles (passenger cars, trucks, buses, construction machinery etc.) and their typical operating conditions.

Various alternator designs which require no special maintainance work during their service life are available to comply with this variety of requirements. In case of malfunction in the vehicle's electrical power-generating system, a check should first of all be carried out directly in the vehicle.

If, in the process, an alternator defect is located, the alternator is either replaced or repaired using the service-information and service-instructions documentation.

Before being installed in the vehicle again, the alternator must be tested on the combination test bench.

Testing directly in the vehicle

A visual inspection is first of all carried out to check the V-belt, the wiring, and the charge-indicator lamp.

Basically, an engine analyzer and a volt-ammeter are required for the electrical tests.

The following tests can be performed using these two testers

- Oscilloscope display of the DC voltage with low harmonic ripple (between D+ and B−),
- Regulator voltage (between D+ and B−),
- Alternator output,
- Charging current under load,
- Current without load,
- Battery voltage,
- Short-circuit of lines to ground or plus (+),
- Line open-circuit, and
- Contact resistance of the lines.

Alternator repair

For the various alternator types, there are specific service instructions available which describe the alternator repairs.

These instructions also contain the relevant test and adjustment values.

A number of different testers (e.g. alternator tester and interturn-short-circuit tester) are used when repairs are carried out on alternators. In addition, in order that defects inside the alternator can

Fig. 1

Combination test bench for starting motors and alternators

1 Operator panel for alternator and starting-motor tests,
2 Adjustable loading resistor (alternator test),
3 Handwheel for adjusting the height of the clamping table (alternator test),
4 Alternator test setup with protective hood,
5 Socket contact for rotational-speed sensor (alternator test),
6 Display unit, 7 Starting-motor test setup, 8 Battery compartment.

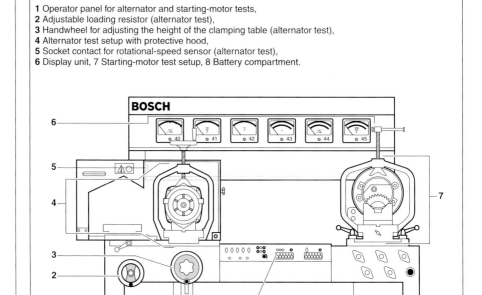

be located and repaired efficiently, special tools are required for each alternator type.

Checking the alternator on the combination test bench

Once the alternator has been repaired, it is clamped in the relevant test set-up on the combination test bench (Fig.1).
Depending upon the version concerned, the alternator can be driven directly up to speeds of approx. 6,000 min^{-1}. At higher speeds, the alternator is driven through a V-belt (Fig. 2).
The alternator is bolted to the clamping device using a swivel arm. The rotational-speed sensor is calibrated after aligning and tensioning the V-belt. The alternator is then connected electrically.

Two points on the power curve are run up to when testing an alternator:
Using an adjustable loading resistor, the alternator is loaded with the maximum attainable current at two different test speeds (e.g. 1,500 and 6,000 min^{-1}). The alternator voltage must remain above the stipulated limit value.
If these desired values are reached, the alternator can be installed in the vehicle.

Fig. 2

Testing the clamped alternator

1 Clamping table,
2 Guide,
3 Clamping device,
4 Swivel arm,
5 Drive,
6 V-belt,
7 Alternator.

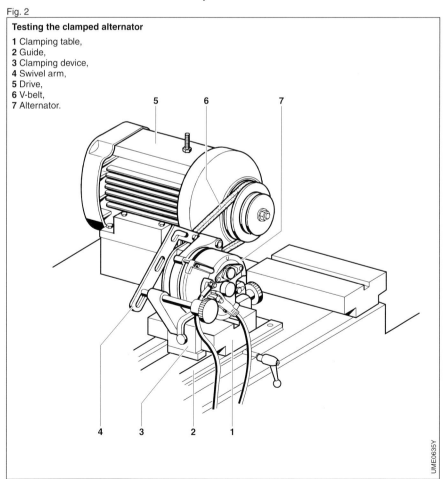

UME0635Y

Starting systems

Basics

Starting

Internal-combustion (IC) engines must be started by a separate system because they cannot self-start like electric motors or steam engines. When starting these engines, considerable resistance resulting from compression, piston friction and bearing friction (static friction) must be overcome. These forces depend greatly on engine type and number of cylinders, as well as on lubricant characteristics and engine temperature. Frictional resistance is highest at low temperatures.

Even under severe conditions, the starter (also known as the "starting motor") must crank the gasoline engine (spark-ignition/SI engine) fast enough so that it can form the combustible air-fuel mixture required for starting, and diesel engines fast enough to reach their self-ignition temperature.

To satisfy these requirements, the starter must rotate the flywheel at a minimum starting speed. It must also continue to support rotation during initial combustion

Fig. 1

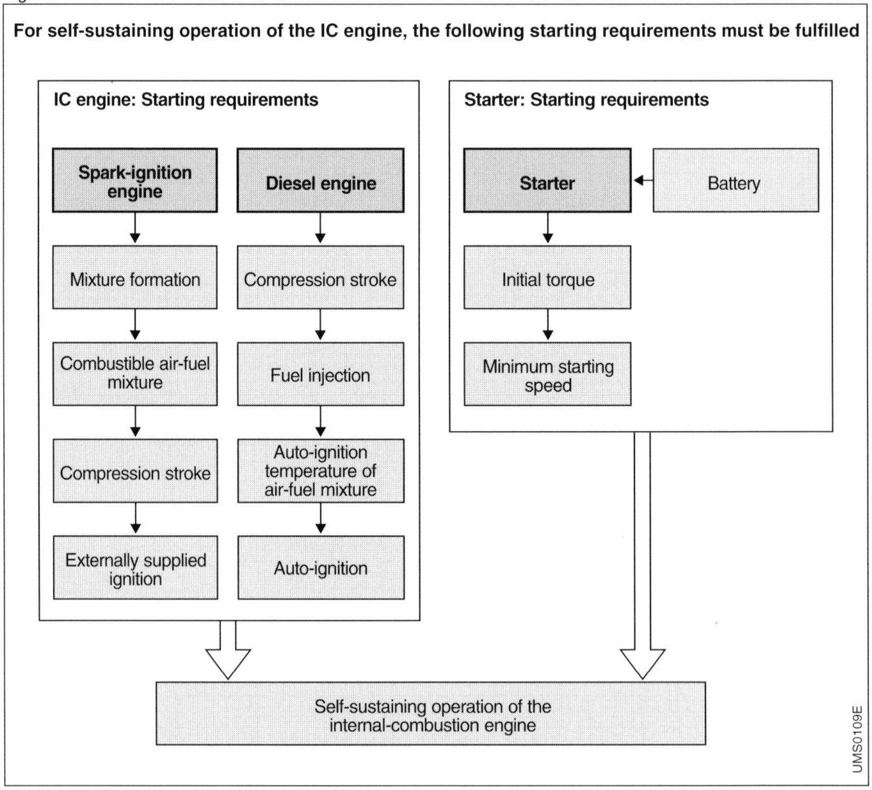

For self-sustaining operation of the IC engine, the following starting requirements must be fulfilled

UMS0109E

164

to maintain momentum until the engine can sustain operation.

Electric motors (DC, AC and three-phase), as well as hydraulic and pneumatic motors are used as starting motors for internal-combustion engines.

The DC series-wound electric motor is particularly well-suited for use as a starting motor, because it generates the high initial torque required to overcome cranking resistance and to accelerate the engine's internal masses. In the majority of cases, starting-motor torque is transmitted to the engine via a starter pinion and a ring gear on the crankshaft-mounted flywheel. However, V-belts, toothed belts and chains are also used, as is direct transmission to the crankshaft.

As a result of the high gear ratio between the starter pinion and the ring gear on the engine flywheel, the "pinion-type starter" can be designed for low torque and high speed, thus allowing small, lightweight starters to be used.

An additional advantage is that the energy required to start the engine can be supplied by the same battery normally used to operate the other electrical devices in the vehicle electrical system (Fig. 1).

For this reason, the starter cannot be viewed as an independent component, but, rather, must be discussed as an integral part of the electrical system.

Both starter and battery are sized such that even under adverse operating conditions cranking power is available long enough to start the engine. Because the starter draws more current than any other vehicle electrical system, it is often decisive in determining the battery specifications.

The starter itself (Fig. 2) must meet the following requirements:
– Continuous readiness for starting,
– Sufficient starting power at different temperatures,
– Long service life for a high number of starts (particularly if the vehicle is used primarily for in-city driving),

– Robust design to withstand meshing, cranking, vibration, corrosion due to dampness and road salt, dirt, temperature cycles within the engine compartment etc.,
– Low weight and small size, and
– Longest possible maintenance-free service life.

Since starting requirements vary considerably, and particularly the effects of temperature are highly significant, the starter must be precisely aligned to the other components of the electrical system, and to the engine with which it is to be used.

Fig. 2

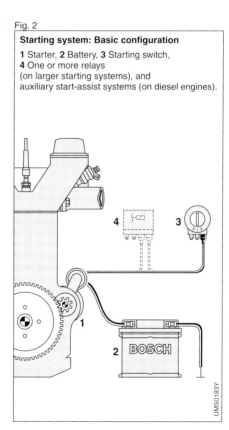

Starting system: Basic configuration

1 Starter, **2** Battery, **3** Starting switch,
4 One or more relays
(on larger starting systems), and
auxiliary start-assist systems (on diesel engines).

UMS0183Y

Starting requirements

In designing a starting system, both engine specifications and starting requirements must be considered. These requirements include:

– Minimum starting temperature. This is the lowest engine and battery temperature at which the system must ensure that the engine starts (Fig. 3),

– The engine's resistance to rotation. This equals the resistance to rotation, as measured at the crankshaft at the minimum starting temperature (thus including permanently-connected ancillaries, Fig. 4)

– Minimum required engine speed at the minimum starting temperature,

– Starter pinion/ring gear ratio,

– Rated voltage of the starting system,

– Specifications/capacity of the starter battery,

– Length and resistance of the cables (Fig. 5) between battery and starter (voltage drop),

– Torque, speed and capacity of the starter (starter characteristic curve, starting process), etc.

Of particular importance in this respect is the minimum starting temperature, i.e. the lowest temperature at which an engine with a given electrical system, a defined state of battery charge, and given oil viscosity can be brought to self-sustaining speed.

An engine's minimum starting temperature is defined by the climatic conditions at the place of use, the conditions under which the engine must operate and economic considerations (the power required of a starting system, as well as its costs, increase rapidly in response to downward definitions of the minimum starting temperature).

Fig. 3

Starting limit temperature (example)

a Starter speed; decreases as the temperature drops due to increased internal resistance of the battery.

b Minimum required initial engine speed; increases as temperature drops due to increased cranking resistance. The intersection of both curves yields the starting limit temperature (here −23 °C).

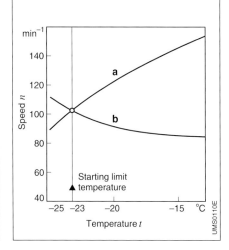

Fig. 4

Engine torques (cranking resistances) and starter torques

M_S Starter torque for various temperatures (referred to the engine crankshaft).

M_M Torque required for starting a 3-liter SI engine at the different temperatures shown. The intersection point of the relevant curves determines the speed at which the engine is cranked at −25 °C, −18 °C, and −10 °C.

The torque curve is referred to a 20 % discharged 55 A·h battery.

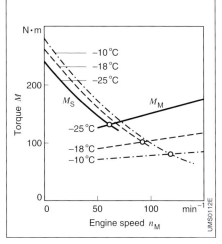

In the example given, a 2.2 kW starter and a 12 V, 90 A·h, 450 A battery are required for a minimum starting temperature of −23 °C. The battery in the example is discharged to 80 % of its rated capacity (Fig. 3). The colder the engine, the higher the rotating speed needed to get it started. Ideally, the starter should compensate for the engine's cold response pattern with higher output speeds.

Unfortunately, since it depends upon the battery for its energy supply, and the battery responds to colder temperatures with a disproportionate increase in internal resistance, the starter turns more slowly. Bosch frequently examines these patterns in starting and cranking tests in the cold-climate chamber at its Technical Center for Automotive Electrical Systems.

Starter systems intended for application within Europe are usually designed with reference to the minimum starting temperatures listed in Table 1.

The starting resistance, that is, the torque required to overcome friction and inertia, is largely a function of engine displacement and oil viscosity (index of the engine's internal friction).

As a general rule, the mean rotational resistance in spark-ignition engines continues to increase as a function of crankshaft speed (on diesels, in contrast, resistance peaks at 80 to 100 min⁻¹ before again falling as the relatively high levels of compression energy are fed back into the system). The intersection of the engine and starter torque curves (Fig. 4) indicates the engine's rotational speed at any given temperature.

Additional factors include: Engine design and number of cylinders, bore/stroke ratio, compression ratio, engine speed, mass of engine's moving parts, crankshaft assembly, etc., along with bearings, additional loads from clutch assembly, transmission, etc.

The minimum starting speed will vary substantially according to the design of the engine and its mixture-formation system. Start-assist systems are another important factor on diesel engines. Table 2 provides some interesting empirical data.

Table 1.

Starting limit temperature	
Engines for	Starting limit temperatures
Passenger cars	−18...−25 °C
Trucks and buses	−15...−20 °C
Tractors	−12...−15 °C
Drive and equipment engines on ships	−5 °C
Diesel locomotives	+5 °C

Table 2.

Empirical values for minimum cranking speeds	
Required cranking speeds at −20 °C	Cranking speed min⁻¹
Reciprocating-piston SI engine	60...90
Rotary-piston SI engine	150...180
Direct-injection diesel engine without start-assist	80...200
with start-assist (e.g. glow plug)	60...140
Pre-chamber and whirl-chamber diesel engines without start-assist	100...200
with glow-plug start-assist	60...100

Fig. 5

Basic circuit diagram of a starting system

R_L Line resistance,
R_B Internal battery resistance,
R_S Internal starter resistance.
1 Battery,
2 Starter cable,
3 Starter.

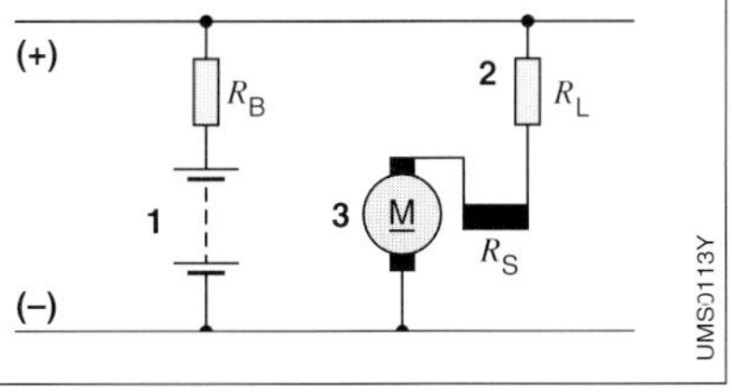

Figure 6 illustrates an actual starting process. The engine's combustion mixture starts to ignite at the minimum starting speed. The torque curve then rises during the transition to self-sustaining operation (Curve 1, simplified illustration showing constant progression).

The engine's torque has been superimposed on the downward curve (Curve 2) for starter torque. In this transitional phase, the engine's speed rises to the levels required for self-sustaining operation. The starter reverts to a supporting function which continues until it is overtaken by the engine.

The sum of the two torque curves provides a theoretical composite curve (Curve 3, broken line). In actual practice, initial fluctuations in the combustion process that started at Point A mean that this theoretical curve is achieved only sporadically. This condition continues until engine operation becomes consistent at Point B. At Point C the starter is switched off and the engine continues to operate without external assistance.

Starting-system voltage ratings

Starting systems are available with various rated voltages:
– Passenger cars today generally have 12 V systems.
– Tractors, small auxiliary power units and marine engines also usually have 12 V systems.
– Systems designed for 24 V are used in some engines of this type as well as in special-purpose vehicles.
– Trucks and buses use 12 V and 24 V systems.
– The starters on large commercial vehicles are generally rated at 24 V, as the higher voltage makes it possible to obtain higher specific output from more compact starters.

Capacity rating

Along with its voltage rating, the starter's rated capacity is also an essential index of performance.

The capacity rating is a precisely defined parameter determined on the test bench. It is referred to the largest permissible battery for the starter in question, with a 20 % discharge at a temperature of –20 °C. It is connected to the starter via a cable with a resistance of 1 mΩ. These criteria guarantee that the starter will operate even under adverse conditions. The torque transmitted through the starter pinion represents its generated torque minus iron, copper and friction losses.

Starter output is therefore highly dependent upon line resistance and internal battery resistance. The lower the internal resistance of the battery, the higher the starter output.

Some of the testing employed to determine starter performance under severe conditions is carried out in the cold-climate chamber.

Fig. 6

Internal-combustion engine: Starting procedure

1 Theoretical engine torque assuming smooth combustion,
2 Starter torque,
3 Theoretical total torque (sum of curves 1 and 2),
4 Actual total torque as a result of irregular combustion.
A Irregular combustion begins,
B Uniform engine speed,
C Self-sustaining engine operation.

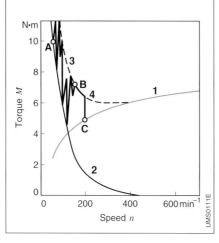

Starting systems for passenger cars

Passenger cars are defined as motor vehicles designed to carry up to 9 persons. Passenger-car starting systems generally have pre-engaged drive starters with a nominal output of approx. 2 kW. The standard rated voltage is 12 V. These systems can start gasoline engines and diesel engines up to a displacement of approx. 7 and 3 litres, respectively. The required cranking power greatly depends upon the type of combustion: A diesel engine requires a more powerful starter than a gasoline engine of equal size.

Passenger-car starter circuits are usually very simple. The engine is located in the vicinity of the driver, who is usually easily able to hear when the engine starts. The driver is therefore not likely to attempt to restart an engine which is already running, thereby possibly damaging the starter pinion as it attempts to engage the ring gear on the engine flywheel. For this reason, passenger cars usually do not have start protection and monitoring de-vices. Many passenger-car models have ignition/starter switches which incorporate additional start repeating blocks to avoid any possibility of accidental starter operation.

Starting systems for passenger cars with gasoline engines

The basic circuit for this starting system is shown in Figure 7. The starting system is usually activated by a multiple-position ignition/starter switch.

The ignition system is switched on before the key reaches the "start" position, because the ignition system must be on for the spark-ignition engine to start, and must remain on for the engine to run.

Ignition continues after the starter is switched off, and allows the gasoline engine to continue running.

In systems with breaker-triggered ignition coils with ballast resistors, starting can be facilitated by increasing the available voltage. This is done by bridging the ignition-coil ballast resistor, and requires starters with an additional terminal (15a).

Fig. 7

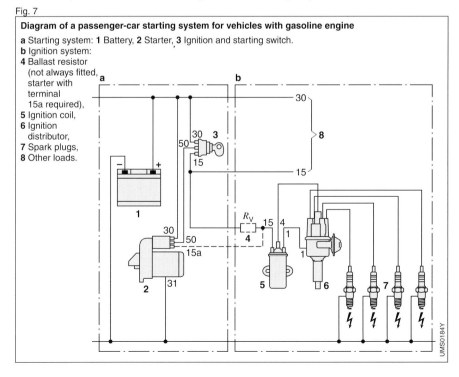

Diagram of a passenger-car starting system for vehicles with gasoline engine

a Starting system: **1** Battery, **2** Starter, **3** Ignition and starting switch.
b Ignition system:
4 Ballast resistor (not always fitted, starter with terminal 15a required),
5 Ignition coil,
6 Ignition distributor,
7 Spark plugs,
8 Other loads.

Starting systems for passenger cars with diesel engines

Before the engine can be started, the preheating system must be switched on. Most late-model passenger-car preheating systems have a combined driving/glow-plug and starter switch which, at the end of the glow duration, can be turned farther to the starting position (Figure 8). In the case of older diesel starting systems, the driving switch and glow-plug and starter switch are still fitted separately. As soon as the surface of the glow-plug becomes hot enough to ignite the diesel fuel, the engine can be started. In contrast to the ignition system of the gasoline engine, the electrical (preheating) system of the diesel engine is switched off together with the starter after the engine has started.

Starting systems for commercial vehicles

Commercial vehicles are vehicles which are designed to carry more than 9 persons, for transporting goods, or for pulling trailers. This category of motor vehicles comprises the following main vehicle groups:

- Buses (e.g. microbuses, public buses, articulated buses),
- Trucks of various sizes,
- Special-purpose vehicles (e.g. tank trucks, fire-department vehicles, tow trucks, sanitation-department vehicles),
- Towing vehicles (road-construction vehicles or tractors and tractor trailers).

Because there are so many different types of commercial vehicles, starting systems must be designed for each specific type of vehicle engine.

Tractors and light-duty commercial vehicles such as delivery trucks and minibuses, are usually equipped with

Fig. 8

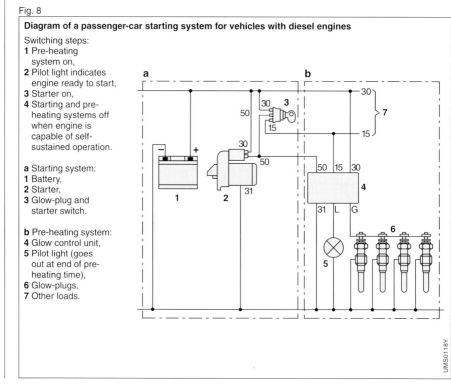

Diagram of a passenger-car starting system for vehicles with diesel engines

Switching steps:
1 Pre-heating system on,
2 Pilot light indicates engine ready to start,
3 Starter on,
4 Starting and pre-heating systems off when engine is capable of self-sustained operation.

a Starting system:
1 Battery,
2 Starter,
3 Glow-plug and starter switch.

b Pre-heating system:
4 Glow control unit,
5 Pilot light (goes out at end of pre-heating time),
6 Glow-plugs,
7 Other loads.

UMS0118Y

simple 12 V starting systems which, apart from higher starter output, are very similar to common passenger-car starting systems. Switching relays and protective relays of the kind found in heavier commercial vehicles are not necessary for trouble-free starting.

Medium-duty commercial vehicles with gasoline engines with displacements of up to approximately 20 l normally have 12 V starting systems, whereas comparable vehicles with diesel engines up to approximately 12 l have 12 V or 24 V starting systems.

Heavy-duty commercial vehicles with diesel engines up to approx. 24 l displacement have only 24 V starting systems fed by two series-connected 12 V batteries.

Starting systems which have a rated voltage of 24 V are the best solution, particularly if the battery and the starter are far apart from each other. Line losses are reduced, so that a given battery provides better starting. Available voltage also determines the amount of power which can be delivered by the starter. For this reason, "hybrid" 12/24 V systems are used in some vehicles. Here, 12 V is used for the vehicle electrical system, and 24 V for the starting system.

Starting systems with start-locking

In some applications, the operator will not necessarily be able to hear the starter or determine if the engine has started (e.g., rear-engined buses). More complex circuitry is then required for effective protection of starter and ring gear.

Figure 9 shows a starting system for a heavy-duty vehicle. The circuit incorporates an electronic start-locking relay designed to provide multiple protection for the starting system:
– Starter deactivates after successful engine start,
– Lockout prevents starter from engaging with engine running,
– Lockout prevents starter from engaging for as long as the crankshaft continues to turn,

Fig. 9

Circuit diagram of a starting system with electronic start-locking relay

1 Battery,
2 Battery main switch,
3 Driving switch,
4 Starter switch,
5 Charge-
indicator lamp,
6 Alternator,
7 Electronic
start-locking relay,
8 Starter.

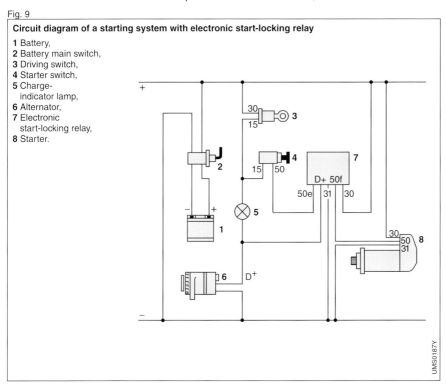

UMS0187Y

– Lockout when starter fails to start the engine.

In the last two cases, a programmed time must elapse before the relay will release the circuit; this prevents premature attempts to re-engage the starter.

Starting systems with 12/24 V battery change-over

A number of heavy-duty vehicles – primarily trucks – have a 12/24 V hybrid system (Fig. 10). In these systems, all electrical devices (except the starter), including the alternator, are designed for a rated voltage of 12 V. The starter, however, is operated at a rated voltage of 24 V. This higher voltage allows the starter to generate the power necessary to start large engines.

For this purpose, the 12/24 V systems incorporate a battery change-over relay. The two 12 V batteries in the vehicle are connected in parallel to supply a voltage of 12 V for the electrical system during normal vehicle operation or with the engine off.

When the starter switch is pressed, the battery change-over relay automatically connects both batteries in series temporarily to provide the starter with a voltage of 24 V. The electrical system continues to supply 12 V to all other electrical devices.

After the starter switch is released, the starter is switched off and the batteries are again connected in parallel. The batteries are recharged by the 12 V alternator (Terminal B+) while the engine is running.

Fig. 10

Circuit diagram of starting system with battery change-over relay

1 12 V battery I, **2** 12 V battery II, **3** Battery change-over relay, **4** Starting switch, **5** 24 V starter, B+ Alternator terminal.

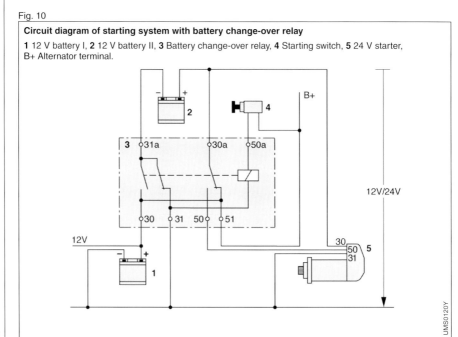

Special starting systems

Special starting systems cannot be limited to a particular type of application or vehicle. They are installed, sometimes in modified form, in large commercial vehicles (e.g. large rear-engine motor coaches or special-purpose vehicles with underfloor engines), diesel rail coaches, ships (in diesel engines used either for propulsion or power generation, depending upon the size of the ship) and in stationary equipment (e.g. drive engines for pumps or standby power units, generator drive units etc.).

The various kinds of operating conditions often require complex starting systems with specially designed protective and supervisory relays in different configurations. These relays control the starting procedure, prevent damage to the starter and ring gear as a result of circuit malfunctions, and also allow two starters to be operated in parallel. In most cases, the engines are so far away from the driver or operator that he cannot see or hear whether or not the engine has started. For this reason, engines in many applications are started either by remote control or fully automatically (e.g. standby power units, heat pumps with diesel engines, etc.). All large electrical systems in commercial vehicles must, for safety reasons, be equipped with an additional battery master switch which disconnects the battery from the vehicle electrical system when the vehicle is parked for a longer period of time or when performing maintenance or in case of damage.

Because there are so many types of special starting systems, only a few distinct circuit types can be discussed here.

Starting systems with start-repeating relays

Starting systems which are remote-controlled or which are characterized by indirect starter operation (e.g. in stationary systems, diesel rail coaches and some rear-engine commercial vehicles) in some cases also include a start-repeating relay, particularly if the operator is so far from the engine that he cannot tell whether or not it has been started. The circuit is designed such that the start-repeating relay is not

Fig. 11

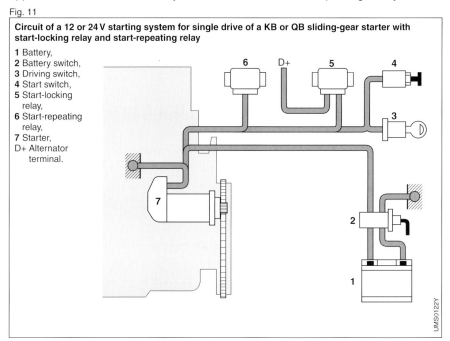

Circuit of a 12 or 24 V starting system for single drive of a KB or QB sliding-gear starter with start-locking relay and start-repeating relay

1 Battery,
2 Battery switch,
3 Driving switch,
4 Start switch,
5 Start-locking relay,
6 Start-repeating relay,
7 Starter,
D+ Alternator terminal.

UMS0122Y

actuated if the starter pinion engages normally. In order to avoid thermally overloading the starter if the engine has not started (starter does not engage), the start-repeating relay interrupts the starting procedure and repeats it automatically. This can occur several times until the starter pinion can engage the ring gear, thus closing the starter current contacts (Fig. 11).

The start-locking relay, which is also part of the circuit, additionally protects the starter by keeping it from operating if the engine is still running.

This type of circuit is used exclusively with sliding-gear starters which are electrically actuated in two steps (K, Q or T starters) and which have an additional terminal 48.

Starting systems (12 V or 24 V) with double starting relay for parallel operation

If single starters were used to start very large engines, they would have to be very big. To save space, two smaller starters are used in place of one large starter. However, both starters must be operated in parallel and must simultaneously engage the ring gear in order to crank the engine at the required speed. Assuming that sufficient power is available, parallel operation of two starters provides roughly double the starting power of a single starter.

In low-voltage parallel starting systems (12 or 24 V), a so-called double starting relay (Fig. 12) is included in the starting circuit along with the start-locking relay and the start-repeating relay described above. The double starting relay controls the following sequence of events: First, one starter after the other meshes with the ring gear of the engine. Full starter current is not applied to the starters until the second starter has completely engaged the ring gear. This allows both starters to simultaneously develop full torque, and neither starter is overloaded. Starters which are suitable for parallel operation have additional terminals for this purpose.

Fig. 12

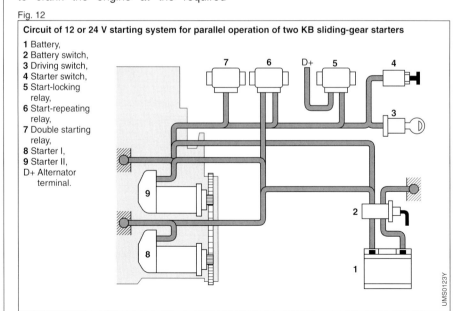

Circuit of 12 or 24 V starting system for parallel operation of two KB sliding-gear starters

1 Battery,
2 Battery switch,
3 Driving switch,
4 Starter switch,
5 Start-locking relay,
6 Start-repeating relay,
7 Double starting relay,
8 Starter I,
9 Starter II,
D+ Alternator terminal.

UMS0123Y

Starting systems (50 to 110 volts) with switching relay for parallel operation

In high-voltage parallel starting systems (50 to 110 volts) a special parallel connection relay is used in addition to a start-repeating relay with control relay and a frequency-controlled start-locking relay.

On the one hand, the parallel connection relay switches the main current to the second starter. On the other hand, it must also ensure that the starters engage one after the other, and must not apply the main current to the starters until they are both completely engaged. Fig. 13 illustrates a starting system with a parallel circuit for vehicles which are started indirectly or automatically, e.g. when a certain oil pressure or temperature is reached. In railroad engines, large stationary engines, etc., monitoring devices are often used for lubricating-oil temperature and water level which may drop briefly, thus interrupting the starter control line. A hold-in relay ensures that the starter is not needlessly switched on and off by these monitoring devices during starting, thereby burning the moving contacts in the solenoid switch.

Whereas the control relay is mounted within the sliding-gear starter in low-voltage systems, high-voltage systems combine this relay with the start-repeating relay in one unit. This design results in greater starting reliability.

Fig. 13

Circuit diagram of a 50 or 110 V starting system for parallel operation of two TB sliding-gear starters

1 Battery,
2 Battery switch,
3 Starter switch,
4 Relay for pump,
5 Motor for oil-pump,
6 Oil-pressure switch,
7 Monitoring system,
8 Relay for hold-in circuit,
9 Start-locking relay (with speed input n_M from alternator or sensor),
10 Start-repeating relay,
11 Parallel connection relay,
12 Starter I,
13 Starter II.

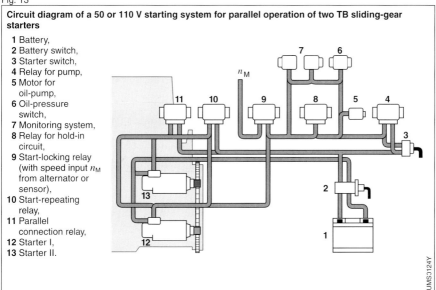

Basic starter design

As a rule, starters incorporate the following subassemblies:
– Electric starter motor,
– Solenoid switch, and
– Pinion-engaging drive (Fig. 1).

Electric starter motor

Operating principle

The electric motor converts electric current into rotary motion. In doing so, it converts electrical energy into mechanical energy.

When current flows through a conductor in a magnetic field, a force is generated which is proportional to the amount of current and the strength of the magnetic field, and is greatest when the current and the magnetic field are perpendicular to each other.

A loop which can freely rotate within a magnetic field is the most efficient design. When current flows through the loop, the loop normally assumes a position which is perpendicular to the magnetic field, and is held in this position by magnetic force. If the direction in which the current flows is reversed at this static neutral point, the loop continues to rotate. The torque of the loop then continues in the same direction of rotation, and allows the loop to rotate continuously. The commutator is responsible for this current reversal, and, in this example, consists of two semicircular segments which are connected to the two ends of the loop and which are insulated from one another. Two carbon brushes transfer the current to the individual loops (Fig. 2a).

In order to achieve uniform torque, the number of loops must be increased. Their additive individual torques produce a much higher, uniform total torque. Fig. 2b also shows three symmetrical loops whose commutator has 6 segments, or bars. In reality, the number of

Fig. 1

Starter subassemblies

1 Electric starter motor, in some cases with reduction gear,
2 Solenoid switch with electrical connections, in some cases with additional control relay,
3 Pinion-engaging drive.

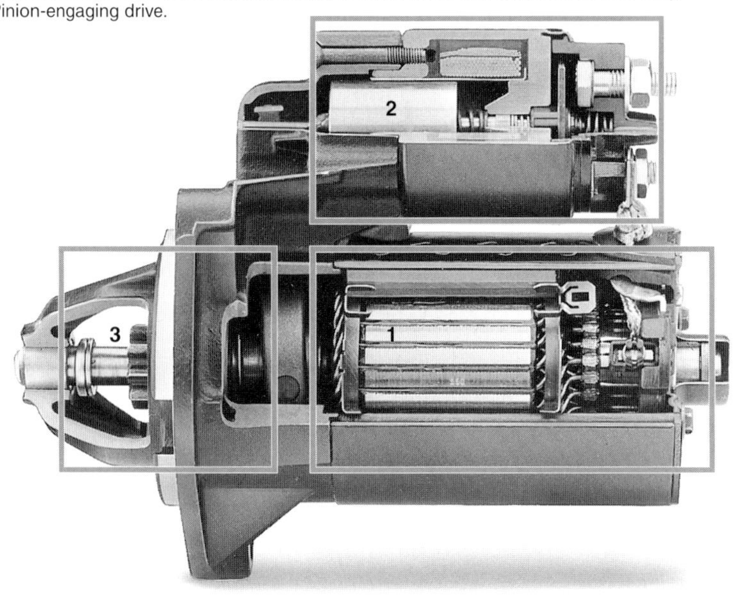

UMS0125Y

loops is considerably higher. The magnetic field can be generated by permanent magnets or by electro-magnets (electromagnetic poles with an excitation winding). We differentiate between shunt-wound, series-wound and compound-wound motors, according to the way in which the excitation winding is connected.

Starter design

In electric starter motors, the electro-magnet consists of a tube-shaped field frame in the interior of which 4 pole shoes (pole magnets) are usually mounted. Apart from DM and DW starters, which use permanent magnets, these pole shoes are provided with an excitation winding through which current flows to produce the magnetic field. The excitation winding is energized with direct current so that the magnetic lines of force always act in the same direction (from each north pole to each south pole). The field frame and pole shoes are made of iron (actually a type of steel with par-ticularly good magnetic characteristics) since the magnetic lines of force are always closed and iron is a particularly good conductor.

The armature acts in the same manner as conductor loops rotating in a magnetic field, however it also has an iron core. When current flows through the armature, a magnetic field with north and south poles is also generated within the armature core. The armature rotates as the like poles of the armature and field frame repel each other when they are in a juxtaposed position. In order to reduce magnetization losses, the iron core of the armature consists of individual disks which are insulated from one another and stacked together to form a "package" on the armature shaft. The grooves in the iron core hold the various armature windings, the ends of which are connected to the corresponding commutator bars. The commutator is mounted directly on the armature shaft and connected in most cases by four carbon brushes (for optimum current

Fig. 2

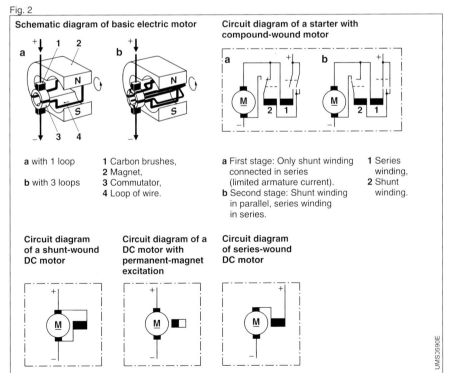

Schematic diagram of basic electric motor

a with 1 loop

b with 3 loops

1 Carbon brushes,
2 Magnet,
3 Commutator,
4 Loop of wire.

Circuit diagram of a starter with compound-wound motor

a First stage: Only shunt winding connected in series (limited armature current).
b Second stage: Shunt winding in parallel, series winding in series.

1 Series winding,
2 Shunt winding.

Circuit diagram of a shunt-wound DC motor

Circuit diagram of a DC motor with permanent-magnet excitation

Circuit diagram of series-wound DC motor

UMSJ590E

transfer) which are connected in pairs to the positive and negative terminals of the battery (or vehicle ground). By continuously reversing the current, the commutator ensures that the polarity in the armature changes at the proper time, whereas the magnetic poles of the field frame maintain the same polarity.

A voltage is induced in the armature of an electric motor which opposes the operating voltage applied to the armature. The faster the motor turns, the greater the countervoltage and the lower the current. If a load is now applied to the motor, the countervoltage drops as the speed decreases, whilst the current increases. The current and, thus, the torque (Fig. 3) is greatest when the motor starts under load. The electric motor automatically adjusts its flow of current to match its mechanical load.

Shunt-wound motors

In shunt-wound motors, the excitation winding is connected in parallel with the armature. When energized with constant voltage, excitation and speed are therefore practically independent of torque; this would not be desirable for starter operation. However, the drop in battery voltage, caused by the high starter current, results in a characteristic suitable for starting – similar to that of series-wound motors.

Motors with permanent-magnet excitation

Motors with permanent-magnet excitation are characterized by simple design and small size. Because the magnetic field is generated by permanent magnets, excitation is the same (permanent) under all operating conditions. Because there is no excitation winding, there is also no excitation current or ohmic resistance in the excitation circuit, and the overall resistance of the electric motor is reduced. The behaviour of motors with permanent-magnet excitation used as battery-operated starter motors is the same as that of shunt-wound motors.

Series-wound motors

In series-wound motors, the excitation and armature windings are connected in series. The excitation current is not diverted, rather the armature current also passes through the excitation winding. The armature current in this type of motor generates a strong magnetic field because it is unusually high when the motor starts under load.

Series-wound motors therefore develop high initial torque which drops sharply as motor speed increases. These characteristics make the series-wound motor a particularly good starter motor. When used in small starters, the series-wound motor is switched on as the starter engages the engine so that its full torque is immediately available.

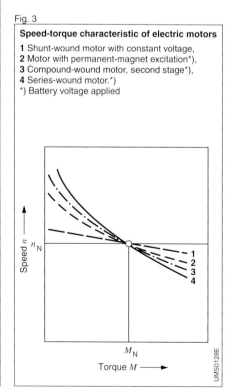

Fig. 3

Speed-torque characteristic of electric motors

1 Shunt-wound motor with constant voltage,
2 Motor with permanent-magnet excitation*),
3 Compound-wound motor, second stage*),
4 Series-wound motor.*)
*) Battery voltage applied

Combined series/shunt-wound motors (Compound motors)

Large starters use compound-wound motors which have a shunt winding and a series winding which act in two stages. In the first stage, the armature current is limited because the shunt winding is connected in series with the armature and acts as a dropping resistance. This keeps the meshing torque of the armature low. In the second stage, the full current is applied to the starter motor which then develops its full torque. The shunt winding is now connected in parallel with the armature and the series winding is additionally connected in series with the armature (Fig. 2, bottom left). When the pinion returns to its initial position, the shunt winding stops the armature quickly.

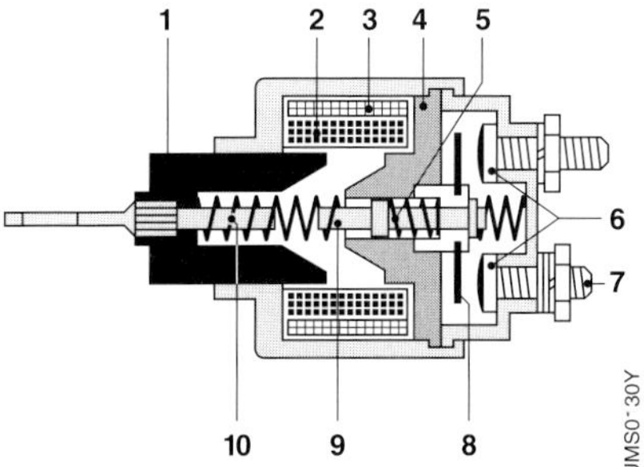

Solenoid switch

Relays are used to switch high currents with relatively low control currents. The starter current in passenger cars, for example, is up to approximately 1,000 A, and approximately 2,600 A in commercial vehicles. The low control current, on the other hand, can be switched using a mechanical switch (starter switch, ignition/starting switch, driving switch).

The solenoid assembly is an integral part of the starter (Fig. 4), and is actually a combined relay and engagement solenoid. It has two functions:

– Pushing the the pinion forward so that it engages in the ring gear of the engine, and
– Closing the moving contact for the main starter current.

The structure of the solenoid switch is shown in Figure 5. The solenoid armature, which is an integral part of the switch housing, enters the solenoid coil from one end and the movable relay armature enters from the other. The distance between the solenoid armature and the relay armature represents the total armature stroke. The solenoid housing, solenoid armature and relay armature together form the magnetic circuit.

In many switch designs, the relay winding comprises a pull-in winding and a hold-in winding. This configuration is very favourable in terms of thermal loadability and the magnetic forces which can be generated. At the beginning of the pull-in phase, increased magnetic force overcomes meshing resistance. When the starter circuit is closed, only the hold-in winding acts; the pull-in winding is shorted. The somewhat lower magnetic force of the hold-in winding is now sufficient to hold the relay armature until the starter switch is again opened.

After the starter is switched on, the magnetic force pulls the relay armature into the winding. This armature movement moves the pinion axially and also

Fig. 4

Starter assemblies "Solenoid switch and pinion-engaging drive"

(on a pre-engaged-drive starter).

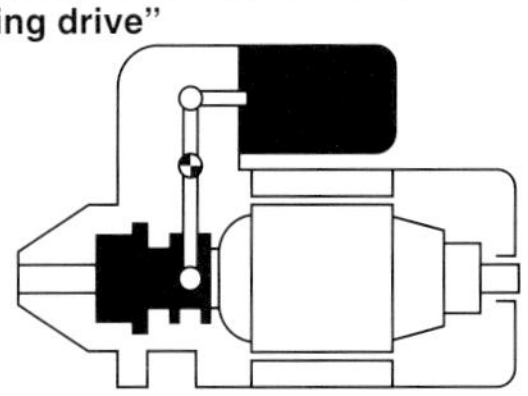

UMS0131Y

Fig. 5

Solenoid switch

1 Armature, 2 Pull-in winding, 3 Hold-in winding, 4 Solenoid armature, 5 Contact spring, 6 Contacts, 7 Terminal, 8 Moving contact, 9 Switching pin (2-part), 10 Return spring.

UMS0·30Y

closes the main current contact. Return springs between the individual components open this main current circuit again when the starter is switched off and return the relay armature to its initial position. The solenoid switch design allows the electrical contacts to be grouped together.

Larger starters do not include a solenoid switch. Instead, the starter solenoid and the control relay for the electrical switching are two separate units.

Pinion-engaging drive

The starter's end-shield assembly consists of the pinion-engaging drive with pinion, overrunning clutch, engagement element (lever or linkage to control engagement travel) and pinion spring. This starter subassembly is responsible for coordinating the thrust motion of the solenoid switch and the rotary motion of the electric starter motor and transferring them to the pinion (Figures 4 and 6).

Pinion

The starter engages the ring gear on the engine flywheel by means of a small sliding gear called the pinion (Fig. 7). A high conversion ratio (normally between 10:1 and 15:1) makes it possible to overcome the high cranking resistance of the internal-combustion engine using a relatively small but high-speed starter motor. Thus, the starter dimensions and weight can be kept small. To ensure perfect meshing between starter pinion and flywheel ring gear, capable of trans-

Fig. 6

Pinion-engaging drive of a pre-engaged-drive starter

1 Drive end shield, **2** Engaging lever, **3** Meshing spring, **4** Driver, **5** Roller-type overrunning clutch, **6** Pinion, **7** Armature shaft.

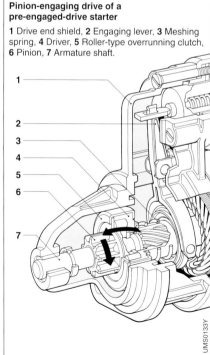

UMS0133Y

Fig. 7

Starter pinion

To facilitate engagement with the ring gear, the pinion teeth are chamfered slightly.

UMS0106Y

mitting the starting torque and then disengaging at the desired moment, special pinion tooth patterns are necessary:

– The pinion gear teeth have an involute shape to promote meshing (engagement-pattern tooth design derived from a special mathematical progression),
– The faces of the pinion gear teeth, and those of the ring gear depending upon starter design, are chamfered,
– By contrast with gears which remain meshed all the time, the center distance between the pinion and ring gear is increased in order to ensure great enough backlash at the tooth flanks,
– The outer face of the pinion in its rest position must be a certain minimum distance away from the face of the ring gear, and
– In order to achieve long service life, pinion and ring gear materials and hardening methods are matched to each other.

As soon as the engine starts and accelerates past the cranking speed, the pinion must automatically demesh in order to protect the starter, i.e. the connection between the starter shaft and the engine flywheel must automatically be broken. For this reason, starters also incorporate an overrunning clutch and a mechanism to mesh and demesh the pinion.

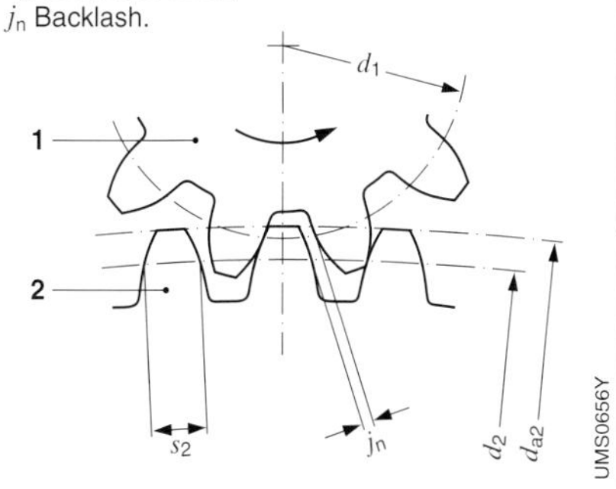

Pinion-engaging drive

The pinion-engaging drive must, in all cases, be designed such that the thrust movement of the solenoid switch and the rotary motion of the electrical starter motor can occur at the same time – but independently – under all meshing conditions. At the same time, individual engagement-mechanism designs vary according to starter size. These differences are reflected in the designations applied to the starter type.

Pre-engaged drive

In pre-engaged-drive starters, the thrust movement of the solenoid switch is transferred to the driver (with pinion) which rides in a helical spline in the armature shaft. This design results in combined axial and rotary motion which greatly facilitates the meshing of the pinion.

Sliding-gear starter, electromotive pinion rotation

In this type of sliding-gear starter, the engagement solenoid switch is mounted along the armature's axis. The pinion is actuated by an engagement rod extending through the hollow armature shaft. The armature simultaneously starts to slowly turn in order to facilitate engagement. Once the gears engage, the motor switches to full-power and turns the engine flywheel.

Sliding-gear drive, mechanical pinion rotation

Sliding-gear starters with external solenoid switch engage the pinion by moving the entire gear assembly forward along its axis. If the gears do not mesh immediately, a second mechanical unit rotates the pinion in a second stage.

Fig. 8

Starter pinion engaged with the ring gear

1 Starter pinion.
d_1 Reference diameter.
2 Ring gear.
d_2 Reference diameter, d_{a2} Tip diameter,
s_2 Tooth face width,
j_n Backlash.

UMS0656Y

Overrunning clutch

In all starter designs the rotary motion is transmitted via an overrunning clutch. The overrunning clutch allows the pinion to be driven by the armature shaft, however it breaks the connection between the pinion and the armature shaft as soon as the accelerating engine spins the pinion faster than the starter.

The overrunning clutch is located between the starter motor and the starter pinion and prevents the starter motor armature being accelerated to an excessive speed when the engine starts.

Roller-type overrunning clutch

Pre-engaged drive starters are equipped with a roller-type overrunning clutch as a protective device (Fig. 9). The most important component of the clutch is the clutch shell with roller race which forms part of the driver and thus communicates with the armature shaft via a helical spline. Rollers which are free to move within the roller race lock the pinion shaft to the clutch shell of the driver.

Fig. 9

Roller-type overrunning clutch

1 Clutch cover, **2** Pinion, **3** Driver with clutch shell, **4** Roller race, **5** Roller, **6** Pinion shaft, **7** Coil spring, **a** Direction of rotation for clutch locking action.

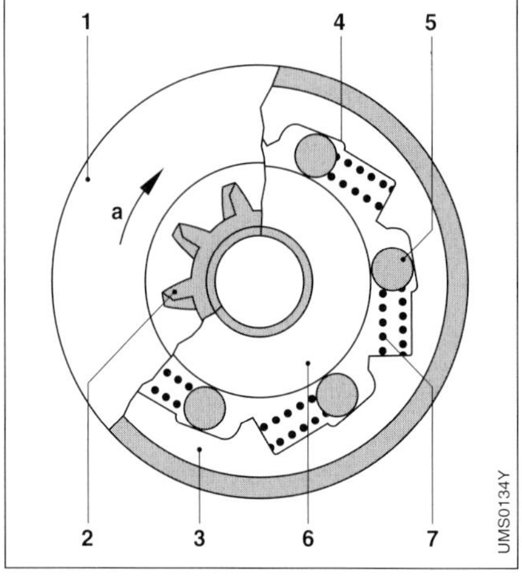

At rest, helical compression springs press the rollers into the narrow areas between the clutch shell race and the cylindrical part of the pinion shaft so that the pinion locks to the armature shaft when the starter is operated.

When the starter armature shaft rotates, the rollers become wedged in the narrow areas.

If the starting engine spins the starter pinion faster than the no-load speed of the starter armature, the rollers become loose and slide – against the force of the helical compression springs – into the wide areas. This unlocks the pinion from the armature shaft. The advantage of this type of overrunning clutch is that only small masses need be accelerated and the effective overrunning torque of the engine is relatively low.

Multiplate overrunning clutch

The multiplate overrunning clutch is used in larger sliding-gear starters. When the engine starts and the flywheel begins to spin the pinion faster than the starter armature was doing, the multiplate overrunning clutch unlocks the starter pinion from the starter armature. This is accomplished by a helical spline on the drive shaft, so that the starter motor is not accelerated to excessive speeds by the flywheel. The multiplate overrunning clutch also acts as an overload clutch by limiting the torque transmitted from the armature shaft to the pinion.

The important design feature of this type of overrunning clutch is that the individual plates must transmit all of the forces involved, and to do so must be able to move axially within the driver flange or on the clutch section, while at the same time resisting radial movement. Alternate plates contact the driver flange radially at the outer diameter (outer plates) and the clutch section at the inner diameter (inner plates). The surrounding driver flange is firmly attached to the armature shaft. The clutch section, on the other hand, rides in the helical spline of the drive spindle (Fig. 10).

Transmission of force

The plates must be <u>pressed</u> together in order to <u>transmit force by friction</u>. At rest, the multiplate stack is compressed by a slight amount of spring pressure such that the resulting friction is sufficient to drlve the clutch sectlon (Hig. 11).

When the pinion reaches its end position after meshing, the entire starter power must be available. With the pinion held in place in the ring gear, the clutch section rides outward in the helical spline against the Belleville spring washer and the armature shaft rotates, thus increasing the pressure on the plates. The pressure continues to increase until the friction between the plates is sufficient to transmit the full starting torque. Force is transmitted along the following path:
– Armature shaft,
– Driver flange,
– Outer plates,
– Inner plates,
– Clutch section,
– Drive spindle,
– Pinion,
(Figure 12).

Fig. 11

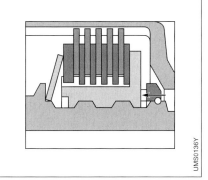

Multiplate overrunning clutch, rest position

Pretensioning spring compresses multiplate stack. Driving of the clutch section is ensured by friction.

UMS0136Y

Fig. 10

Pinion-engaging drive with multiplate overrunning clutch

1 Drive end shield,
2 Stop collar,
3 Belleville spring washer,
4 Pressure plate,
5 Outer and inner plates,
6 Drive flange,
7 Armature,
8 Pinion,
9 Drive spindle,
10 Helical spline,
11 Stop ring,
12 Field frame.

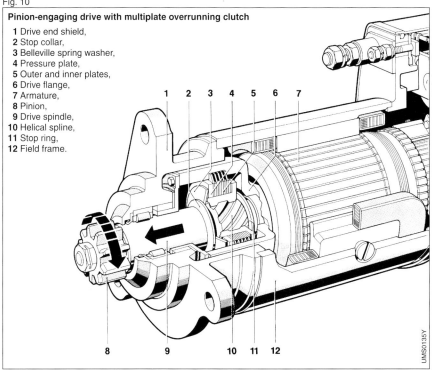

UMS0135Y

**Multiplate overrunning clutch,
transmission of force**

Pinion is engaged. Clutch section presses
against Belleville spring washer. Pressure
increases. Plate transmits total force.

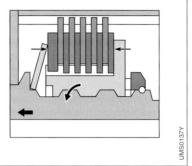

UMS0137Y

Fig. 13

**Multiplate overrunning clutch,
torque limiting**

Clutch section fully compresses
Belleville spring washer. Equilibrium of
forces at maximum setting. Plates slip.

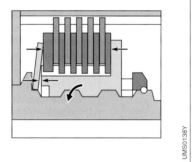

UMS0138Y

Fig. 14

Overrunning

Change in direction of force. Clutch section
runs up against stop ring and interrupts
plate compression or transmission of force.

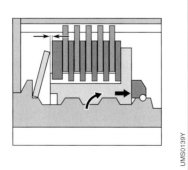

UMS0139Y

Fig. 12

Torque limitation

Increasing pressure on the plates caused by the screw action of the clutch section, and thus the transmitted torque, is limited when the clutch section runs up against the inside surface of the spring washer under maximum permissible load. The face of the clutch section depresses the spring washer against the stop collar of the drive spindle (Fig. 13). The forces are now at equilibrium. Pressure on the plates can no longer be increased. In this case, the multiplate overrunning clutch acts as an overload clutch because the plates begin to slip when the set maximum force and resulting maximum torque are reached.

Overrunning

When the flywheel is accelerated by ignition pulses or the engine starts, the pinion "overruns" the starter motor. This change in the direction of applied force causes the clutch section on the helical spline to run up against the stop ring at the starter motor. When this happens, the Belleville spring washer is completely relieved and can exert no more pressure. The pressure on the plates is relieved, and they sit loosely in the clutch section. This overrunning clutch interrupts the transmission of force so that the starter armature is not accelerated to a dangerous level (Fig. 14).

Radial-tooth overrunning clutch

The radial-tooth overrunning clutch is a special clutch used in type KE sliding-gear starters in conjunction with the two-stage mechanical pinion-engaging drive. As soon as overrunning starts, the ring gear of the engine as shown in Figs. 15 and 16 drives the pinion (1) which is engaged with the clutch section (4) by means of radial teeth. The clutch section is pushed backward on the helical spline of the drive spindle in the direction of the starter motor. It compresses the spring (5) which later will return the clutch section to its normal position. The separation of the radial-tooth components of the radial tooth overrunning clutch (pinion and clutch section) is aided by several flyweights (2) which exert a longitudinal force via a ring (3) with a conical bore. A rubber bumper (6) cushions the shock of the clutching action when the clutch components are again driven.

Armature braking

Sometimes a second attempt must be made to start the engine. The starter armature, however, must first be quickly stopped. In the case of pre-engaged drive starters this is accomplished simply by the return spring which presses the pinion-engaging drive or armature against a friction washer or brake disc

after the starter is switched off. The resulting friction stops the armature. In the case of permanent-magnet excitation, magnetic force additionally acts to brake the armature as it runs down. In sliding-gear starters, the shunt field limits starter speed so that the armature rapidly stops. On the other hand, other types of starters have a specially connected brake winding which is connected in parallel with the rotating starter armature after the starter is switched off, thereby acting as an electric brake.

Fig. 15

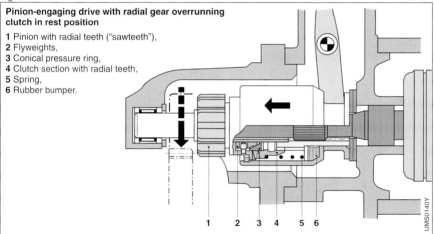

Pinion-engaging drive with radial gear overrunning clutch in rest position

1 Pinion with radial teeth ("sawteeth"),
2 Flyweights,
3 Conical pressure ring,
4 Clutch section with radial teeth,
5 Spring,
6 Rubber bumper.

1 2 3 4 5 6

UMS0140Y

Fig. 16

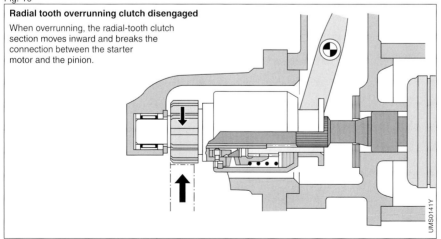

Radial tooth overrunning clutch disengaged

When overrunning, the radial-tooth clutch section moves inward and breaks the connection between the starter motor and the pinion.

UMS0141Y

Starter types

Summary

There are many different kinds of internal-combustion engines and vehicle electrical systems, and there are therefore just as many different operating conditions which determine the design of electrical starting systems and suitable starters. A broad range of starter types must therefore be available. The most important starter characteristics are:
– Rated voltage,
– Rated output,
– Direction of rotation,
– Starter size (diameter of starter-motor field frame),
– Type, and
– Design.

Rated voltage is determined by the type of starter used. Small starters are designed for 12 V, medium-sized starters for 12 and 24 V, and large starters for various rated voltages between 24 and 110 V, depending upon application. The starter's performance specifications are defined according to whether the unit will be used on a diesel or spark-ignition engine (starters for diesel engines must be more powerful) and the engine's displacement. The starter pinion's rotating direction is determined by the unit's installation position and the engine's normal operating direction. The starter's size is a function of the required power rating.

The basic design is determined by the pinion-engagement concept being used which, in turn, is largely determined by the starter's power rating and the resulting dimensions. The unit's mechanical construction features will depend upon space requirements, mounting type and operating conditions (Figures 2 and 3).

Type designation
The type designation provides pertinent initial information and is given together with the part number in the technical starter documents (Fig. 1).

Starter labelling
Starter labels (stamped into the housing) are a combination of part number, direction of rotation and rated voltage.

Example:
0 001 314 002 → 12 V.

Fig. 1

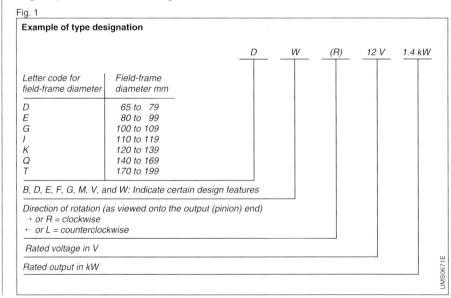

Example of type designation

		D	W	(R)	12 V	1.4 kW

Letter code for field-frame diameter	Field-frame diameter mm
D	65 to 79
E	80 to 99
G	100 to 109
I	110 to 119
K	120 to 139
Q	140 to 169
T	170 to 199

B, D, E, F, G, M, V, and W: Indicate certain design features

*Direction of rotation (as viewed onto the output (pinion) end)
→ or R = clockwise
← or L = counterclockwise*

Rated voltage in V

Rated output in kW

UMS0671E

Chart of starter types

Pinion-engaging drive, function	Reduction gear	Design E Pinion-engaging drive M Motor, R Relay	Based on design	Similar types	Starter motors
Pre-engaged drive Pinion moves forward with screw action until it meets ring gear, and is meshed by solenoid switch. Meshing is facilitated by spiral spline. Full starter current is switched on at the end of solenoid travel.	without		IF	ID	Series-wound motor
	with		EV	–	Motor with permanent-magnet excitation
			DW	–	
	without		DM	–	
Sliding-gear drive with mechanical pinion rotation Pinion moves straight forward until it meets the ring gear, and is meshed by solenoid switch. Two-stage mechanical pinion-engaging drive facilitates meshing. Full starter current is switched on after complete meshing.	without		KE	–	Series-wound motor
Sliding-gear drive with electromotive pinion rotation Pinion moves straight forward until it meets the ring gear, and is meshed by engagement solenoid. Simultaneous slow motor start-up to facilitate meshing (electrical first stage). Full starter current is switched on just before end of pinion travel (second stage).	without		KB	QB	Compound-wound motor
			TB	–	
	with		TF	On TB basis	

UAS0965E

Fig. 3

Fig. 2

Examples of pre-engaged-drive starters

1 Type IF,
2 Type EV,
3 Type DW,
4 Type DM.

1   2   3   4

UMS0681Y

Pre-engaged-drive starters without reduction gear

The main features of pre-engaged starters without reduction gear are the electric motor with direct drive, the attached solenoid switch, the pinion-engaging drive for axial and rotary pinion movement and the roller-type overrunning clutch.

Type IF with series-wound motor

Design
The design and internal circuitry of a pre-engaged drive starter without reduction gear are illustrated in Figures 4 and 5.

Starter motor:
These starters have DC series-wound motors in which the excitation and armature windings are connected in series. The motor drives the pinion-engaging drive directly with a ratio of 1:1. The extension of the armature shaft has

a helical spline which holds the driver of the pinion-engaging drive (Fig. 6).

Solenoid switch:
Pre-engaged-drive starters without reduction gear are actuated by an integral solenoid switch with pull-in and hold-in windings. The solenoid armature has a slot in its protruding end in which the end of the engaging lever fits with a certain amount of free play. This free play allows the return spring to pull the solenoid armature back to its initial position to switch off the starter, and quickly pull the moving contact away from the fixed contacts. This is necessary so that the starter can be switched off quickly in the event of the engine not being started.

Pinion-engaging drive:
The driver which rides in the helical spline of the armature shaft is coupled to the pinion via a roller-type overrunning clutch. The direction of the helical spline is selected so that the pinion, which cannot turn, is pushed into the

Fig. 4

Sectional drawing of type IF pre-engaged-drive starter

1 Hold-in winding, **2** Pull-in winding, **3** Return spring, **4** Engaging lever, **5** Meshing spring, **6** Roller-type overrunning clutch, **7** Pinion, **8** Armature shaft, **9** Stop ring, **10** Terminal, **11** Contact, **12** Moving contact, **13** Solenoid switch, **14** Commutator end shield, **15** Commutator, **16** Brush holder, **17** Pole shoe, **18** Armature, **19** Field frame, **20** Excitation winding.

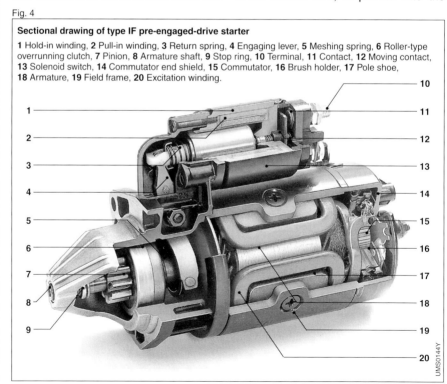

UMS0144Y

ring gear when the armature shaft rotates. The driver has two guide rings or discs which are engaged by the forked end of the engaging lever which moves the driver axially. The meshing spring sits between the guide ring and driver to allow the engaging lever to move against its stop so that the starter current is always switched, even if the pinion meets but does not engage the ring gear (the contacts close shortly before the engaging lever reaches its end position).

The driver and pinion are moved axially by the engaging lever while they are simultaneously augered forward by the helical spline until the pinion reaches its stop. The helical spline thus prevents torque from being imparted to the engine until the pinion is fully meshed. The overrunning clutch transmits the force of the starter armature to the engine flywheel after the pinion is fully meshed, and breaks this connection as soon as the engine speed exceeds the speed of the starter.

Fig. 6

Internal connections of pre-engaged-drive starters

a Basic circuit
b Terminal 15a for connection of ignition-coil ballast resistor
E Pull-in winding, H Hold-in winding.
1 Solenoid switch,
2 Excitation winding.

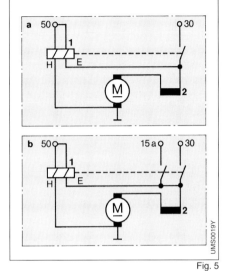

Fig. 5

The most important components of the starter's electric motor

1 Armature shaft, **2** Armature winding, **3** Armature stack, **4** Commutator, **5** Pole shoes, **6** Excitation winding, **7** Carbon brushes, **8** Brush holder.

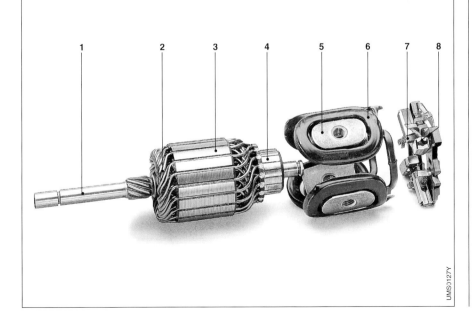

Operation

In pre-engaged-drive starters, the total meshing travel is the sum of the axial travel and the helical travel.

Axial travel:
When the starter switch or the ignition/starting switch is operated, the pull-in and hold-in windings of the solenoid switch are also energized. The solenoid armature pulls in the engaging lever against the force of a return spring. The engaging lever, via guide rings and a meshing spring, pushes the driver with pinion against the ring gear on the engine flywheel; the driver and pinion simultaneously rotate due to the action of the helical spline. The starter motor armature does not yet turn in this phase, because the main current for the excitation and armature windings has not yet been switched on.

If the pinion can immediately engage the ring gear, the pinion moves forward until it reaches the end of its travel and the moving contact in the solenoid switch meets the solenoid contacts (Fig. 7, Pos. 2). The starter motor is now switched on.

If a pinion tooth meets a ring-gear tooth, the pinion cannot immediately mesh with the ring gear. As a result, the meshing spring is compressed via the engaging lever and the guide rings until the moving contact in the solenoid switch meets the solenoid contacts (Fig. 7, Pos. 3).

The starter motor is now switched on and begins to turn. Initially, the pinion turns along the ring-gear surface. At the first opening, the pinion gears respond to the pressure exerted by the engagement spring and, especially, the helical motion along the engagement axis, by meshing with the teeth in the ring gear.

Helical travel:
At the end of the solenoid travel, the solenoid switch contacts close – independent of the pinion position – and switch on the starter current. The starter armature now begins to rotate, and the helical spline forces the pinion, which is prevented from turning by the ring gear with which it has meshed, even further into the ring gear until it contacts the stop ring of the armature shaft.

When the starter circuit is closed, the pull-in winding is simultaneously shorted. Now only the hold-in winding acts. However, its magnetic force is sufficient to hold the solenoid armature in its pulled-in position until the engine is started (Fig. 7, Pos. 4).

Demeshing:
After the engine starts and the speed of the starter pinion exceeds the no-load speed of the starter motor, the roller-type overrunning clutch described earlier breaks the connection between the pinion and the armature shaft. This keeps the armature from being rotated too fast and damaged. The pinion remains meshed as long as the engaging lever is held in the engaged position. The engaging lever, driver and pinion are returned to their initial positions by the return spring only when the starter is switched off. The return spring also ensures that the pinion remains in its rest position in spite of engine vibration until the starter is again operated.

Fig. 7

Schematic diagram of the most important working phases of a pre-engaged-drive starter

① **Rest position**

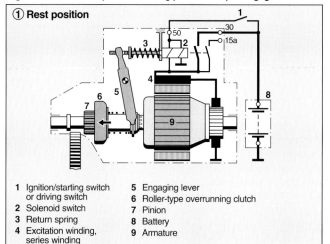

1 Ignition/starting switch or driving switch	**5** Engaging lever
2 Solenoid switch	**6** Roller-type overrunning clutch
3 Return spring	**7** Pinion
4 Excitation winding, series winding	**8** Battery
	9 Armature

② **Tooth meets gap**

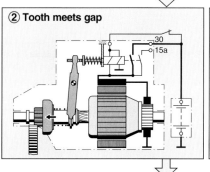

③ **Tooth meets tooth**

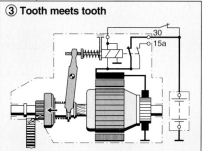

④ **Engine is cranked**

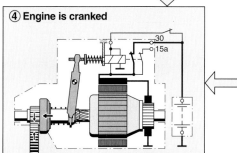

① **Rest position**
No current supplied to starter, pinion demeshed.

② **Tooth meets gap**
Favorable meshing position. Pull-in and hold-in windings are energized. A pinion tooth meets a gap in the ring gear, and the pinion meshes immediately. The starter position just before the main current is switched on is shown.

③ **Tooth meets tooth**
Unfavorable meshing position. Pinion tooth meets ring-gear tooth. Engaging lever in end position, meshing spring compressed, pull-in winding not energized. Main current flows, armature rotates. Pinion attempts to mesh with ring gear.

④ **Engine is cranked**
End position. Engaging lever in end position, pull-in winding not energized. Main current flows, pinion is fully meshed. Engine is cranked.

UMS0682E

191

Type DM with permanent-magnet motor

Starter Type DM with a permanent-magnet field is designed for use in passenger vehicles powered by spark-ignition engines with swept volumes of up to 1.9 liters. It is up to 15% lighter, and is smaller than customary starter types designed for the same operating conditions. The reduction in weight helps to cut fuel consumption. The compact dimensions are important since, owing to the increase in the number of new systems and the streamlined, low design of passenger cars, less and less space is available in the engine compartment.

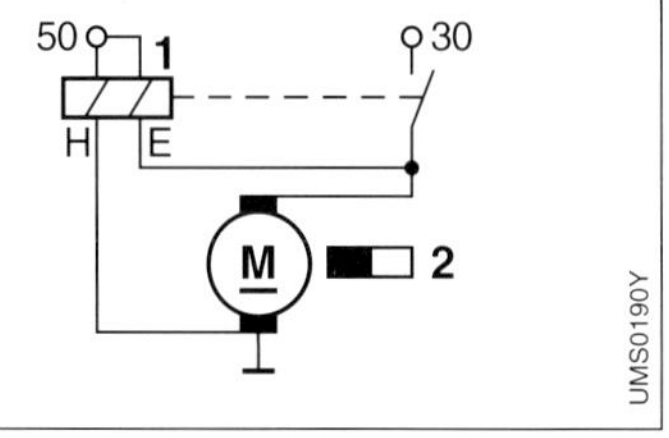

Design
The design of pre-engaged-drive starter type DM is shown in Figs. 8 to 10.

Starter motor: A DC motor with permanent-magnet excitation is used as the starter motor. Only permanent magnets with flux-concentrating pieces (Fig. 11) are used in place of the electromagnets (pole shoes with excitation winding) in the excitation circuit. The lengths of the armature and permanent magnets are graded depending upon the rated output of the starter. The permanent magnets used on this model feature long-term stability and are insensitive to demagnetizing influences.

Fig. 8

Internal connections of DM starter

1 Solenoid switch, **2** Permanent magnets.
E Pull-in winding, H Hold-in winding.

50 1   30

H   E

M   2

UMS0190Y

Fig. 9

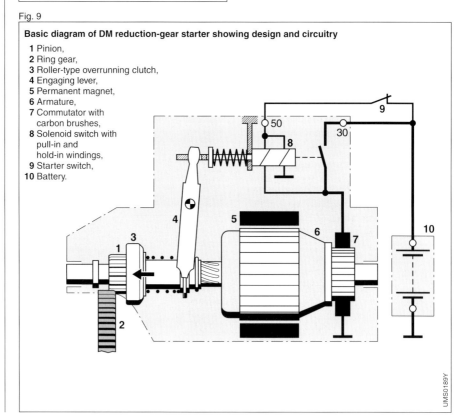

Basic diagram of DM reduction-gear starter showing design and circuitry

1 Pinion,
2 Ring gear,
3 Roller-type overrunning clutch,
4 Engaging lever,
5 Permanent magnet,
6 Armature,
7 Commutator with carbon brushes,
8 Solenoid switch with pull-in and hold-in windings,
9 Starter switch,
10 Battery.

UMS0189Y

The special, two-component carbon brushes comprise a power zone with a high-percentage copper content and a commutating zone with a high-percentage graphite content.

Solenoid switch:
As in all pre-engaged-drive starters, the solenoid switch which actuates the driver and closes the starter circuit is mounted on the starter and transfers movement to the components on the armature shaft via the engaging lever.

Pinion-engaging drive:
The pinion-engaging drive with roller-type overrunning clutch is of the same design and operates in the same way as on other pre-engaged-drive starters.

Operation
Starter type DM does not differ greatly from type IF. The only minor difference is in the electrical circuitry where perma-

nent magnets take the place of the excitation winding. When the starter circuit is closed, current flows directly to the carbon brushes and the armature.

Fig. 11

Field frame of DM starter with permanent magnets M and flux-concentrating pieces F

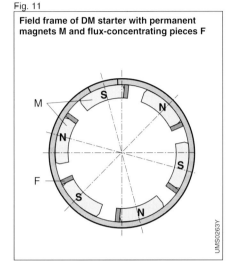

UMS0263Y

Fig. 10

Cutaway view of DM starter with permanent-magnet motor

1 Drive end shield, **2** Pinion, **3** Solenoid switch, **4** Terminal, **5** Commutator end shield, **6** Brush plate with carbon brushes, **7** Commutator, **8** Armature, **9** Permanent magnet, **10** Field frame, **11** Engaging lever, **12** Pinion-engaging drive.

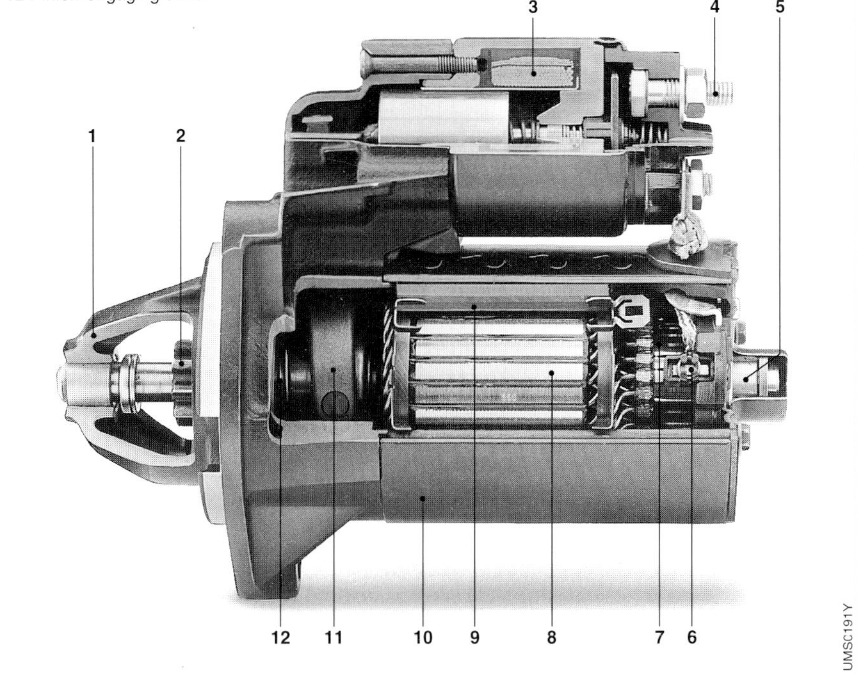

UMSC191Y

193

Pre-engaged-drive starters with reduction gear

In their design and function, reduction-gear starters are much the same as the pre-engaged-drive starters without reduction gear which transfer the motor torque directly to the pinion-engaging drive in a conventional manner.

Special feature

The main difference between conventional starters and this new generation of starters is a planetary gear added between the field frame and the drive end shield. It transfers the armature torque to the pinion, free of any transverse forces. Whereas the planet gears are steel, the internal gear is a high-grade polyamide compound with mineral additives for increased strength and wear resistance.

This new design allows use of smaller and lighter starters; the savings in weight over traditional starter/drive combinations is approx. 35 to 40%. An added benefit of the lower weight is an increase in vehicle fuel economy.

Type EV with series-wound motor

Starter type EV is designed for use in motor vehicles powered by diesel engines with displacements of 1.8 to 3 litres.

Design

The design of pre-engaged-drive starter EV is shown in Figs. 12 and 13.

Starter motor with reduction gear:
The starter motor is a DC series-wound motor, in which the excitation and armature windings are connected in series. The speed of the high-speed electric motor is reduced by the planetary gear (reduction gear) and transferred to the pinion-engaging drive. The torque is increased by the same ratio. The shaft of the internal gear bears a helical spline on which the driver of the pinion-engaging drive rides.

Fig. 12

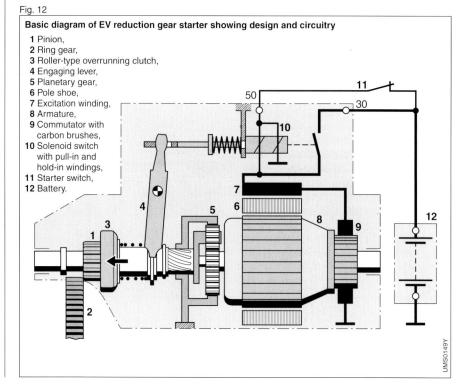

Basic diagram of EV reduction gear starter showing design and circuitry

1 Pinion,
2 Ring gear,
3 Roller-type overrunning clutch,
4 Engaging lever,
5 Planetary gear,
6 Pole shoe,
7 Excitation winding,
8 Armature,
9 Commutator with carbon brushes,
10 Solenoid switch with pull-in and hold-in windings,
11 Starter switch,
12 Battery.

UMS0149Y

Solenoid switch:
Pre-engaged-drive starters with reduction gear are operated in the same manner as starters without, namely via a solenoid switch with pull-in and hold-in windings. The switch is built onto the starter. Axial force is transmitted to the pinion in the same manner by the engaging lever.

Pinion-engaging drive:
The pinion-engaging drives of these starters do not differ materially from those of pre-engaged-drive starters without reduction gear described above.

Operation
The switching positions and switching functions are the same as those described in the section entitled "Pre-engaged-drive starters without reduction gear".

Type DW with permanent-magnet field

Reduction-gear starter type DW with permanent-magnet field is suitable for use in passenger vehicles powered by gasoline engines with displacements of up to 5 litres or diesel engines with displacements of up to 1.6 litres. While offering equal or increased starting power, it is 40 % lower in weight and has considerably smaller dimensions than customary starter types designed for the same operating conditions.

Design
The design of pre-engaged-drive starter type DW is shown in Figs. 14 to 17.

Starter motor with reduction gear:
A DC motor with permanent-magnet excitation is used as the starter motor. Permanent magnets take the place of the pole shoes with excitation winding. The length of armature and permanent magnets determines the starter rated output. This type of motor design allows

Fig. 13

Cutaway view of EV reduction gear starter with series-wound motor

1 Drive end shield, **2** Pinion, **3** Solenoid switch, **4** Terminal, **5** Commutator end shield,
6 Brush plate with carbon brushes, **7** Excitation winding, **8** Field frame, **9** Armature, **10** Pole shoe,
11 Planetary gear (reduction gear), **12** Engaging lever, **13** Pinion-engaging drive.

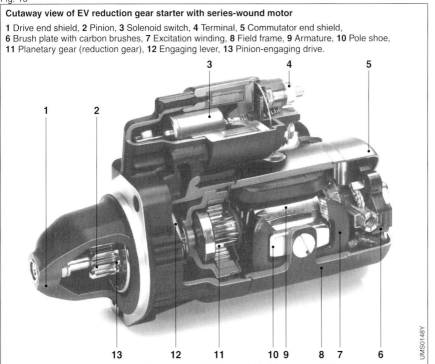

UMS0148Y

the starter-motor size and thus the size of the entire starter to be drastically reduced, thus significantly reducing weight. As in the case of type EV, the high motor speed is reduced to appropriate starter speed by a reduction gear, thus simultaneously achieving the required high starter torque.

Solenoid switch:
As in all other pre-engaged-drive starters, the solenoid switch which actuates the driver and closes the starter circuit is mounted on the starter and transfers movement to the components on the armature shaft via the engaging lever. All versions of starter type DW have the same solenoid switch.

Pinion-engaging drive:
The pinion-engaging drive with roller-type overrunning clutch is of the same design and operates the same as the drive described above for other pre-engaged-drive starters, and is used for all starter versions.

Operation
Reduction-gear starter type DW operates no differently from the other pre-engaged-drive starters. The only difference is in the electrical circuitry which does not include the excitation winding normally connected in series. When the starter circuit is closed, current flows directly to the carbon brushes and the armature.

Fig. 15

Internal connections of DW starter

1 Solenoid switch, **2** Permanent magnets.
E Pull-in winding, H Hold-in winding.

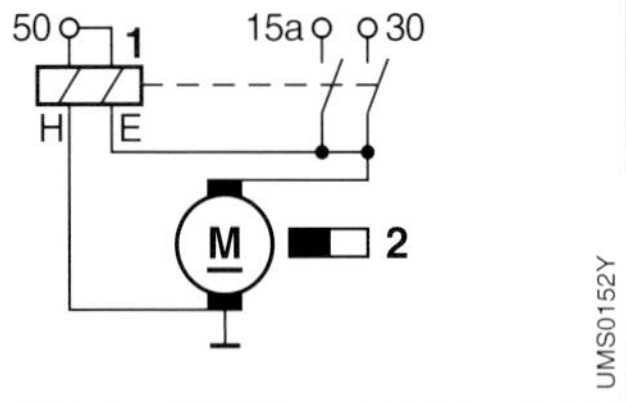

Fig. 14

Basic diagram of DW reduction-gear starter showing design and circuitry

1 Pinion,
2 Ring gear,
3 Roller-type overrunning clutch,
4 Engaging lever,
5 Planetary gear,
6 Permanent magnet,
7 Armature,
8 Commutator with carbon brushes,
9 Solenoid switch with pull-in and hold-in windings,
10 Starter switch,
11 Battery.

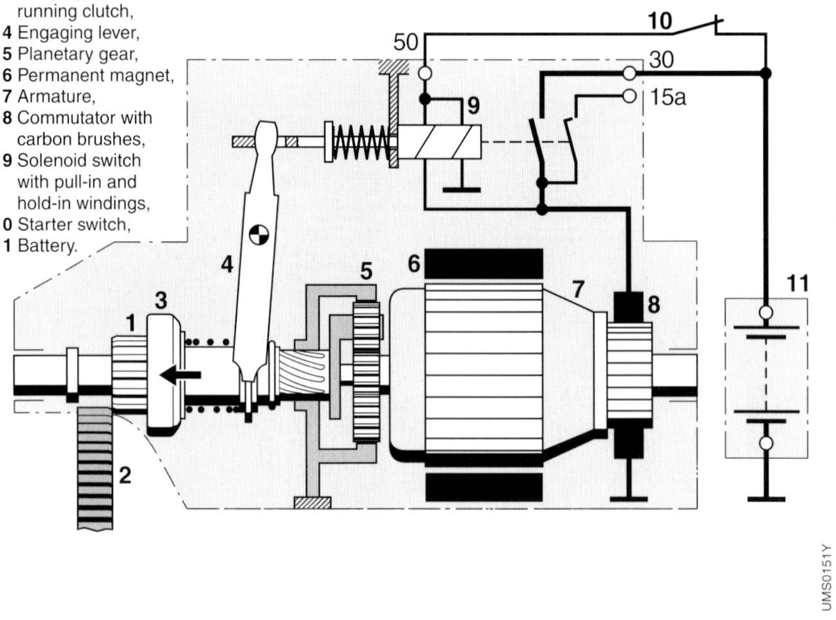

Armature and planetary gear (reduction gear) of a DW reduction-gear starter

1 Planetary-gear carrier shaft with helical spline, **2** Internal gear (ring gear), also serves as intermediate bearing, **3** Planet gears, **4** Sun gear on armature shaft, **5** Armature, **6** Commutator.

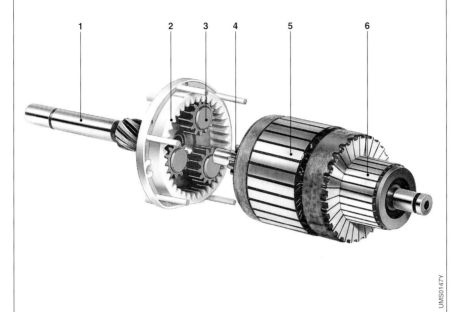

Fig. 16

Fig. 17

DW reduction-gear starter with permanent-magnet motor. Cutaway view

1 Drive end shield, **2** Pinion, **3** Solenoid switch, **4** Terminal, **5** Commutator end shield, **6** Brush plate with carbon brushes, **7** Commutator, **8** Armature, **9** Permanent magnet, **10** Field frame, **11** Planetary gear (reduction gear), **12** Engaging lever, **13** Pinion-engaging drive.

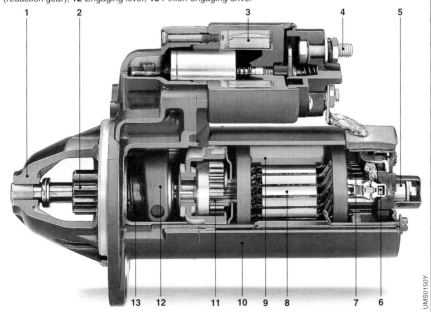

Sliding-gear starters with mechanical pinion rotation

A characteristic feature of this starter is the piggyback solenoid switch, the two-stage pinion-engaging drive and the radial-tooth overrunning clutch.

Type KE with series-wound motor

The various versions of type KE are designed for use in heavy-duty commercial vehicles powered by diesel engines with displacements of up to 21 litres which are operated under the most adverse conditions. Special features are:
- Maintenance-free for up to 800,000 km,
- Resistant to extremely severe vibration,
- Watertight and resistant to hydraulic-oil penetration at the drive end for engines equipped with oil-type clutches or oil-bath torque converters,
- Highly resistant to thermal overload due to use of insulation material which has good high-temperature characteristics.

Design
The design and internal connections of these starters are shown In Figs. 18 and 19.

Starter motor: Excitation and armature winding of the DC series-wound motor are connected in series. The extended armature shaft carries a drive for the pinion-engaging drive.

Solenoid switch: The solenoid switch which is mounted on the starter pushes the pinion-engaging drive with the pinion forward via the engaging lever. The protruding end of the solenoid-switch armature holds a "spool" which is engaged by the engaging lever fork with a certain amount of free play. Additional free play between the sliders of the engaging lever and the guides of the pinion-engaging drive – termed the free-play distance – enables the return spring on the solenoid switch to move the solenoid armature the distance of

Fig. 18

Cutaway view of KE sliding-gear starter with mechanical pinion rotation

1 Pinion, 2 Engaging lever, 3 "Spool", 4 Cutoff spring, 5 Return spring, 6 Solenoid switch,
7 Hold-in winding, 8 Pull-in winding, 9 Moving contact, 10 Terminal, 11 Contact, 12 Brush spring,
13 Commutator, 14 Carbon brushes, 15 Pole shoe, 16 Field frame, 17 Armature, 18 Excitation winding,
19 Brake disc, 20 Helical spline, 21 Meshing spring, 22 Drive end shield.

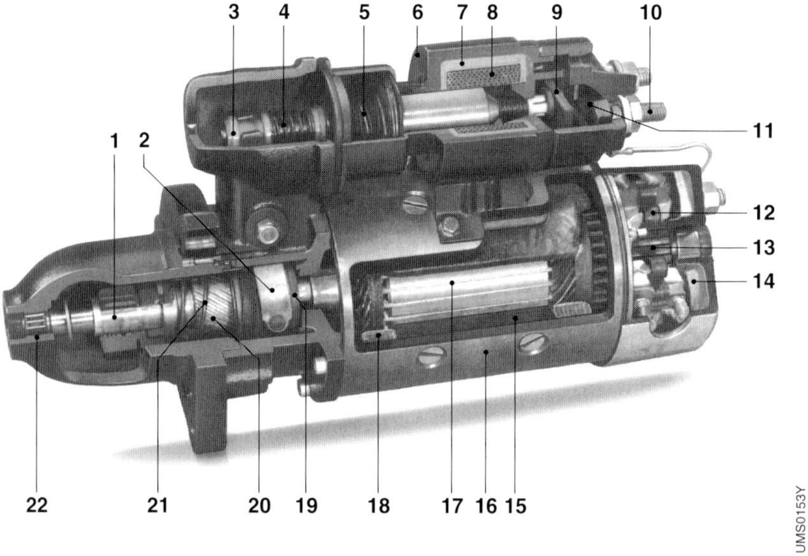

UMS0153Y

the free play toward its initial position, and thus to pull the moving contact quickly away from the fixed contacts. Since, under certain conditions, e.g. if the pinion attempts to mesh with a blocked ring gear, this free play can be zero, an additional cut-off spring has been installed. The spring characteristics of the cutoff spring and the return spring are such that the force of the cutoff spring is greater in the rest position, while the force of the return spring predominates in the meshed position. The free play and the cut-off spring guarantee that the starter motor is reliably switched off.

Pinion-engaging drive:
Type KE sliding-gear starters have a two-stage mechanical pinion-engaging drive in order to protect the pinion and the ring gear. The driver of the pinion-engaging drive which is connected to the pinion via a clutch section and the "sawteeth" of the integral radial-tooth overrunning clutch slides on the straight teeth of the armature shaft. The engaging lever slides the pinion-engaging drive axially in the direction of the ring gear.

Operation
First meshing stage: When the starter switch is operated, the solenoid switch first moves the engaging lever against the return spring before the excitation and armature winding are energized at all. The engaging lever slides the entire drive along the straight-tooth guide against the ring gear. If the pinion finds a gap in the ring gear, it is able to mesh with the ring gear as far as the pivoting range of the engaging lever allows. In this case, the pinion has moved the full axial distance.

Second meshing stage: If the pinion meets a tooth of the ring gear as it moves forward, the other drive components continue to be pushed straight forward in the direction of the ring gear. The helical spline of the crown-tooth clutch section rotates the pinion in the

cranking direction and, at the same time, compresses the spring of the pinion-engaging drive. The pinion tooth slides past the ring gear tooth until it finds the next gap; the pinion then engages completely due to the pressure exerted by the compressed spring. In the meantime, the radial-gear clutch section rotates in the overspeed direction. It is also possible that the pinion may meet a damaged or notched ring gear tooth and is therefore prevented from rotating. In this case, while the pinion-engaging drive is pushed forward by the engaging lever, the starter armature rotates in the direction opposite to the cranking direction due to the helical spline of the radial-tooth clutch section, and compresses the spring. Drive and solenoid travel are matched such that the movement of these components in this phase does not yet switch on the main current; the starting attempt must be aborted due to failure of the pinion to mesh. The spring relaxes again when the starter switch is switched off, and the pinion rotates in the overspeed direction and is then able to mesh with the ring gear when the starter is again operated.

Starting phase:
When the pinion meshes completely, the main current is applied to terminal 30 via the solenoid switch travel. Only the hold-in winding is energized in this position. The electric motor of the starter can now transfer its full torque to the ring gear.

Fig. 19

Internal connections of KE sliding-gear starters with mechanical pinion rotation

1 Solenoid switch, **2** Excitation winding.
E Pull-in winding, H Hold-in winding.

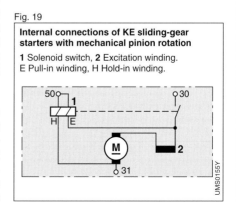

Overspeeding and demeshing:

As soon as the ring gear of the engine begins to turn the pinion faster than the starter, the clutch section connected to the pinion by radial teeth is moved toward the rear via the helical spline, simultaneously compressing the over-running clutch spring. Flyweights also aid in breaking the connection between the radial-tooth overrunning clutch components. The flyweights apply longitudinal force via a conical ring. This prevents the starter from being rotated at a speed higher than its maximum permissible speed. The engaging lever and the pinion-engaging drive are moved back to their rest positions via return-spring pressure only after switching off the starter. A mechanical brake disc rapidly stops the armature which is still rotating. The return spring holds the pinion-engaging drive firmly in its rest position.

Sliding-gear starters with electromotive pinion rotation

Sliding-gear starters with electromotive pinion rotation are used for starting large internal-combustion engines; they have a two-stage electrical pinion-engaging drive which protects the pinion and the ring gear. In the first switching stage, only the starter pinion is meshed. The pinion does not yet, however, crank the engine. The full excitation and armature current is not switched on until the second stage directly before the pinion reaches the end of its travel.

KB/QB and TB/TF starter design is characterized by the engagement solenoid and other subassemblies being mounted coaxially, i.e. on one axis.

Fig. 20

Sectional drawing of a KB sliding-gear starter with two-stage electrical pinion-engaging drive

1 Drive spindle, **2** Drive housing, **3** Multiplate overrunning clutch, **4** Armature, **5** Terminal, **6** Commutator end shield, **7** Control relay, **8** Moving contact, **9** Stop, **10** Tripping lever, **11** Release lever, **12** Engagement solenoid, **13** End cover, **14** Commutator, **15** Carbon brush, **16** Brush holder, **17** Pole shoe, **18** Excitation winding, **19** Field frame, **20** Pinion.

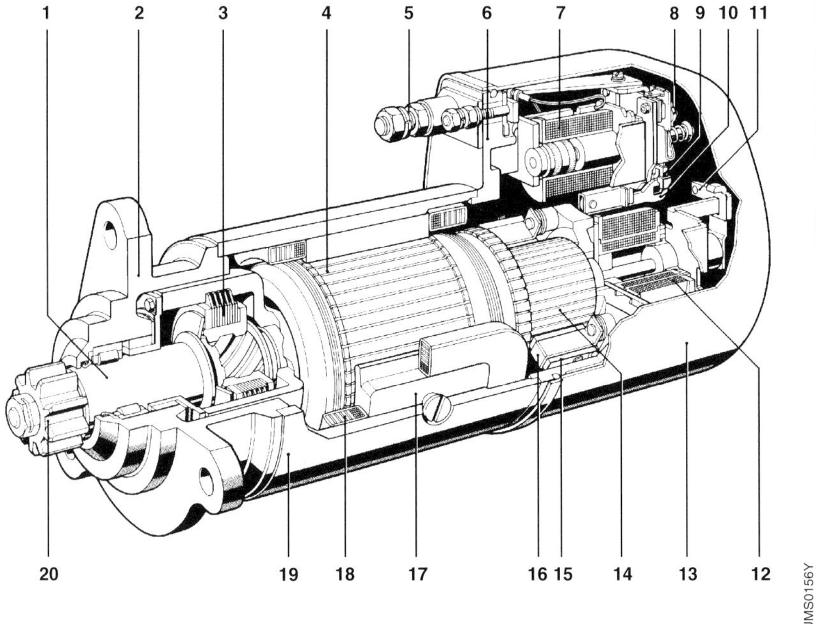

UMS0156Y

Type KB/QB with compound-wound motor

Design
KB starter design is shown in the sectional drawing (Fig. 20)

Starter motor:
The armature of the starter motor is held in the drive housing and commutator bearing; it has a hollow shaft which has a driving flange at the drive housing end for the multiplate overrunning clutch. This driving flange is closed by a cover which holds a plain bearing for the starter armature in the drive housing. The starter armature is held in a plain bearing at the commutator end. The schematic diagram of the internal connections (Fig. 21) shows that a shunt winding for field excitation is present in addition to the series winding. This shunt winding remains connected in parallel with the starter motor in both switching stages in the various versions of starter type KB. In other versions, on the other hand, the shunt winding is connected as a dropping resistor in the first stage, and is connected in series with the starter motor in order to promote slow armature rotation by limiting the armature current. In the main phase the shunt winding is connected in parallel with the starter motor and limits the maximum starter speed. On QB starters, an additional auxiliary winding serves to increase the torque during the first stage (Fig. 22).

Engagement solenoid and control relay:
An engagement solenoid for the pinion and a control relay for the two switching stages are flanged to the commutator end shield. Owing to this engagement solenoid configuration, it is necessary to actuate the pinion via an engagement rod which passes through the hollow armature shaft. The engagement solenoid also has the function of releasing the moving contact of the control relay via a release lever, tripping lever and stop plate.

Pinion-engaging drive:
The drive spindle which has a helical spline to hold the multiplate overrunning clutch is held in a roller bearing in the drive housing and in an armature shaft needle bearing. A parallel key connects the pinion to the drive spindle. The multiplate overrunning clutch either transmits the force from the starter armature to the pinion or interrupts it.

Operation:
The illustrations and the descriptions of the meshing and demeshing procedures refer to type KB.

First (preliminary) switching stage (Fig. 23):
After the starter switch is turned on, current flows through the control relay winding and through the hold-in winding of the engagement solenoid. As a result, the control relay also immediately closes

Fig. 21

Internal connections of a KB starter

1 Control relay, **1a** Tripping lever,
2 Engagement solenoid, **3** Series winding,
4 Shunt winding, **5** Changeover switch for shunt winding.
E Pull-in winding, H Hold-in winding.

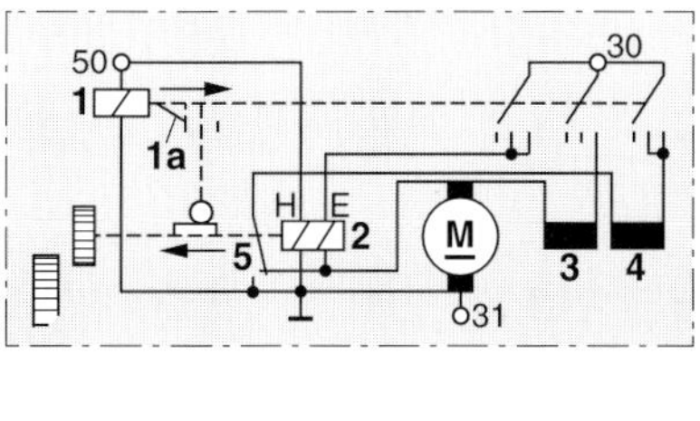

Fig. 22

Internal connections of a QB starter

1 Control relay, **1a** Tripping lever,
2 Engagement solenoid, **3** Auxiliary winding,
4 Series winding, **5** Shunt winding.
E Pull-in winding, H Hold-in winding.

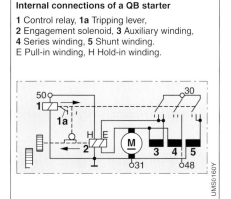

the pull-in winding circuit of the engagement solenoid. The solenoid armature now pushes the pinion against the engine ring gear via the engagement rod and the drive spindle. The shunt winding (which is still connected in series with the starter armature) is energized at the same time. Together with the pull-in winding of the engagement solenoid, it acts as a dropping resistor for the starter armature winding (additionally for the auxiliary winding in QB starters). This circuit limits the armature current to such an extent that the starter armature is only able to develop very low torque, and therefore rotates very slowly. In the first switching stage the starter pinion is therefore pushed axially and at the same time slowly rotated to provide smooth meshing. The pinion, however, does not yet crank the engine because the low torque of the starter is not sufficient. If, due to its position, the starter pinion cannot immediately mesh, it rotates on the face of the flywheel ring gear until the pinion tooth engages the next gap. In a tooth-to-tooth or corner-to-corner position, the pinion is unable to mesh. In this case the start attempt must be immediately aborted and then repeated.

Second (main) switching stage:
Just before the pinion reaches the end of its meshing travel, a release lever lifts a tripping lever and releases the moving contact of the control relay. The snap-action of a tensioned spring can thus push the moving contact against the fixed contacts. This switches on the main current which then flows through the series winding and the armature. In several starter versions a changeover switch on the engagement solenoid additionally switches the shunt winding in parallel (Fig. 21). The starter motor now receives its full current and cranks the engine at full torque via the multiplate overrunning clutch.

Overspeeding and demeshing:
Force is applied to the pinion in the reverse direction when the engine starts and spins the starter pinion faster than the no-load speed of the starter armature. The multiplate overrunning clutch breaks the connection between the starter pinion and the starter armature via the helical spline, thus preventing the starter motor from overspeeding. The pinion itself remains engaged as long as the starter switch is actuated.

When the starter switch is released, the starter is switched off and the current to the control-relay winding and the hold-in winding of the engagement solenoid is interrupted. The control relay then interrupts the main circuit, whereby a return spring inside the hollow armature shaft returns the drive and pinion to their initial position. The pinion demeshes and returns to its initial position. The above-mentioned return spring also has the task of holding the drive spindle in its rest position despite vibrations caused by the running engine, until the next start. When the pinion demeshes, the spring-loaded tripping lever of the control relay returns to its locked position and the starter is ready for the next two-stage starting operation.

Type TB/TF with compound-wound motor

Design
The basic design of starters in the T series is largely the same as that of KB/QB starters. There are minor differences in the housing shape, the bearings and in the electrical range of the starters. Thus, for instance, the driving flange of the starter armature is not held in a plain bearing, but rather in a rolling bearing in the drive housing. Various components have special seals to keep oil, dirt and dust out of the interior of the starter.

Fig. 23

Schematic diagram of the most important working phases of a sliding-gear starter with electromotive pinion rotation, type KB

1 Ignition/starter switch or driving switch, **2** Tripping lever, **3** Release lever, **4** Control relay, **5** Moving contact, **6** Stop, **7** Engagement solenoid, E Pull-in winding, H Hold-in winding, **8** Armature, **9** Excitation winding, N Shunt winding, R Series winding, **10** Pinion, **11** Battery, **12** Multiplate overrunning clutch.

1 Rest position
No voltage applied.

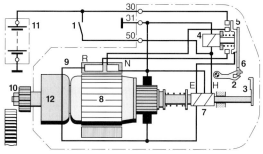

2 Tooth meets gap
Favorable meshing position (switching stage 1).

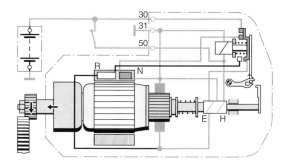

3 Tooth meets tooth or corner
Meshing not possible (switching stage 1). Start attempt must be repeated.

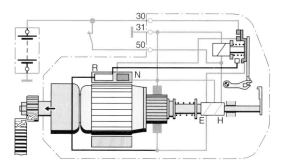

4 Engine is cranked
End position (switching stage 2). Starter torque is maximum.

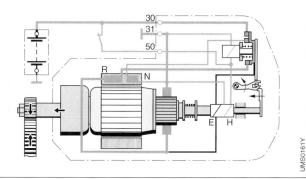

UMS0161Y

Starter motor:
Similar to type KB/QB, the armature of the starter motor has a hollow shaft which serves as a driving flange at the drive-housing end for the multiplate overrunning clutch (Fig. 24). In versions for rated voltages of up to 36 V, the pole shoes incorporate a brake winding in addition to the series winding. The brake winding has no effect as long as the starter is operating. When the starter is switched off, the brake winding is switched in parallel to the rotating starter armature via a contact on the control relay, and acts as an electric brake which quickly stops the armature (Fig. 25). Versions for rated voltages of 50 V or more have a shunt winding instead of a brake winding.

Engagement solenoid and control relay:
The terminals, control relay and engagement solenoid are located together in a cylindrical commutator housing, i.e. not in a recess in the housing as is the case with types KB/QB. However, the engagement solenoid also pushes the pinion forward via an engagement rod

Internal connections of a TB starter 24 V

1 Control relay, 2 Engagement solenoid, E Pull-in winding, G Opposing winding, H Hold-in winding, 2a Tripping lever, 3 Series winding, 4 Brake winding, 5 Thermostatic switch.

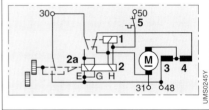

Fig. 25

inside the hollow armature shaft. The engagement solenoid also has an opposing winding which acts as a dropping resistor for controlling starter torque during meshing.

Thermostatic switch:
Starting systems which are expected to operate for long periods of time and which will be expected to perform a high number of repeated starts (e.g. in the

Fig. 24

Sectional drawing of a TB sliding-gear starter (without reduction gear)

1 Pilot for engine flange, 2 Multiplate clutch, 3 Engaging spindle, 4 Pole shoe, 5 Armature, 6 Excitation winding, 7 Carbon brush, 8 Brush holder, 9 Terminal stud, 10 Control relay (24 and 36 V), 11 Solenoid switch, 12 Cutoff spring, 13 Commutator, 14 Field frame, 15 Helical spline, 16 Pinion.

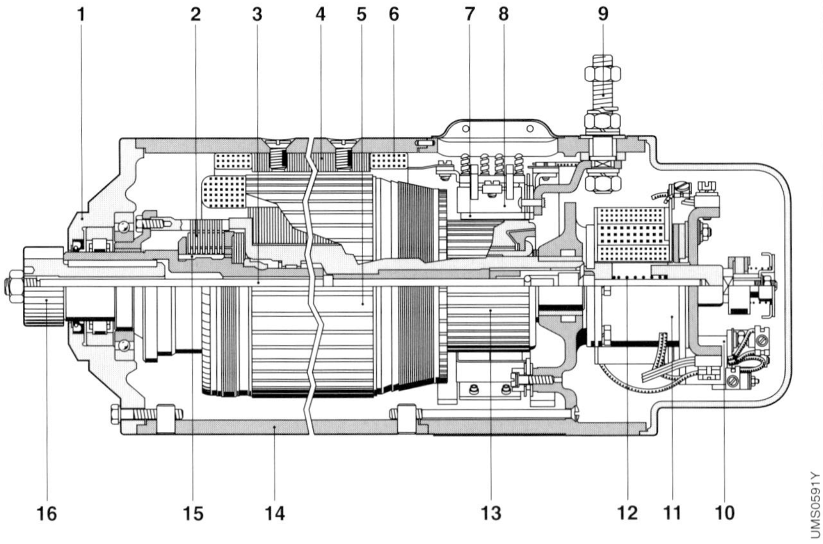

case of low battery voltage, damaged ring-gear teeth or engine malfunction) incorporate T-series starters with two built-in thermostatic switches for protection against thermal switch overload of the starter during the first and second switching stages. These thermostatic switches (which are connected in series) are installed in the carbon brushes or in joining bars. If the temperature of the engagement-solenoid windings exceeds certain values as a result of failure of the starter to mesh or due to radiated heat from other live components, the thermostatic switches interrupt starter cable 50 and switch off the starter. The starter can be operated again after a cooling period of approx. 20 minutes.

The internal connections of such starters may be slightly different according to rated voltage. High-voltage starters have a spark-suppressor capacitor connected in parallel with the thermostatic switches.

Pinion-engaging drive:
As in type KB/QB, the pinion-engaging drive consists of a multiplate overrunning clutch, drive spindle with helical spline and pinion. Primarily, the TF starter differs from the TB starter in having a reduction gear, i.e. its pinion is offset with respect to the armature axis (Fig. 26). The eccentric drive thus achieved frequently improves the installation conditions when attaching the starter on the engine. The intermediate shaft with pinion is mounted in the intermediate-shaft bearing so that it can rotate and move axially. An engaging lever located in the intermediate housing transfers the pushing motion of the engagement rod to the intermediate shaft with pinion. Force is transferred from the engagement solenoid to the pinion as follows:

– Engagement-solenoid armature,
– Engagement rod,
– Guide ring,
– Upper fork,
– Engaging lever,
– Lower fork,
– Clutch sleeve,
– Intermediate shaft,
– Pinion.

Fig. 26

Sectional drawing of a TF sliding-gear starter (with reduction gear)

1 Reduction-gear housing, **2** Guide disc, **3** Multiplate clutch, **4** Pole shoe, **5** Excitation winding, **6** Carbon brush, **7** Brush holder, **8** Terminal stud, **9** Control relay (24 and 36 V), **10** Solenoid switch, **11** Cutoff spring, **12** Commutator, **13** Armature, **14** Engaging spindle, **15** Helical spline, **16** Field frame, **17** Reduction-gear, **18** Drive housing, **19** Pilot for engine flange, **20** Pinion.

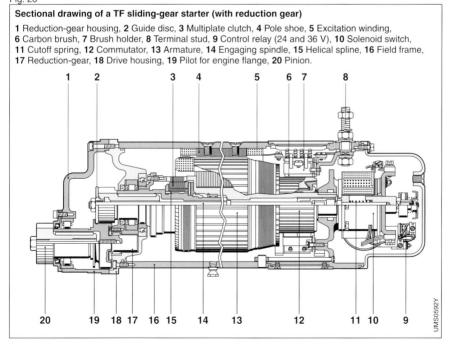

UMS0592Y

Since a single starter does not suffice for starting very large engines, some starters models are provided with additional terminals for parallel operation.

Operation
Operation of types TB/TF and KB/QB is nearly identical.

First (preliminary) switching stage:
When the starter switch is actuated, the current flows via terminal 50 through the hold-in winding of the control relay, thus causing the control relay to open the contacts for the brake winding and energize the pull-in and opposing windings of the engagement solenoid. The solenoid armature pushes the drive spindle with pinion against the ring gear via the engagement rod (Fig. 27).
Simultaneously, the series winding (main winding) is supplied with low current via the pull-in and opposing winding which acts as a dropping resistor. This causes the starter armature to rotate slowly.
In this first switching stage the starter does not yet develop its full torque. Even before the end of the meshing travel is reached, a stop plate contacts a tripping lever which stops the moving contact of the engagement solenoid. However, the solenoid armature continues to move. Thus the pinion is simultaneously pushed forward and slowly rotated so that it smoothly meshes with the ring gear.
If a pinion tooth meets a ring-gear tooth, the pinion continues to rotate on the face of the ring gear until it finds the next gap. Even if the pinion easily engages the ring gear, it does not yet crank the engine since the torque developed by the starter motor in this switching phase is still inadequate. If the pinion comes into corner-to-corner contact with the ring gear and cannot mesh, a restart attempt must be made.

Second (main) switching stage:
A release lever lifts the tripping lever just before the end of the meshing travel to initiate the second switching stage. The tripping lever releases the stop plate and a spring which was compressed during meshing is again relieved. This causes the moving contact to press with snap-action against the contact bar, and prevents the moving contact from burning if meshing is not immediate, as well as considerably lengthening the service life of the contacts. Although the pull-in and opposing windings of the engagement solenoid are simultaneously shorted, the energized hold-in winding holds the solenoid armature in the "on" position. The full current is now supplied to the starter; it develops its full torque and cranks the engine via the multiplate overrunning clutch which is now engaged.

Overspeeding and demeshing:
As soon as the speed of the engine increases, the overrunning clutch disengages to prevent the starter armature from being accelerated excessively. The pinion, however, remains engaged. After the starter is turned off, the control-relay winding and the hold-in winding of the engagement solenoid are de-energized. The control relay interrupts the pull-in winding circuit and energizes the brake winding. A cutoff spring returns the solenoid armature to its rest position. This also interrupts the main circuit. The pinion demeshes and moves, together with the drive, to the rest position under the action of a return spring inside the hollow armature shaft. When the pinion demeshes, the spring-loaded tripping lever of the control relay is also returned to its locked position, and the starter is ready for the next two-stage starting operation.

Fig. 27

Schematic drawing of the most important working phases of a TB 24 V sliding-gear starter

1 Ignition/starter switch or driving switch, **2** Tripping lever, **3** Release lever, **4** Control relay,
5 Moving contact, **6** Stop, **7** Engagement solenoid, **8** Armature, **9** Excitation winding (brake winding and series winding), **10** Pinion, **11** Battery, **12** Multiplate overrunning clutch.

I Rest position
No voltage applied.

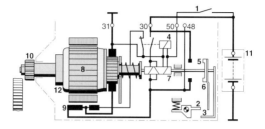

2 Tooth meets gap
Favorable meshing position
(switching stage 1).

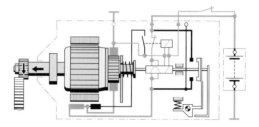

3 Tooth meets tooth or corner
Meshing not possible
(switching stage 1).
Start attempt must
be repeated.

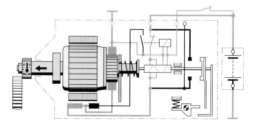

4 Engine is cranked
End position
(switching stage 2).
Starter torque
is maximum.

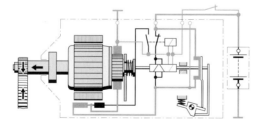

5 Demeshing and braking

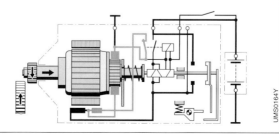

UMS0164Y

Starting-system installation

Installing the starter

Starters are mounted either ahead of the flywheel next to the crankcase or behind the flywheel next to the transmission. Depending upon design, starters are either flange-mounted or are mounted in a cradle such that they make good electrical contact with the engine.

Small and medium-sized flange-mounted starters usually use a two-hole flange (Fig. 1), whereas larger starters use an SAE flange (named after the Society of Automotive Engineers). In various models an additional support is provided in order to reduce the effect of vibration on the starter (Fig. 2).

Cradle-mounted starters are held in place by strong hold-down clamps (Fig. 3).

Starters are generally mounted horizontally with the terminals and the solenoid switch on top. Starters whose bearings must be lubricated at frequent intervals due to adverse operating conditions (dust, dirt) must have freely accessible lubrication points.

A pilot on the starter allows the starter to be centered and the proper gear backlash to be maintained.

Main starter cable

A look under the hood of a passenger car will reveal that the cable from the battery to the starter has an unusually large cross section. The distance between the battery and the starter, and thus the cable length, are also kept short. This indicates how important the starter cable is. The cross section of an electrical conductor always depends upon the electrical devices connected to it. The largest load in the vehicle – if only briefly for starting the engine – is always the starter. For this reason, the size of the battery and the routing and cross section of the starter cable are determined by the starter itself.

When starting the engine, extremely high current flows between the battery and the starter. When the starter pinion is engaged and the starter is stalled (starter speed "0"), brief short-circuit currents between 335 A (DM starter) and 3250 A (TB/TF starter) can flow. The starter cable must therefore have as low a resistance as possible in order to keep the voltage drop to a minimum. The resistance of the supply and return line together should not exceed $1\,\text{m}\Omega$, and

Fig. 1

Flange-mounted starter

UMS0165Y

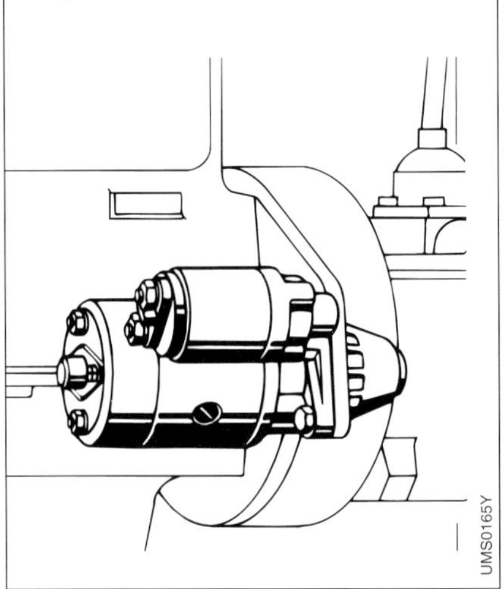

Fig. 2

Starter with additional support
1 Flange mount,
2 Support.

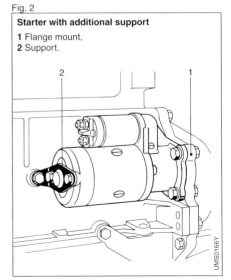

UMS0166Y

the maximum permissible voltage drop for a rated voltage of 12 V and 24 V is limited to 0.5 V and 1.0 V respectively. The starter cable must therefore be as short as possible and have an adequate minimum cross section.

Example:
Starter Type DW 12 V 1.4 kW for gasoline engines with swept volumes of up to 3 liters draws a short-circuit current of 427 A when connected to a battery with a nominal capacity of 66 Ah. Taking into consideration the increase in the cable temperature and the voltage drop, a 1.9 m long starter cable should theoretically have a minimum cross section of approx. 30 mm^2. Referring to the standard cross sections prescribed for automotive engineering, the next larger cross section of 35 mm^2 is then actually installed in practice.

In most cases, the starter is grounded to the engine by its housing. A ground cable then returns from the engine to the battery. If the starter has an insulated ground cable, the starter need not be physically grounded to the engine. The starter terminals are protected by rubber sleeves or caps.

The following parameters determine the cross section of a starter cable:

– Current consumption of the starter under short-circuit conditions (zero speed) and the brief permissible cable loading in terms of temperature,
– Starter-cable material and its specific resistance (copper cables are usually used due to their favourable material characteristics),
– Cable length, and
– Rated voltage of the starting system and permissible voltage drop under short-circuit conditions.

Starter switch

Switches which are normally used in starting systems are, in most cases, manually operated mechanical switches. They are used for either directly switching small starters or indirectly switching large starters via additional relays.

Single-purpose starter switch
The simplest type of starter switch is the pushbutton which is a standard single-purpose switch with "on" and "off" positions. The pushbutton returns automatically to its initial position.

Ignition/starting switches
Ignition/starting switches with built-in locks are multi-purpose switches for battery ignition systems. They are used to switch current from a central point to most of the vehicle loads including the ignition system and the starter.

Glow-plug and starter switches for diesel-engine vehicles
Glow-plug and starter switches in the form of rotary switches, push-pull switches, or key-operated switches are used to start diesel engines. The latter are a combination of starter switch and glow-plug and starter switch, thus making an additional driving switch unnecessary.

Fig. 3

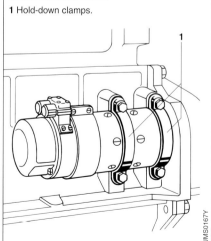

Cradle-mounted starter
1 Hold-down clamps.

UMS0167Y

Relays

Relays are normally installed to control large starters. These relays perform different functions, depending upon the specific application:

– Switching the high starter currents,
– Preventing damage to the starter or ring gear,
– Controlling the start-repeating relay in case the engine fails to start,
– Switching electrical circuits in the vehicle, and
– Parallel switching of starters.

Battery change-over relay

Various heavy-duty commercial vehicles feature a combined 12/24 V electrical system. This type of 12/24 system (12 V standard system voltage, 24 V for starting) must include a battery relay (Fig. 4). It switches the contacts in such a way that the two 12 V batteries previously connected in parallel are temporarily connected in series to supply the starter with a voltage of 24 V.

Start-locking relay

The start-locking relay is used in cases where it is difficult to monitor the starting procedure. It protects the starter, the pinion and the engine ring gear in com-

Fig. 4

Battery change-over relay

UMS0186Y

mercial vehicles with underfloor or rear engines, remotely controlled starting systems and fully automatic starting systems (e.g., standby power units). The following functions must be fulfilled in all cases:

– Shut-off after successful start,
– Blocked restart with engine running,
– Blocked restart with engine running down, and
– Blocked restart after failed start.

In the last two cases, a restart is only possible after the integrated blocking time has elapsed. The start-locking relay depends on the voltage from the alternator or from a rotational-speed sensor for its operation.

Start-repeating relay

The purpose of the start-repeating relay is to protect the starter solenoid switch in vehicles in which the driver is unable to hear the engine start (e.g. rear-engine commercial vehicles, diesel railroad engines), during parallel operation of two starters as well as in remote-controlled stationary systems. The relay is designed exclusively for use with starters which operate electrically or mechanically in two stages (IE, K, Q and T starters) which require the additional terminal 48. The start-repeating relay does not respond if the starter meshes normally. If the pinion is not able to mesh with the ring gear, however, the main current contacts are not closed even if the solenoid switch is actuated. In order to keep the pull-in winding of the solenoid switch from being overloaded and burning if the starting switch is operated too long, the start-repeating relay interrupts the starting procedure automatically and initiates a restart. This uses an NC time-lag relay, and is repeated until the pinion meshes with the ring gear and the main current contacts are closed.

Start-repeating relay with control relay

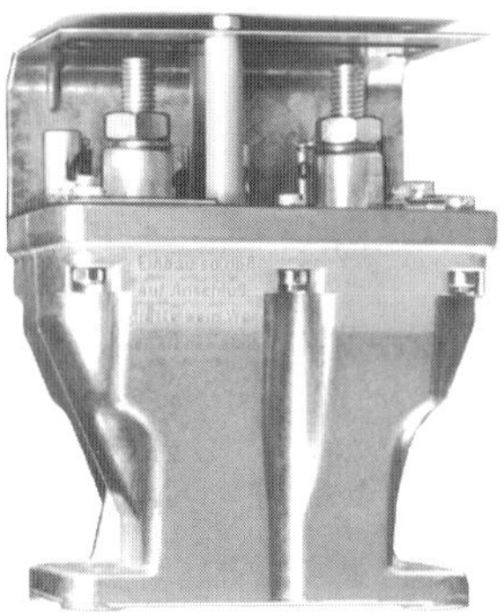

Whereas the control relay for the main starter current is located in the starter in low-voltage systems (up to 36 V), it is combined with the start-repeating relay in higher-voltage systems (50 to 110 V). This improves switching reliability.

Hold-in relay

Starting systems in railroad engines, locomotives and large stationary engines with lubricating-oil, temperature and water-level monitoring systems frequently contain monitoring devices which can suddenly respond and interrupt the flow of current in the starter control line.

The hold-in relay prevents these monitoring devices unnecessarily switching the starter on and off during starting. This could lead to burning of the moving contact in the solenoid switch.

Double-starting relay

For starting, very large engines require two parallel-connected sliding-gear starting motors which operate simultaneously. These parallel starting systems develop roughly double the starting power of single-starter systems if battery capacity is increased accordingly.

Low-voltage parallel starting systems (up to 36 V) require a double-starting relay. This relay supplies main current to both starters simultaneously, but only when both starter pinions are fully meshed. The engine is then cranked by both starters simultaneously.

Switching relay for parallel operation

High-voltage parallel starting systems (50 to 110 V) contain a switching relay for parallel operation, besides the start-repeating relay with control relay. The start-repeating relay with control relay switches the main current for starter I and, by contrast, the circuit for parallel operation switches the main current for starter II; this relay also meshes both starters.

Battery relay (battery main switch)

A main switch is prescribed by law in the electrical systems of buses, railroad engines, tank trucks etc., and disconnects the vehicle electrical system from the battery. This avoids short circuits (e.g. during repairs or as the result of accidents) as well as deterioration of live parts due to leakage currents brought about by the effects of salty splash-water (winter operation).

An electromagnetic battery main switch must be used in systems with alternators. This switch prevents the alternator from being electrically disconnected from the battery while the engine is running.

Fig. 5

Double-starting relay

Fig. 6

Switching relay for parallel operation

Workshop testing techniques

Bosch customer Service

Customer service quality is also a measure of product quality. The car driver has more than 10,000 Bosch Service Agents at his disposal in 125 countries all over the world. These neutral workshops are not tied to any particular make of vehicle. Even in sparsely populated and remote areas of Africa and South America the driver can rely on getting help quickly. Help which is based on the same quality standards as in Germany, and which is backed of course by the same guarantees which apply to Bosch customer-service work all over the world.

The data and the performance specs for the Bosch systems and assemblies of equipment are precisely matched to the engine and the vehicle. In order that they can be checked in the workshop, Bosch developed the appropriate measurement techniques, test equipment, and special tools and equipped all its Service Agents accordingly.

Testing techniques for starters

Today's passenger-car starting systems are designed to last for the average service life of the vehicle's engine, and therefore need no special servicing.

On the other hand, when commercial use is made of vehicles with short-distance trips or other unusual loading, and this applies particularly to taxis and parcel services etc., it is advisable that the starting system is checked regularly.

This is also applicable to trucks which, compared to passenger cars, are designed for far higher mileages.

If trouble occurs with the starting system, a check should first of all be made directly in the vehicle. If a fault is located on the starter, this is either replaced or repaired using the Service Information and Service Instructions documentation. Before being installed in the vehicle again, the starter must be tested on the combination test bench.

Testing directly in the vehicle

Before the starter is checked, the battery voltage under load, the electroylte level, and the electrolyte specific gravity must be measured to prove the correct functioning of the battery.

By means of an acoustic check, the following malfunctions can be pinpointed:
– Unusual starter noises,
– Starter meshes, but the engine only

Fig. 7

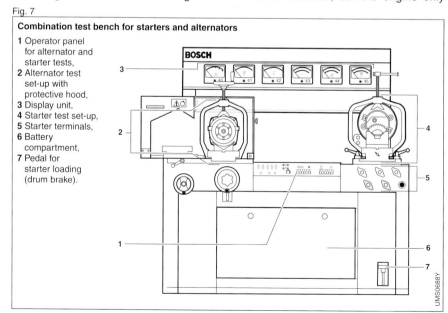

Combination test bench for starters and alternators

1 Operator panel for alternator and starter tests,
2 Alternator test set-up with protective hood,
3 Display unit,
4 Starter test set-up,
5 Starter terminals,
6 Battery compartment,
7 Pedal for starter loading (drum brake).

BOSCH

UMS0688Y

turns slowly or not at all,
- No meshing sound,
- Starter fails to mesh or only de-meshes very slowly.

Unusual starter noises can be attributed to starter faults, to the starter installation, or to the engine's ring gear. Specific electrical testing of the starting system (with an engine analyzer for instance) is required for the other faults. The following tests are carried out with the starter switched off:
- Short-circuit to ground, or short-circuit to + of the cables and lines,
- Voltage at Terminal 30,
- Open-circuit of the cables and lines,
- Contact resistance of the cables and lines.

The following tests are carried out when starting the engine:
- Voltage at Terminal 50,
- Voltage at the relay output, and
- Starter current (up to 1,000 A).

Starter repair

When repairing a starter with the help of the Service Instructions, a number of different testers are used (e.g. alternator tester and interturn short-circuit tester). In addition, in order that the defects inside the starter can be located and re-paired efficiently, special tools are required for each starter type.

Checking the starter on the combination test bench

Once the starter has been repaired, it is clamped in the starter test set-up on the combination test bench (Fig. 7). Depending upon starter model, the starter is clamped using a flange (Fig. 8) or a V-block and hold-down clamp. Pinion backlash and pinion spacing are adjusted using the handwheel and clamping table, and then the speed sensor is calibrated. Finally the starter is connected electrically. Basically, the check of the starter comprises two test schedules:
- As the name implies, in the no-load test the starter is driven without load. The rotational speed must reach a minimum value, and the starter current must not exceed a given limit.
- With the short-circuit test, the starter is braked to standstill using the test-bench drum brake, whereby the starter may only remain blocked for a maximum of 2 secs.

The starter current and voltage are measured during short-circuit operation. The measurement results must comply with the specified values.

Fig. 8

Starter clamped for testing

1 Ring gear,
2 Starter,
3 Protective cover,
4 Speed sensor,
5 Handwheel,
6 Clamping bracket,
7 Clamping flange,
8 Clamping table.

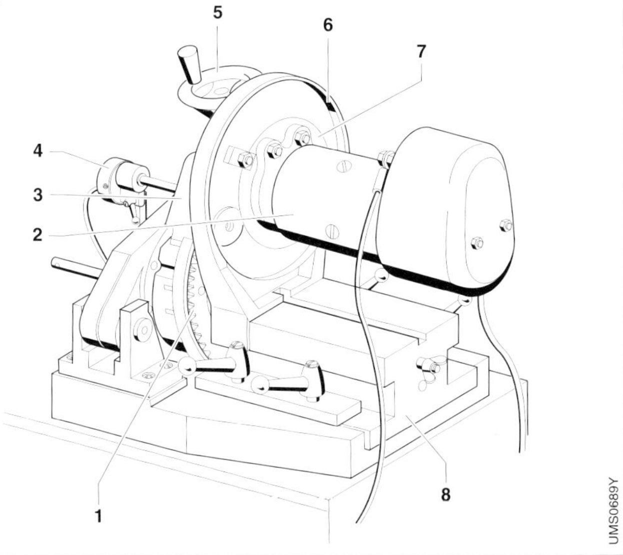

UMS0689Y

Lighting technology

Technical demands

Increasing traffic density is combining with higher vehicle speeds and a wide spectrum of driving conditions to confront headlamp and lighting-system designers with an increasingly exacting array of challenges.

The challenge of mastering an extensive range of potential driving situations is the decisive factor behind the engineering and layout of automotive lighting equipment.

Effective illumination of the road surface under a variety of conditions not only serves as a vital reinforcement for human vision; in some cases sight would be completely impossible without the help of the vehicle's lighting system. Priorities include brightness, color and spatial perception, while also embracing shape and motion along with recognition of luminance and contrasting hues. This is why automotive headlamps are subject to such stringent technical demands.

High-performance headlamps and other lamps at the front and rear of the vehicle form the foundation of an effective "see and be seen" policy.

Requirements in the realm of illumination technology are now being joined by the proliferating stylistic demands that headlamps and vehicle lights are expected to satisfy. One essential condition for meeting these demands is partial or total deletion of optical profile elements from the external lens to produce a "glass-smooth" finish. The inside of the headlamp or lamp unit can then be designed to reflect stylistic requirements. The ultimate result is powerful headlamps and attractively styled lamps.

Front-end lighting

The primary function of the front headlamps (Figure 1) is to illuminate the road, making it possible for the driver to monitor road conditions and recognize obstructions in time to react. The image resulting from switched-on headlamps also serves as a vehicle-recognition signal for oncoming traffic. The turn signals alert other drivers to changes of direction and, when they flash together, to potentially critical situations.

The headlamps and other lamps installed on the front end of the vehicle include:

– High-beam/low-beam headlamps,
– Fog lamps,

Figure 1
In combination: Low beam, high beam, fog lamp and turn signal form a single unit for harmonious integration in the vehicle's front end (example)

UKB0281Y

- Auxiliary driving lamps,
- Turn-signal lamps,
- Parking lamps,
- Side-marker and clearance lamps (wide vehicles), and
- Daytime running lamps (as prescribed in some countries).

Rear-end lighting

Rear-mounted lamps (Figure 2) are switched on in response to inclement weather and darkness to indicate the vehicle's position, and also to signal the vehicle's current and intended direction. The stop lamps show whether a vehicle's brakes are being applied or not. The turn-signal lamps indicate an intended change of direction, and when both flash at once they warn of a hazardous situation. The backup lamps provide illumination when the vehicle is reversing.

The lamps and lights on the vehicle's rear end include:

Figure 2
Rear lamps and backup lamp integrated to form a compact and efficient assembly at the rear of the vehicle (example)

- Stop (brake) lamps,
- Tail lamps,
- Fog warning lamps,
- Turn-signal lamps,
- Parking lamps,
- Clearance lamps (wide vehicles),
- Backup lamps, and
- License-plate lamp.

Vehicle interior

Regarding the inside of the vehicle, the vital factors include safe operation of controls and switchgear along with an adequate flow of information on operating conditions (all with minimal driver distraction). These priorities dictate effectively illuminated instrument panels and discrete lighting for various control clusters (such as the sound and navigation systems), where they satisfy a prime requirement for relaxed and safe vehicle operation. Visual and acoustic signals should be transmitted to the driver in accordance with their priority.

Legal requirements

Overview

Numerous regulations govern highway traffic both within Germany and in the EU as a whole.

These legal mandates regulate such areas as traffic safety, driver conduct and homologation approval for vehicles intended for operation on the public highways.

Automotive lighting equipment is defined in the Design and Operation Regulations, while official certification is subject to yet other stipulations.

Approval codes and symbols

Lighting equipment on motor vehicles is subject to national and international design and operation regulations that govern both manufacturing processes and test procedures.

A special approval code is specified for each kind of lighting device. This code or the corresponding symbol must be clearly visible on the lens of the headlamp or other lighting device (Figure 3). This

Figure 3
Headlamp lens with ECE and SAE approval-code symbols (examples)

provision also applies to replacement lamps and lights.

The presence of a code or symbol attests that the light unit has absolved testing with an official technical appraisal institute to earn official government type approval (i.e., from the Kraftfahrzeugbundesamt in Germany). The specifications of all units from series production that display a particular code or symbol must correspond to those of the test specimens. Examples of certifications symbols are:

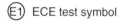 K German national test symbol

Ⓔ1 ECE test symbol

e1 EU test symbol.

The number 1 following each letter indicates that type approval has been granted in Germany (Figure 3).

Moving beyond the individual national ordinances, installation of all automotive lighting and visual signalling equipment within Europe as a whole is governed by European regulations (ECE: All of Europe with the exception of Albania, EU: 15 countries); these pan-European ordinances assume legal priority over national laws. Continuing steps toward European unity have been accompanied by harmonization of regulations and ordinances; this, in turn, has allowed progressive relaxation of export restrictions. Because headlamps from Bosch comply with the applicable ECE and EU regulations, they can be used in all ECE and EU nations, regardless of the country of acquisition.

A completely different regulatory framework applies in the US. The self-certification principle compels each manufacturer, in its role as importer of technical lighting equipment, to verify (and demonstrate when required) 100% compliance with the specifications defined in FMVSS 108 as entered in the Federal Register. As this implies, there is no specific individual homologation certification in the European sense. The specifications in FMVSS 108 are partly based on SAE industrial standards.

This means that vehicles being reimported to Europe must be modified to comply with the European standards.

Development of lighting technology

Introduction of lighting equipment

When the first motor vehicles appeared at the end of the 19th century, lighting technology played only a minor role; the dubious vehicle safety of the period meant that night driving was too dangerous in any case to become common. The first lamps were candle-powered hurricane lanterns, followed by petroleum lamps and, finally, acetylene lamps.

In 1908 the only electric lamps on the vehicle were auxiliary devices incorporated in the rear and side lights, as no means was available for recharging the batteries while the vehicle was actually being driven. The introduction of the "dynamo," or Lichtmaschine (the eponymous German term) fostered the gradual rise of electric headlamps. The year 1908 also witnessed the first deliveries of Bosch lighting equipment, consisting of the following components:
- Headlamps and side lanterns (Figure 5),
- Number-plate and splash-wall lanterns,
- Generator,
- Battery, and
- Switch unit (Figure 5)

Figure 4 (next page)
Advances in headlamp design as reflected in vehicles from different periods

Figure 4

UKB0362...69

The march toward mass motorization, higher traffic density and stylistic demands have all combined to propel lighting technology to its current level (Figure 4).

Enhanced road-surface illumination

Reflectors designed as parabolic "mirrors" made an early appearance as a means of directing the beam from the light source toward the road surface for maximum illumination. Focusing equipment was employed to adjust the filament to the focal point. These headlamps usually consisted of lanterns in all conceivable – and sometimes completely arbitrary – configurations. Essential elements were the clear lens protecting the reflector and the bulb against dirt and damage. The beam they projected produced nothing more than a simple patch of light on the road surface. As early as 1911, the realization that enhanced driving safety goes hand in hand with even

and consistent illumination of the road surface became the basis upon which Bosch proceeded with its headlamp development and manufacturing activities. Major developments included highly polished metal reflectors, suitable bulbs and a new bulb socket. The latter, adopted as an industrial standard in 1917, made the complex focusing mechanisms a thing of the past.

Reduced glare

Bulbs continued to get brighter until around 1919. When two vehicles met, both drivers were blinded (especially if the road was wet), and at least one would be compelled to stop.

Arrangements designed for switching on either of two different headlamp pairs, either high beams for long-distance illumination or low-intensity "approach lamps," proved to be a substantial improvement.

The costs incurred by the additional headlamp pair though, combined with the basic idea that a low-beam concept could be implemented by alternating between two different intensities and projection paths, resulted in the shift of the adjustment device for luminous intensity and projection angle into the bulb itself.

With Bosch participation, the "Bilux" bulb with individual filaments for low and high

(Figure 5)
Bosch lighting system from 1913

1 Headlamp,
2 Side lamp,
3 Generator,
4 Control and switch box.
Battery, number plate and splash-wall lamps not shown.

beam appeared in 1925. A metal shield blocked part of the light from the low-beam filaments, which then illuminated only the top half of the reflector to project a beam focused immediately forward of the vehicle.

Improved peripheral illumination

Better dispersion of the light patterns to the side allowed headlamps to provide still higher levels of safety. Newly developed pressed-glass lenses featuring vertically-oriented cylindrical lens technology extended the beam sideways without increasing low-beam glare. This concept permitted these lights to satisfy demands for illumination to the side, as well as for illumination of the road surface directly forward of the vehicle.

In 1931 the range was supplemented by a fog lamp projecting a light beam that was aimed both downward and offset to the side in order to inhibit backglare (an effective strategy because fog density is generally lowest just above ground level).

(Figure 6)
Light distribution patterns with symmetrical (a) and asymmetrical low beam (b)

1 Vehicle,
2 Road surface,
3 Symmetrical low beam (no longer common),
4 Asymmetrical low beam (current standard).

This improved the driver-side visibility along the roadway shoulder.

Extended range

With the construction of Autobahnen and other limited-access high-speed roadways, the projection ranges provided by the period's headlamps became inadequate. The high beams, and even more so the low beams, did not satisfy the requirements for extended range which went hand in hand with higher vehicle speeds.

Supplementary driving lamps (with a tilting arrangement for "dimming") were installed with the aim of extending visual ranges from the 200 m provided by contemporary high beams to between 800 and 1,000 metres.

Asymmetrical projection patterns

The period after 1945 witnessed the emergence of new automotive styling trends, and one characteristic development was to incorporate the previously free-standing headlamps within the bodywork.

This period also was marked by increasing reliance on the low beams, as the continued rise in traffic density meant that high beams could only rarely be used, even on the open road. The introduction of headlamps with an asym-

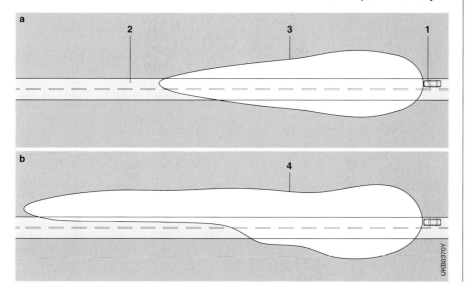

metrical light/dark cutoff – characterized by an ascending cutoff line on the right side – was a major advance over the symmetrical distribution patterns with horizontal beam limits used till then. The new lamps allowed substantial extensions in low-beam range along the shoulder of the road without producing glare that could blind approaching traffic (Figure 6).

Enhanced illumination

One disadvantage of conventional light bulbs lies in their tendency to become opaque. The opacity arises as vaporized tungsten particles from the filament gradually accumulate on the inside of the bulb. In all respects, this process reduces the lighting efficiency, while at the same time shortening service life. The introduction of the halogen bulb meant enhanced luminous density from the filament as well as extended operating life. The halogen lamps available from 1964 onward had only one filament and were used initially in fog lamps and in separate low and high-beam headlamp assemblies (H1, H2 and H3 bulbs). The dual-filament H4 halogen units with their combined high and low beams became available in 1971 and soon advanced to become practically an industry standard. The H7 lamp, available since 1992, is seeing increased application in modern systems with reflectors featuring computer-aided design technology and clear outer lenses. The prime characteristics of this bulb are high luminous density and precise tolerances in the filament.

Since 1991 the contemporary state-of-the-art has been the Litronic gaseous-discharge (or HID, high-intensity discharge) lighting system. In addition to its extended service life, its prime distinction is its considerably improved road illumination, achieved despite its substantially smaller projection surface. The system provides a luminous flux two to three times more powerful than that of a standard halogen bulb throughout a service life that extends up to 2,000 hours.

Effective signal image to the rear

In terms of size, features and illumination, rear-lamp developments have proceeded on roughly the same lines as those for headlamps.

Early rear lighting tended to consist of various individual units mounted separately on the body. Tail lamps, stop (brake) lamps, reflectors, and, finally, license-plate lamps and backup lamps were all independent units. In the course of time the individual components, now improved for better lighting performance, were incorporated in combined assemblies.

Current passenger cars rely on the assemblies that are now available for total integration within the bodywork. These incorporate such equipment as turn signals, stop lamps, tail lamps, reflectors, parking lamps, backup lamps and fog warning lamps, all in a single integrated assembly. License-plate illumination usually stems from a separate source. On heavy commercial vehicles the tail-lamp assemblies often consist of two or three-chamber units in which it is frequently possible to integrate the license-plate lamp.

Tail lamps can rely on LEDs as well as conventional bulbs as their illumination source.

Clear signaling of direction

During the period in which open cars were the norm, a hand signal gave sufficient warning of intended changes in direction. Bosch responded to the widespread introduction of enclosed vehicles and higher traffic density in major urban centers by starting production of semaphore blinkers in 1928. The driver could then signal an intended change in direction by pressing a blinker switch to activate a solenoid and extend a semaphore lever mounted on the left or right side of the vehicle. A bulb within the unit lit up at the same time. It was only in the period following 1949 that these semaphore turn signals started losing ground to electric turn-signal lamps (later combined with hazard-warning flashers) with their considerably improved effectiveness.

Visual perception

The human eye

The eye relies on various mechanisms to adapt to different lighting conditions. One of these is an adjustment in the sensitivity of the retina, while another is the adaptive response of the iris. The iris, which encircles the lens like a shutter, reacts to changes in light intensity by expanding and contracting to enlarge or reduce the effective aperture of the pupil. The eye reacts to sudden glare with contraction, marked by muscular tension and lid closure. This leads to a temporary but substantial loss of vision.

This is why the avoidance of glare is one of the essential priorities of headlamp technology: The aim being to ensure consistent visual performance.

Evaluating visual acuity

Sensitivity

The human visual apparatus assumes a vital role as a source of sensory information, of which it supplies roughly 90% of the total. The primary perceptual classifications for all visual information are brightness, color and shape, while sensitivity to brightness, in particular, is also marked by pronounced spectral sensitivity.

In 1924, efforts to quantify "perceived" light based on technically reliable data led to the definition of a spectral brightness sensitivity function $V(\lambda)$ for a standard observer. The corresponding spectral sensitivity peaks at a wavelength of $\lambda = 555$ nm (in the yellow-green range).

The boundary separating UV radiation and visible light (violet) lies at the short-wave end of the spectrum, at 380 nm. In the long-wave range the red sector extends to 780 nm, which marks the start of the "invisible" infrared range.

Visual and signal identification (ID) ranges

The visual range is the maximum distance at which objects (such as vehicles) and persons on the road or on the road shoulder remain visible.

Among the factors which influence visual range are the shape, size and reflectivity of the person or object, the road-surface material, the technical characteristics and cleanliness of windshields and headlamps, and the visual acuity of the driver.

This extensive array of influences renders it impossible to define visual range in purely numerical terms. Under extremely favorable conditions it can extend forward beyond 100 m (along the verge of the road), while negative factors (the other side of the road being wet) can reduce it to below 20 m.

The signal identification range is the distance at which a light signal (such as that generated by the brake lamps on another vehicle) can be perceived in fog, mist or other inclement weather conditions.

Safety

Light sensitivity is particularly significant for "perceptibilty efficiency," a factor of vital importance in road safety.

Satisfying the demands of perceptibility efficiency entails total exploitation of all available means for ensuring that the driver retains the unobstructed ability of perceiving and recognizing all visual and acoustic impressions and signals generated or transmitted by road traffic.

Yet another significant factor is "control efficiency," which focuses on fostering road safety by minimizing certain demands on the driver. Here the object is to position all instruments and control elements in highly visible locations within convenient reach of the driver, thereby facilitating their operation and use. Lighting for instruments and displays joins dashboard control configuration and general interior illumination as an essential contributor to achieving this end.

Visual impairment factors

The lighting system can support the driver's perceptiblity and control efficiency while driving under restricted visibility (dawn, dusk, darkness, driving in a tunnel or on roads surrounded by dense forest). Even the surface of a dry, brightly illuminated road absorbs 70% of the incident light, leaving only 30% for visible roadway illumination.

Inclement weather and the accompanying effects (fog, rain, snow, etc.) also impinge upon the driver's clear field of sight. When wet, a dark road surface absorbs 85% of the light directed toward it. Under these conditions perceptibility efficiency can be substantially enhanced by lighting equipment such as headlamps on low or high beam as well as by front fog lamps.

Among other factors with the potential to directly limit visual perception are condensation, contamination and damage (scratches and fissures) on the windshield. Yet another danger source is sudden glare (from approaching traffic at night, the setting sun, etc.).

Indirect impairment of the driver's sight may arise from dirty headlamps, on which dirt prevents more than a fraction of the available light from reaching the road surface.

Various countermeasures are available for responding to these negative influences:

– Wipe/wash systems for windshields and headlamps,
– Sun visors,
– Variable light distribution, and

<u>Luminous intensity</u>
The "brightness" of light sources can vary. Luminous intensity serves as an index for comparing them. It is the visible light radiation that a light source projects in a specific direction. The unit for defining levels of luminous intensity is the candela (cd), roughly equivalent to the illumination emitted by one candle. The "brightness" of an illuminated surface varies according to its reflective properties, the luminous intensity and the distance separating it from the light source.

Samples of approved intensities (in cd):

Stop (brake) lamp (single)	60...185
Tail lamp (single)	4...12
Fog warning lamp (single)	50...300
High beam (max. total)	225,000

– Optimal road illumination (Figure 7, good low-level illumination, extended side illumination, concentration on specific road sectors).

Light sources

Thermal radiation devices

Thermal radiators generate light from heat energy. This means that luminous-intensity levels in these devices are proportional to the heat that can be generated at the source.

Figure 7
Road-surface illumination from two vehicles meeting at night

1 Wide-ranging side illumination,
2 Targeted road-surface illumination,
3 Effective illumination depth.

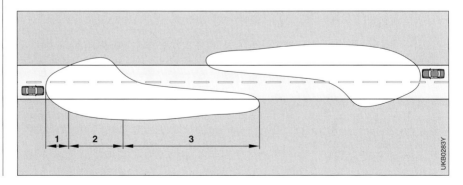

UKB0283Y

The major liability of the thermal radiator is its low working efficiency (below 10 %), which leads to very low potential for luminous efficiency relative to the gaseous-discharge lamp.

Incandescent (vacuum) bulb

The bulb is a thermal radiation device featuring a tungsten filament that glows when electrical energy flows through it.

The luminous power of a standard bulb is modest, and service life is limited by vaporized tungsten particles from the filament, which deposit on the inner surface of the bulb. These considerations led to the virtual extinction of the incandescent bulb in headlamps, where it has been replaced by the halogen bulb. Other lamps, including reversing lamps, are still equipped with incandescent bulbs (Figure 8).

Halogen bulbs

H1, H3, H7, HB3 and HB4 halogen bulbs have only a single filament. These and similar bulbs are employed in low-beam headlamps and fog lamps. The H4 halogen bulb (Figs. 9, 10) is a dual-filament unit capable of alternating between low and high beam. A cap covers that portion of the low beam with the

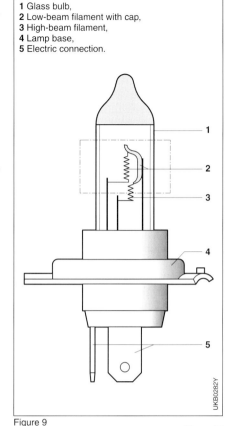

H4 Halogen lamp
1 Glass bulb,
2 Low-beam filament with cap,
3 High-beam filament,
4 Lamp base,
5 Electric connection.

UKB0282Y

Figure 9

Figure 8

Figure 10

Incandescent bulb

1 Glass bulb,
2 Filament,
3 Lamp socket base,
4 Electric connection.

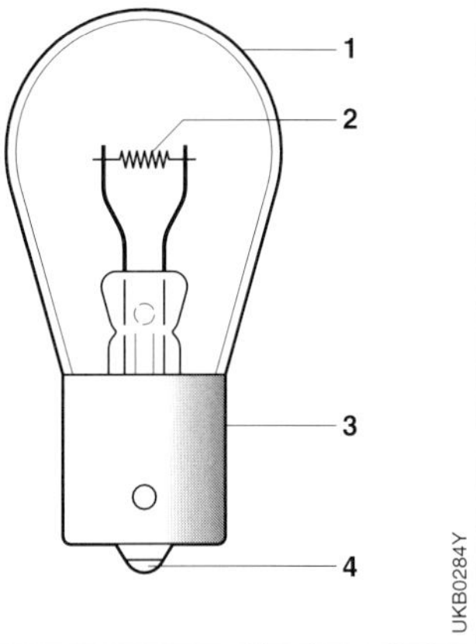

UKB0284Y

H4 halogen bulb (sectional view)

1 Tungsten filament,
2 Halogen charge (iodine or bromine),
3 Tungsten evaporate,
4 Tungsten halogenide,
5 Tungsten deposits.

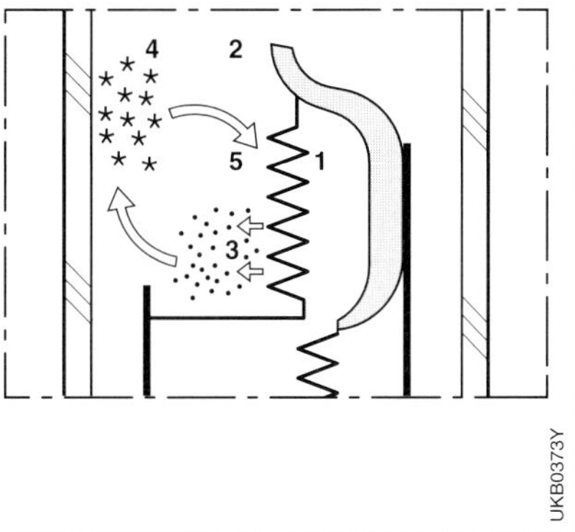

UKB0373Y

223

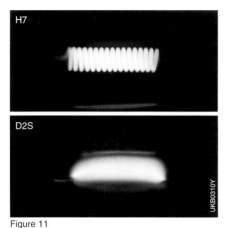

Figure 11
Comparison of light generation
H7 Halogen bulb, **D2S** Gaseous-discharge lamp
(filament compared to high-intensity discharge arc)

Figure 12

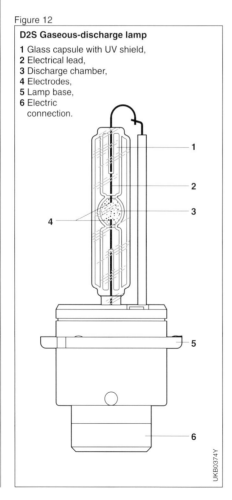

D2S Gaseous-discharge lamp
1 Glass capsule with UV shield,
2 Electrical lead,
3 Discharge chamber,
4 Electrodes,
5 Lamp base,
6 Electric
 connection.

highest glare potential to form a light/dark cutoff line within the light beam. A halogen unit rated at 60/55 W radiates approximately twice as much light as a comparable 45/40 W Bilux unit, and the inner surface of the bulb resists clouding, remaining clear throughout the H4's service life. The halogen-gas charge (iodine or bromine) makes it possible to use filament temperatures approaching tungsten's melting point (around 3,400 °C) for commensurately high levels of luminous power. The tungsten vapor adjacent to the hot bulb walls combines with the atmospheric halogen to form a translucent gas (halogenated tungsten) that remains stable through a temperature range of roughly 200...1,400 °C. Tungsten particles approaching the filament respond to the high local temperatures by dispersing to form a consistent tungsten layer (Figure 10). As external bulb temperatures of approximately 300 °C are needed to maintain this cycle, clearances between quartz bulb and filament must be minimal. A further advantage of this layout is the fact that higher fill pressures can be used which contribute to inhibiting tungsten's inherent evaporative tendency.

Even minute traces of oils and grease on the bulb's surface, of the kind caused by exposure to bare fingers, lead to harmful desposits that can attack and destroy the glass during high-temperature operation.

Gaseous-discharge lamps

Gaseous discharge (high-intensity discharge, arc discharge) describes the electrical discharge that occurs when an electrical current flows through a gas and causes it to emit radiation (Examples: sodium-vapor streetlamps and fluorescent interior lamps).

Light sources relying on the gaseous-discharge concept acquired new significance for automotive applications with the advent of the "Litronic" electronic lighting system. This kind of system is better adapted to human vision, as well as providing for longer ranges and brighter and more consistently distributed illumination of the road surface.

Durability is such that the units usually last for the entire life of the vehicle, with no need for replacement. The concept also makes it possible to design more compact headlamps for smooth front-end styling. The gaseous-discharge lamp is filled with the noble gas xenon and a mixture of metal halides, and an electronic ballast unit is required for ignition and operation. Application of an ignition voltage in the 10...20 kV range ionizes the gas between the electrodes, producing an electrically conductive path in the form of a luminous arc (Figure 11). A regulated supply of alternating current (400 Hz) heats the lamp body to vaporize the metallic charge and radiate light.

Under normal circumstances the lamp requires several seconds to ionize all of the particles and generate full illumination. This process is accelerated by feeding a high "start-up current" into the unit until it reaches maximum luminous power, at which point limitation of lamp current commences. A sustained operating voltage of only 85 V is then sufficient to maintain the arc.

This concept features several decisive assets compared with conventional bulbs:
– Extended service life. Because no solid metal evaporates, the lamp is not subject to mechanical wear,
– High levels of luminous efficiency owing to the extreme temperature of the gaseous mixture (hotter than 4,000K),
– Higher overall efficiency levels, due to high luminous efficiency combining with the low power consumption resulting from the lower operating temperatures.

Gaseous-discharge bulbs for automotive applications are designed around a socket and a glass element that serves as a shield against ultraviolet radiation:
– For PES headlamps, the D2S bulb (Figure 12),
– For reflection headlamps, the D2R bulb with an integral light shield designed to generate a pronounced

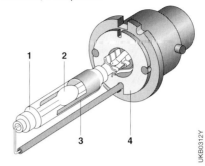

D2R gaseous-discharge lamp
1 Glass capsule, **2** Discharge chamber, **3** Shutter, **4** Lamp base.

UKB0312Y

Figure 13

light/dark cutoff (comparable to the low-beam bulb shield found in H4 units (Figure 13).

Automotive lamps
ECE-R37 defines lamps for vehicle lighting systems in 6 V, 12 V and 24 V versions. Various socket configurations are employed to ensure that the correct bulb types are installed at individual locations, while the operating voltages are also given on the bulb units to further distinguish bulbs of different voltage ratings that share a single socket design. The correct bulb type must be indicated on the assembly.

The luminous efficiency (in lumens per watt) indicates photometric efficiency as a function of the electrical power supplied to the unit. Thus the luminous efficiency of incandescent vacuum bulbs is 10...18 lm/W, while halogen bulbs operate at higher levels in the 22...26 lm/W range, primarily due to higher filament temperatures. At 85 lm/W, the D2S and D2R gaseous-discharge lamps (Litronic) make substantial contributions to improving low-beam illumination.

Luminous flux
Luminous flux is that light emitted by a light source that falls within the visible wavelength range. It is quantified in lumens (lm). For comparitive data refer to Table 1: Automotive lamps and bulbs.

Table 1

Specifications for motor-vehicle bulbs (not including motorcycles etc.)

Application	Designation	Voltage rating V	Wattage W	Luminous flux lm	Base type IEC	Illustration
High/ low beam (not on new vehicles)	R2	6 12 24	45/40[1] 45/40 55/50	600 min/ 400-550[1]	P 45 t-41	
Fog-, high/low beam, driving lamps in quad system	H1	6 12 24	55 55 70	1,350[2] 1,550 1,900	P14,5 e	
Fog lamp, auxiliary driving lamp	H3	6 12 24	55 55 70	1,050[2] 1,450 1,750	PK 22s	
High/low beam	H4	12 24	60/55 75/70	1,650/ 1,000[1], [2] 1,900/1,200	P 43 t – 38	
Side-marker lamp	H6W	12	6	125[3]	BAX9s	
High/low beam, fog lamp	H7	12	55	1,500[2]	PX 26 d	
High/low beam/ fog lamp (E vehicles)	H8	12	35	800[2]	PGJ19	
High beam	H9	12	65	2,100[2]	PGJ19-5	
Low beam/ fog lamp	H11	12 24	50 70	1,350[2] 1,750[2]	PGJ19-2	
Turn signal/ stop lamps	H21W	12	21	600[2]	BAY 9s	
Low beam in quad system	HB4	12	55	1,100	P 22 d	

Table 2, cont.

Application	Designation	Voltage rating V	Wattage W	Luminous flux lm	Base type IEC	Illustration
High beam in quad sytem	HB3	12	60	1,900	P 20 d	
Stop, turn signal, fog warning lamp, backup lamp	P 21 W PY 21 W[7]	6 12 24	21	460[3]	BA 15 s	
Stop lamp/ tail lamp	P 21/5 W PY 21 W[7]	6 12 24	21/5[4] 21/5 21/5	440/35[3] 440/35 440/40[3]	BAY 15d	
Side-marker lamp, tail lamp, tail lamp	R 5 W R 10 W	6 12 24 6 12 24	5 10	50[3] 125[3]	BA 15 s	
License-plate lamp, tail lamp, backup lamp	C 5 W C 21 W	6 12 24 12	5 21	45[3] 460[3]	SV 8,5 X SV 8,5	
Position lamp	T 4 W	6 12 24	4	35[3]	BA 9 s	
Side-marker lamp, license-plate lamp	W 5 W/ W 3 W	6 12 24	5/3	50/22[3]	W 2,1 x 9,5 d	
Low/high beam Bi-Litronic low beam (since 1991)	D1S[5]	85 12[6]	35 ca. 40[6]	3,200	PK 32 d-2	
Low/high beam Bi-Litronic low beam (since 1994)	D2S[5]	85 12[6]	35 ca. 40[6]	3,200	P 32 d-2	
Low/high beam Bi-Litronic low beam (since 1996)	D2R[5]	85 12[6]	35 ca. 40[6]	2,800	P 32 d-3	

[1] High/low beam, [2] Specs. at test voltage of 6.3 V; 13.2 and 28.0 V, [3] Specs. at test voltage of 6.75 V; 13.5 and 28.0 V, [4] Primary/secondary filament, [5] Gaseous-discharge (HID) lamp. Definition of standards in progress. [6] With ballast unit, [7] Amber version.

Headlamp elements

The headlamps installed in most vehicles still consist of the bulb, reflector and lens as their primary elements.

The bulb serves as light source by emitting a concentric beam (actual geometry may vary according to configuration). That portion of the light beam which is not emitted directly along the target path impacts against the reflector, which concentrates it to form a roughly parallel projection pattern (Figure 14).

A refraction lens then redirects the light waves toward the desired target region on the road surface (Figure 15).

In a number of newer headlamp designs it is the reflector that redirects the light toward the target area, thus assuming the function earlier performed by the lens. In these units the lens merely seals and protects the interior of the headlamp against external influences and contamination.

Reflectors

Purpose

The purpose of the reflectors installed in automotive headlamps is to collect as much of the bulb's light as possible to maximize beam range. As a basic rule, a headlamp's potential illumination range is proportional to its lens aperture, while luminous efficiency grows as the included angle embraced by the reflector increases (reflector depth).

Further demands on headlamp design arise from such requirements as dictated by vehicular styling considerations (for installation in flat-surfaced front ends or for general adaptation to body shape, etc.).

Earlier reflectors were almost always parabolic. Current designs respond to various imperatives by using a variety of configurations, as found in stepped (graduated) reflectors, transitionless free-form reflectors and headlamp designs based on optical imaging technology (PES, or Poly-Ellipsoid System).

Reflection

High levels of reflection are encountered whenever a beam of light impacts against a mirrored surface. The light's angles of incidence and reflection are identical. As this also applies to irregular mirror surfaces, parabolic headlamp reflectors are designed so that light beams leaving the focal point are emitted parallel to the reflector axis.

Figure 14

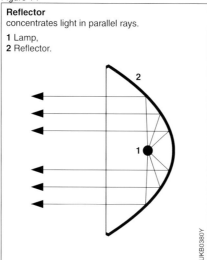

Reflector
concentrates light in parallel rays.

1 Lamp,
2 Reflector.

UKB0380Y

Figure 15

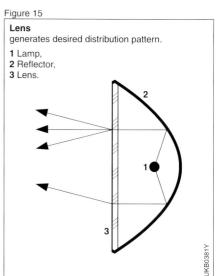

Lens
generates desired distribution pattern.

1 Lamp,
2 Reflector,
3 Lens.

UKB0381Y

The intensity of the emerging light relative to the incident illumination impacting against the reflector (reflection factor) is determined by the angle of incidence and the refraction indices of the contiguous materials. Reflectors featuring aluminum applied in an evaporative coating process exhibit reflection factors in the 90 % range. This figure can drop to below 50 % for a reflector surface that has been damaged by corrosion, which is why rust protection is a major consideration with headlamp reflectors. Reflector-surface quality and corrosion protection are prime determinants of headlamp quality.

Reflector focal length

Reflectors with shorter mean focal lengths promote effective exploitation of the bulb to achieve high efficiency levels. On these units the reflector encloses the lamp bulb to a large extent, making it possible to redirect a high proportion of the incident light into the ultimate beam pattern.

The focal length (interval between the vertex of the parabola and the focal point) is between 15 and 40 mm. A lens redistributes the light to provide the desired pattern on the road surface (Figure 15).

The geometries of modern reflectors display – sometimes quite substantial – departures from the purely parabolic form. Special mathematical (HNS, or Homogenous Numerically Calculated Surface) processes are employed to calculate optimal configurations. The mean focal length, as defined relative to the distance between the reflector's vertex and the center of the filament, is a given figure. The actual value is between 15 and 25 mm. Reduced focal lengths make it possible to install three separate reflectors (Figure 16) – for low-beam, high-beam and fog lamp (H1 and H2 bulbs) – within the same space needed by a conventional H4 parabolic reflector, all with a simultaneous increase in luminous efficiency.

Reflector materials

Bosch reflectors are precision-manufactured in sheet-metal or plastic.

The first step in the production process for sheet-metal reflectors is deep-draw molding to produce a paraboloid or one of the more complex geometrical configurations described above. This is followed by galvanization or application of a powder coating to protect against corrosion. The reflector is then painted to produce a smooth and consistent surface, after which evaporative or sputter-spray techniques are employed to apply the aluminum reflective layer and the special protective coating.

Figure 16

HNS Reflectors (example):
Using numerically-calculated reflector surfaces to modify light distribution for

1 Fog lamp,
2 Low beam (2a Side-marker lamp aperture),
3 High beam.

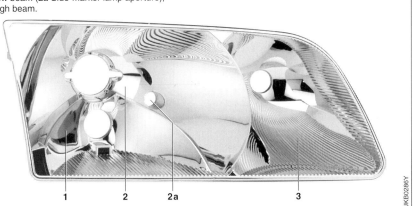

UKB0286Y

This process hermetically seals the sheet-metal while providing an extremely smooth and consistent surface featuring residual unevenness of no more than 1/10,000 mm.

Plastic reflectors are also components of extremely high quality, and high-precision spray (Thermoplast) or compression molding (Duroplast) techniques are employed to endow them with their ultimate geometry. This manufacturing strategy facilitates production of reflectors with special graduation patterns as well as multi-chamber units. Although no special anti-corrosion treatment is needed to protect the base material, the reflective layer is guarded by a special coating.

Lenses
(outer lens with light-dispersion optics)

Purpose

The purpose of the lens is to produce the desired illumination pattern on the road surface by refracting, dispersing and collecting the light emitted by the reflector. During the lens molding process high priority is assigned to surface quality. Flaws must be avoided, as a lens that emits stray light beams could pose a glare hazard to oncoming traffic.

Refraction and reflection

Headlamps and lamps in general rely on transparent substances such as air, glass, plastic, etc., to project light. Light rays entering the transition zone between air and the transparent lens material split into refracted and (relatively minute) reflected elements. Refraction results from variations in light-wave propagation velocities in different media such as air, glass and plastic. Both the transition from air (refractive index n of roughly 1) to glass (refractive index n of roughly 1.52) and the passage from glass to air produce reflection rates of around 4.3% at a 90% angle of incidence. The orientation of the reflected component is defined by the fact that the angle of incidence is equal to the projection (reflection) angle.

"Total" reflection, in which all light rays are reflected with no loss, occurs once the angle of incidence in a high-density substance exceeds a specified level. This effect is exploited in devices such as reflectors, roadside guard-post light refractors and optical-fiber cables (fiber-

Figure 17

Lens sectors with optical elements

1 Lens elements,
2 Prismatic elements,
3 Combined elements.

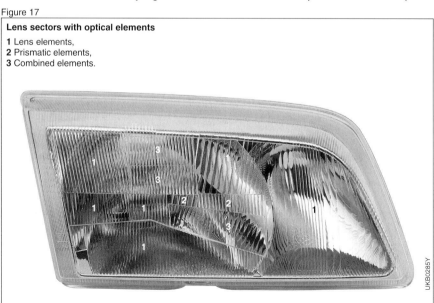

UKB0285Y

glass or plastic) as a means of relaying light.

On the inside of the lens there is a precisely defined arrangement of focal and prismatic elements, which combine with planar sectors to provide an extended high-beam range as well as good luminous distribution when the headlamps are dimmed. The specific type and configuration of these optical elements on the projection surface of the lens varies according to the reflector's focal length and the desired light pattern (Figs. 17 and 18).

The outside of the lens is always smooth to prevent the accumulation of dirt.

Optical elements (Figure 18).
Light refraction disperses parallel rays impacting against a focal (refractive) lens sector at a right angle to the axis of the cylindrical lens. The actual scatter effect is inversely proportional to the radius of the lens.

Light rays encountering a prismatic lens sector at a given angle are redirected at an angle defined by the geometry of the prismatic element. Parallel incident rays remain parallel as they emerge from the prism.

Combination sectors unite focal and prismatic elements in a single field.

"Clear" outer lenses
Modern multidimensional reflectors can be developed specifically for use in combination with "clear" lenses, in which no optical elements of any kind are required.

Lens materials
Conventional lenses are manufactured from a high-purity glass that must be absolutely free of bubbles or streaks. Today plastic is often specified as the lens' material in order to reduce weight. Within Europe the legal framework allowing official certification of plastic lenses is in place, and the first plastic lens entered series production in 1992. A varnish coat is applied to the exterior surface of the lens to protect it against aging and scratches. While plastic lenses can be specified as a weight-reduction measure, they also open new vistas in headlamp design by placing practically no limits on geometrical configurations. Yet another significant factor in technical automotive applications is the fact that plastic units are less expensive.

Figure 18

Optical elements in lens

1 Lens elements,
2 Prismatic elements,
3 Combined elements.

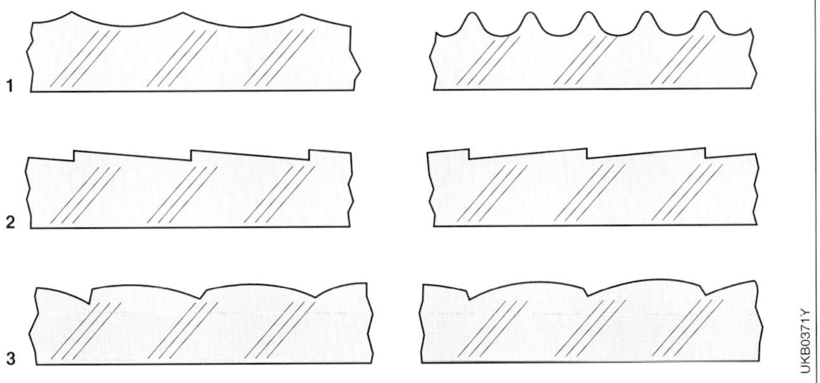

UKB0371Y

Lights and lamps

Purpose
Lights and lamps are intended to enable recognition of the vehicle's presence and its current or intended state of motion.

Regulations
Minimum and maximum luminous intensity levels are prescribed for the light emerging along the central horizontal axes of all lamps. The intent of the regulations is to ensure highly-visible and conspicuous signal images while simultaneously preventing dazzle for other road users. Luminous intensity may be lower in light emerging from above, below and to the side of the horizontal axis.

Versions
All vehicle-mounted lamps can be designed based on either one of two basic concepts, with the final choice being dictated by such factors as available space:

Lamps with reflector optics
A reflector in any of a wide variety of different shapes (often parabolic) redirects the light emitted by the bulb to produce a roughly axial projection angle. A lens featuring optical dispersion elements can then define the ultimate projection pattern as required (Figure 19).

Lamps with fresnel optics
The light is projected directly against the lens, with no refraction contributed by the reflector. A lens based on fresnel refraction technology then redirects the beam in the desired direction (Figure 20). Fresnel optics devices are usually less efficient than reflector-optics lamps.

Lamps combining fresnel and reflector optics
Combination lamps incorporating elements from both concepts are applied with success. As an example, the GP (Gedrehte Parabel, or rotated parabola) reflector can be employed to derive undiminished luminous flux from volumetrically compact units with both shallower reflectors and smaller lenses. This design employs a specially designed (axially offset) reflector to capture the light beam emanating from the bulb at the largest possible peripheral angle. The fresnel optics then homogenize the beam for projection along the specified path. Free-form lamps with fresnel cap (Figure 22) combine good phototechnical efficiency with substantial scope for stylistic variation. The reflector refracts and redirects the rays from the bulb while simultaneously contributing all or part of the required diffusion to the final distribution pattern. As a result, the cover lens can be clear and unfluted, or it can alternatively incorporate cylindrical dispersion elements to generate vertical or

Figure 19

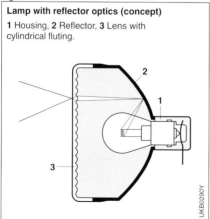

Lamp with reflector optics (concept)
1 Housing, **2** Reflector, **3** Lens with cylindrical fluting.

UKB0290Y

Figure 20

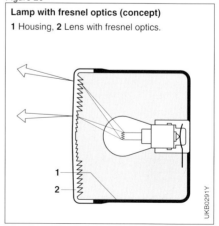

Lamp with fresnel optics (concept)
1 Housing, **2** Lens with fresnel optics.

UKB0291Y

horizontal scatter. The fresnel lens provides greater photometric efficiency by capturing light that would otherwise be wasted and reprojecting it in the desired direction.

The primary application for both designs is in front turn signals. Here and in other applications the ultimate considerations are vehicle shape (which defines space availability), stylistic requirements and the required level of light intensity.

Color filters

Specific red and yellow-based color tones have been assigned to automotive lamps according to their intended function (brakes, turn signals, fog warning lamps, etc.). These colors are defined on the basis of a standard color spectrum (color location). Because white light is actually composed of numerous colors, it is possible to filter luminous radiation to partially or completely remove undesired spectral elements from an emerging beam. The color filter can be in the form of a dyed lens or a colored layer on the bulb's glass surface (e.g., amber bulbs in turn signals with neutral lens tones).

This filter technology can also be employed to produce color-coordinated lamp lenses. These adapt to the vehicle's finish when not in operation, but still comply with the certification regulations when switched on.

The color locations specified within the jurisdiction of the ECE include "amber" with a wavelength of 592 nm for turn signals, and "red" at 625 nm for stop (brake) lamps and tail lamps.

Designs

The term "grouped design configuration" refers to components that share a single housing but still have individual lenses and bulbs (example: multichamber tail lamp assembly containing different individual light units).

"Combined design configuration" in-dicates a component assembly in which a single housing and bulb are employed together with more than one lens (example: combined tail lamp and license-plate lamp in commercial-vehicle applications).

The term "nested design configuration" is applied to component assemblies that share a common housing and lens, but with individual bulbs (example: headlamp unit with built in side-marker lamp). A frequent form of nested design features a bulb in which several functions are combined (example: tail and stop (brake) lamp both relying on a single dual-filament bulb).

Figure 21

Reflector with rotated parabola

1 Fresnel lens, **2** Rotated parabola reflector.

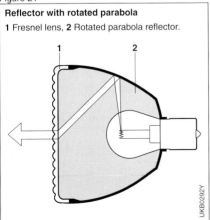

Figure 22

Open design lamp with fresnel blind

1 Fresnel cap, **2** Reflector.
3 Lens with cylindrical fluting.

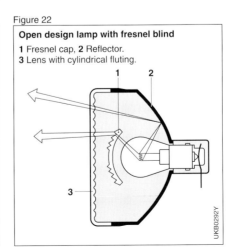

Lighting devices
Vehicle front

Main headlamps (Europe)

Purpose

The vehicle's main headlamps must satisfy several – sometimes mutually antagonistic – imperatives. On the one hand, they must provide maximum visual range while at the same time ensuring that light distribution immediately in front of the vehicle remains commensurate with the requirements of safe operation. Although blinding glare for approaching road users must be minimal, it is vital to provide the lateral illumination needed to safely negotiate curves, i.e., the light must extend outward to embrace the verge of the road. Although it is impossible to achieve absolutely consistent luminance across the entire road surface, it remains possible and necessary to avoid blatant contrasts in light density.

High beam

The high beam is usually generated by a light source located at the reflector's focal point, causing the light to be reflected outward along a plane extending along the reflector's axis (Figure 23). The intensity of the luminous intensity which is available during high-beam operation is largely a function of the reflector's mirrored surface area.

In four and six-headlamp systems, in particular, purely paraboloid high-beam reflectors can be replaced by units with complex geometrical configurations for simultaneous activation of high and low beams.

In these systems the high-beam component is designed to join with the low-beam's light (simultaneous operation) to produce a harmonious overall high-beam distribution pattern. This strategy abolishes the annoying overlapping sector that would otherwise be present at the front of the distribution pattern.

Low beam

Because the high traffic density encountered on modern roads severely restricts the use of high beams, the low beams serve as the primary and exclusive light source under normal conditions. Basic design modifications implemented within recent years have fostered substantial improvements in low-beam performance. Developments have included:

– Introduction of the asymmetrical low-beam pattern, affording extended visual range along the road shoulder,

– Official certification for various types of halogen bulbs capable of raising luminous intensity on the road surface by 50...80 %,

Figure 23

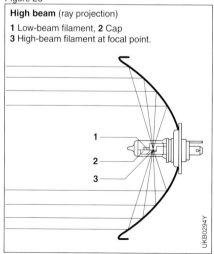

High beam (ray projection)
1 Low-beam filament, **2** Cap
3 High-beam filament at focal point.

UKB0294Y

Illuminance and range

Illuminance is defined as the luminous flux impacting upon a given surface. It is proportional to luminous intensity and decreases at a rate defined by the square of the distance. Illuminance is measured in lux (lx):

$1 \text{ lx} = 1 \text{ lm/m}^2$

The illumination range is the maximum distance at which a light beam's illuminance continues to reach a specified level (e.g., 1 lx). The geometrical range is the distance at which the horizontal element of the low-beam's light/dark cutoff remains visible on the road surface.

– Introduction of innovative new head-lamp concepts with complex geometries (PNS, HNS) with the potential to improve efficiency levels by up to 50 %.

– Headlight leveling control (also known as vertical aim control) devices adapt the attitude of the headlamps to avoid projecting blinding glare toward oncoming traffic when the rear of the vehicle is heavily laden. Headlamp washing systems are mandatory equipment on motor vehicles fitted with headlight leveling control.

– "Litronic" gaseous-discharge head-lamps supply more than twice as much light as conventional halogen units.

Method of operation

Virtually all of the headlamp designs employed up to 1988 (with H4 bulbs) used a low-beam light source mounted forward of the parabolic reflector's focal point which, following reflection, endowed the beam with a post-reflective inclination toward the reflector's axis (Figure 24).

A shield prevents light from the bottom of the bulb impacting against the lower section of the reflector and being projected back up along the vertical plane. The edge of the shield is projected onto the road surface as the light/dark cutoff. The resulting "dark above/bright below" distribution pattern provides acceptable visual ranges under all driving conditions. This layout holds the glare projected

Figure 24

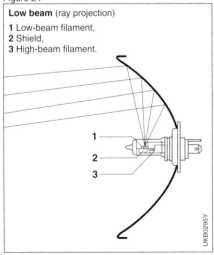

Low beam (ray projection)
1 Low-beam filament,
2 Shield,
3 High-beam filament.

UKB0295Y

toward approaching road users within reasonable limits while simultaneously supplying relatively high levels of illuminance in the sector below the light/dark cutoff line (Figure 25).

Automotive headlamp performance is subject to technical assessment to verify compliance with the regulations governing official certification. Precise test points are specified for measuring light under high and low-beam operation. Among the requirements to be satisfied at these locations are maximum illuminance levels – to ensure adequate road-surface visibility – and maximum intensity levels, to prevent glare.

Regulations

The following ordinances apply for main headlamps, their installation and use.

76/761/EEC and ECE-R1 and R2: high and low-beam headlamps and their bulbs.

ECE-R8: for headlamps with H1, H2, H3, H7, HB3 and HB4 bulbs.

ECE-R20: for headlamps with H4 bulbs.

StVZO §50: high and low-beam head-lamps.

76/756/EEC and ECE-R48-01: mounting requirements.

ECE-R 98-99: headlamps with gaseous-discharge bulb units.

High beams, installation

A minimum of two and a maximum of four headlights are prescribed for the high-beam mode.

Grouped and nested design configurations including other front lighting devices are approved. Combined-design assemblies with other front lighting devices are not approved.

Prescribed color for high-beam indicator lamp in instrument cluster: blue or yellow.

High beams, illumination technology

High-beam light distribution is defined in the regulations and guidelines together with stipulations governing low beams.

The most important specifications are: symmetrical distribution relative to the central vertical plane and maximum illumination along the headlamp's central axis.

Figure 25
Testing low-beam illuminance and light/dark cutoff in the light tunnel

The maximum approved luminous intensity, a composite of the intensity ratings for all high-beam headlamps installed on a vehicle, is 225,000 cd. The corresponding rating is indicated by a reference code located adjacent to the official certification code on each headlamp. The code for 225,000 cd is 75. The luminous intensity of the high beam is indicated by a figure, such as 20 (as an example), stamped next to the round ECE test symbol.

If these are the only headlamps on the vehicle (no auxiliary driving lamps) then this sample system's composite luminous intensity would be in the range of 40/75 of 225,000, viz., 120,000 cd.

Low beams, installation
Regulations prescribe 2 white-light low-beam headlamps for all mutiple-track vehicles.

Grouped and nested design configurations including other front lighting devices are approved. Combined-design assemblies with other front lighting devices are not approved.

Low beams, illumination technology
The relevant ordinances governing symmetrical low-beam headlamps within Germany are contained in the Technical Specifications (TA) of the German StVZO (FMVSS/CUR). The only applicable regulations for asymmetrical low beams are the international regulations and guidelines; these contain precise definitions of the photometric test procedures prescribed for use on various low-beam units (with incandescent, halogen or gaseous-discharge bulbs).

Certification testing is carried out using test lamps manufactured to more precise tolerances than units stemming from standard series production.

Headlamp glare is assessed based on StVZO § 50 (6). Glare is considered to be eliminated when the illuminance at a height equal to that of the headlamp's center does not exceed 1 lx at a distance of 25 m. This test is conducted with the engine running at moderate rpm.

Low/high beam control
Selection of the low-beam mode must simultaneously extinguish all high-beam illumination. Gradual dimming is permissible within a maximum period of 5 seconds. A 2-second response delay is required to prevent the dimming mode from activating when the high-beam flashers are used. The low beams may continue in operation when the high-beam mode is selected (simultaneous operation). The bulbs are generally designed for limited use with both filaments in operation.

Design configurations

Lens (variable position with respect to the bodywork)

This design concept, now superseded, featured a lens and reflector that combined to form a single headlamp module, with beam adjustment being performed by pivoting the complete assembly. Under unfavorable circumstances this type of layout can result in the lens being slightly tilted relative to the vehicle's body. The headlamp assembly was usually equipped with seals in the area immediately surrounding the bulb as well as a special ventilation system.

Lens (fixed with respect to the bodywork)

There is no direct connection between the lens and the reflector, which is mounted in a housing that moves relative to the lens during adjustment (housing concept). Because the lens is not attached to the housing it remains stationary in one invariable position relative to the body, an asset allowing total integration within the front end. The complete headlamp is sealed or provided with ventilation elements.

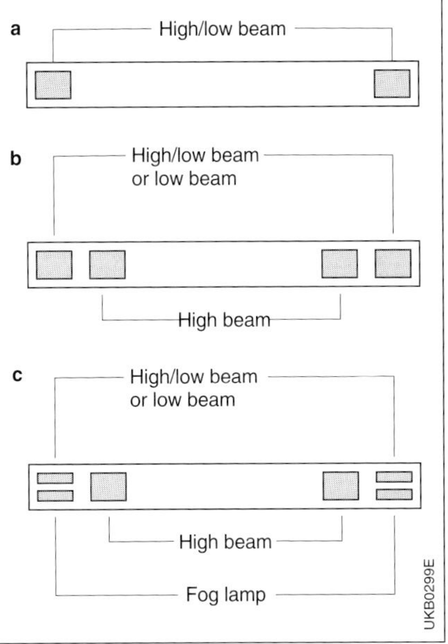

Headlamp systems

Dual-headlamp systems rely on a single shared reflector for low and high-beam operation, e.g., in combination with a dual-filament H4 bulb (Figure 26a).

In quad headlamp systems one pair of headlamps may be switched on in both modes or during low-beam operation only, while the other pair is operated exclusively for high-beam use (Figure 26b).

Six-headlamp systems differ from the quad configuration by incorporating a supplementary fog lamp within the main headlamp assembly (Figure 26c).

Main headlamps (North America)

High beam

As in the European system, the light source is usually located at the focal point of a parabolic reflector (Figure 27, Pos. 3).

Low beam

Headlamps with a light/dark cutoff that rely on visual/optical adjustment procedures have been approved in the US since 1 May, 1997. These units reflect the ECE guidelines in effect for Europe, thus making it possible to equip vehicles for Europe and the US with the same headlamps.

The "classical" sealed-beam design in use prior to this date featured a low-beam light source located just above (Figure 27, Pos. 1) and somewhat to the left of the reflector's focal point (as viewed looking in the direction of vehicle travel). This arrangement directs virtually the entire luminous flux downward onto the road surface. This is accompanied by a rightward offset producing an asymmetrical beam. Because there is no shield beneath the light source, the entire reflector can be used. However, the absence of a clearly defined light/dark cutoff limits visual ranges on the driver's side of the lane, and

Figure 26

Headlamp systems

a Dual-headlamp system,
b Quad headlamp system,
c Six-headlamp system.

a — High/low beam

b — High/low beam or low beam — High beam

c — High/low beam or low beam — High beam — Fog lamp

UKB0299E

also produces higher glare for oncoming traffic than the European system.

Legal regulations

Federal Motor Vehicle Safety Standard (FMVSS) No. 108 and the SAE Lighting Equipment and Photometric Tests (Standards and Recommended Practices).

The regulatory framework governing installation and control arrangements is comparable to that in Europe.

As mentioned above, headlamps with a light/dark cutoff as defined in ECE Directives have been approved in the US since 1 May, 1997.

Prior to this date there were major differences in the respective headlamp systems. Until 1983 only sealed-beam units in the following dimensions were approved in the US:

Dual-headlamp systems:
– 178 mm diameter (round),
– 200 x 142 mm (rectangular).

Quad headlamp systems:
– 146 mm diameter (round),
– 165 x 100 mm (rectangular).

An amendment to FMVSS 108 that entered effect in 1983 made it possible to start using headlamp modules of various shapes and sizes with replaceable bulbs. These were known as the RBH, or Replaceable Bulb Headlamps.

Headlamps adjusted using the visual/optical method: The position of the light/

Figure 27

American sealed-beam headlamp system

a Low beam, 1 Filament for low beam,
b High beam. 2 Focal point,
                    3 High-beam filament
                     (at focal point).

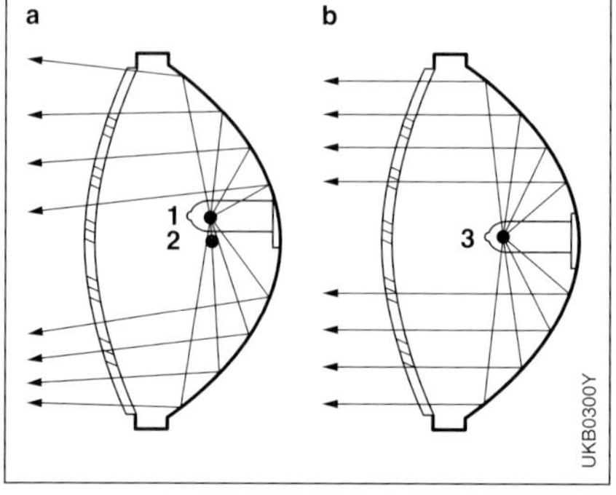

dark cutoff and its sharpness are defined using the following formula:
$G(a) - \log E(a) - \log E(a + 0.1)$ with "E(a)" representing the illuminance along a vertical plane intersecting the light/dark cutoff and "a" as the vertical angle.

The maximum value for "G" indicates the position "a" on the light/dark cutoff, while the absolute value for G at position "a" is its sharpness.

Design configurations

Sealed beam

This design concept has disappeared from common use. Because the light source is exposed, the glass reflector with its vapor-coated aluminum reflective layer must be hermetically sealed with the lens. After sealing, the entire unit is filled with an inert gas. A ruptured filament necessitates replacement of the entire unit. Units relying on halogen light bulbs are also available.

The limited range of sealed-beam headlamps on the market severely restricted the designers' latitude in front-end styling.

Replaceable Bulb Headlamp (RBH)

Starting in 1983, European advances in the realm of replaceable bulb technology began making inroads into American system technology. The RBH headlamp's geometry and dimensions can be adapted for improved design (styling); these units usually feature plastic lenses and housings.

Vehicle Headlamp Aiming Device (VHAD)

This is a mechanical device used with RBH headlamps. The integral mechanism relies on a concept used spirit levels for vertical adjustment, while a system using a pointer and scale is used for horizontal adjustment in an "on-board aiming" process.

Headlamp systems

North America mirrors European practice in employing dual, quad and six-headlamp systems.

Headlamp versions

Conventional headlamps

In conventional headlamp systems the quality of the light obtained from the low beams improves as the reflectors get larger, while geometrical range is proportional to installation height. A contrary consideration is that vehicle front ends should remain low in response to the dictates of aerodynamics.

Headlamps with stepped reflectors

Stepped, or graduated, reflectors are segmented reflectors consisting of paraboloid and/or parelliptical (combined parabola and ellipse) sectors designed to provide various focal lengths. These units retain the advantages of a deep reflector in a more compact unit suitable for installation in shallower apertures (Figure 28).

Homofocal reflector

The homofocal reflector is composed of base and supplementary reflector units (Figure 28, Pos. 1a and 1b). Because the homofocal reflector's supplemental sectors share a single focal point to achieve a shorter focal length than can be derived from a base reflector alone, they make a substantial contribution to effective luminous flux. While the light from the supplementary reflectors improves close-range and side illumination, it does not enhance projection range. A dual-filament (low and high beam) H4 bulb is suitable for use in this type of unit.

Multifocal reflector

The basic concept behind the multifocal reflector is similar to that employed in its homofocal counterpart. Based on mathematical definitions, parelliptic reflector sections designed for horizontal light diffusion provide a large number of focal points.

Stepless reflectors

Specially-developed computer programs (CAL, Computer Aided Lighting) assist in designing stepless VFR (Variable-Focus Reflectors) with non-parabolic sectors.

Headlamps without focal lenses (clear outer lens)

Today's expanded HNS (Homogeneous Numerically Calculated Surface) technology makes it possible to achieve headlamp efficiency levels of up to 50%. In addition, the entire pattern of light distribution can be defined by the reflector, consigning fluted lens patterns to redundancy. Headlamps with clear, unfluted lens covers are also expanding the horizons of headlamp styling (Figure 4, BMW 3 Series).

Figure 28

Stepped (graduated) reflector (example)

1 Homofocal reflector, **1a** Base reflector, **1b** Supplementary reflector surfaces,
2 High-beam reflector, **3** Fog-lamp reflector.

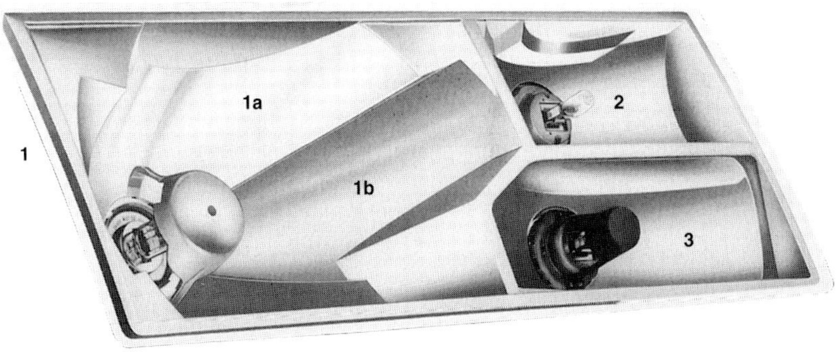

Headlamps with complex-surface (segmented) reflectors

In these headlamps the entire surface of the reflector is divided into numerous segments, each of which can be optimized with the aid of the CAL program. The salient factor in segmented reflectors is that graduations and inconsistencies can be accepted on all four boundaries. This makes it possible to concentrate on generating light with an optimal distribution pattern in a unit that also satisfies the vehicle manufacturer's styling demands.

PES headlamps

The headlamp designated PES (Poly-Ellipsoid System) as shown in Figure 4 (VW New Beetle) is distinguished by reliance on imaging optics to improve on the phototechnical performance provided by conventional headlamps. This system celebrated its world premiere with the start of series production at Bosch in 1985.

While the conventional headlamp relies on a diffusion lens for light distribution, the PES light-distribution pattern is defined by the reflector and then projected through the lens and onto the road surface. The underlying concept is related to that used for overhead slide projectors, with optical reproduction of an image being the basic objective in both cases. While this image (or object) is the slide itself in an overhead projector, in the headlamp it is the light-distribution pattern generated by the reflector and modified by the screen that produces the light/dark cutoff needed for low-beam operation (Figure 29).

Thanks to the elliptical (CAL designed) reflector and to optical projection technology, the PES needs a light-emission surface of no more than 28 cm² to project the light-distribution patterns formerly available only from large-surface conventional lamps. An imaging screen projects precisely defined light/dark cutoffs. This concept offers tremendous latitude for defining these transition areas, with either high or low contrast, and there is also virtually unlimited range for accomodating special geometries (Figure 30a).

The PES PLUS concept relies on light projection directed toward a section of the reflector beneath the imaging screen to enhance close-range illumination (Figure 30b). The signal image is also enlarged to reduce psychological glare. This effect is enhanced by using a supplementary ring-shaped reflector element (Figure 30c) to provide special benefits in case of oncoming traffic.

Litronic

Assignments and requirements

The Litronic (Light Electronics) headlamp system from Bosch features a xenon gaseous-discharge bulb to satisfy the most exacting of recent phototechnical performance demands. The catalog of new requirements affects the type of light and its intensity while also embracing a call for compact designs.

Figure 29

Optical-imaging concept of projection headlamp
1 Screen image,
2 Rear-reflector focal point,
3 Front reflector and lens focal point,
4 Screen.

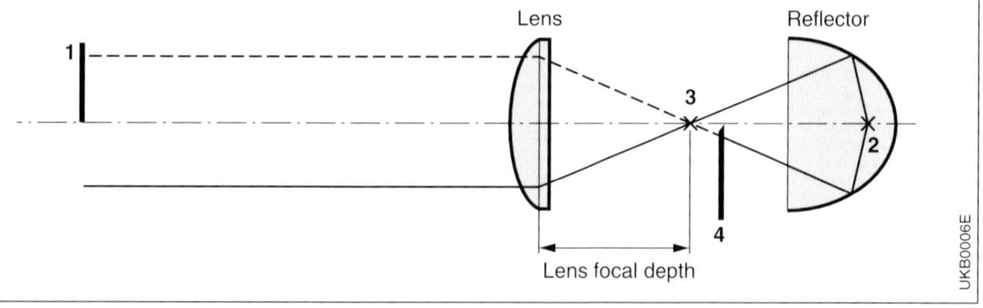

A service life in excess of 1,500 hours means the unit can be be expected to last for as long as the passenger car itself. The illumination furnished by Litronic headlamps also represents a substantial improvement over that provided by conventional halogen units (Figure 31).

Design
The components of the Litronic headlamp system are:
- Optical unit with gaseous-discharge lamp,
- Electronic ballast unit with ignition unit and controller.

Method of operation
Relative to halogen units, Litronic head-lamps produce a higher luminous flux with specifically adapted light distribution. This results in well-illuminated road shoulders, while significant improvements in both visibility and general orientation are valu-able in potentially hazardous situations and in bad weather. Litronic headlamps comply with ECE-R48 by always in-corporating automatic headlight leveling

Figure 30
PES low-beam unit
Projection patterns

a PES,	**1** Lens,
b PES PLUS,	**2** Shield,
c PES PLUS with ring-shaped reflector.	**3** Reflector,
	4 Bulb.

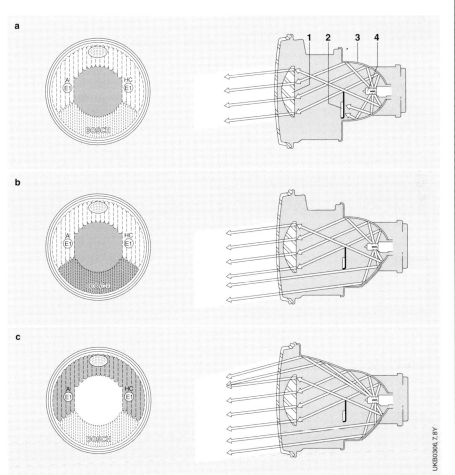

UKB0306, 7.8Y

and a headlamp washer. These features combine to ensure consistently optimal exploitation of the unit's inherently extended range by maintaining unobscured light projection.

An electronic ballast unit – comprising the ignition unit and the controller – is responsible for igniting, operating and monitoring the gaseous-discharge lamp. The ignition unit generates the high voltage required to initiate arcing in the gaseous-discharge lamp. The controller regulates the current supply in the warm-up phase before reverting to maintenance of a consistent 35 W for static operation. The current flowing to the lamp is raised in the initial seconds after ignition to accelerate the lamp's progress to 100 % illumination.

The system also compensates for fluctuations in the vehicle's battery voltage to maintain highly consistent levels of luminous flux.

Should the lamp go out owing to a momentary lapse or collapse in the vehicle's on-board voltage supply, re-ignition is automatic. The electronic ballast unit responds to defects (such as a damaged lamp) by deactivating the power supply to prevent personal injury in the event of inadvertent contact.

Versions

Headlamps with gaseous-discharge lamps are installed in quad systems in combination with conventional high-beam lamps (Figure 32).

Two different optical systems are available:

PES projection headlamps

The world's first Litronic headlamp debuted in 1991 in a unit combining PES projection headlamps with D1 gaseous-discharge units. Current new vehicles equipped with PES projection headlamps are supplied exclusively with DS2 units (Figs. 29 and 33).

Reflection headlamps

Provided that larger areas are available for the light-emission surface, Litronic can also assume the form of a reflection headlamp. Either a lens with integral optical focal elements or a clear lens cover may be found within this substantially larger projection surface.

The low beam is generated by a D2R gaseous-discharge unit equipped with shutter strips to produce the light/dark

Figure 31
Light-distribution pattern (at road level)
a PES H1 lamp, **b** Litronic PES D2S lamp.

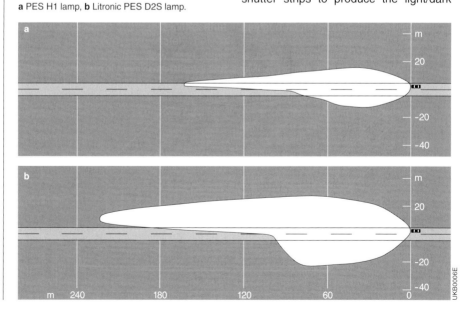

cutoff (Figure 34). Because virtually the entire circumference of the D2S unit provides effective lighting, it can also serve as the basis for an extremely efficient high-beam headlamp.

Bi-Litronic "Reflection"

Bosch sponsored the world premiere of the Bi-Litronic in 1998. This system is unique in having just a single gaseous-discharge bulb to generate the light for both high and low beam within a dual-headlamp system. The concept relies on an electromechanical positioner that responds to dimmer-switch activation by varying the attitude of the gaseous-discharge unit within the reflector. It alternates between two different positions (Figure 35) to generate separate projection patterns for low and high beam.

Quad headlamp system with Litronic

1 Vehicle electrical system,
2 ECU,
3 Ignition unit with lamp connection,
4 Headlamp optics with gaseous-discharge lamp,
5 Halogen high beam.

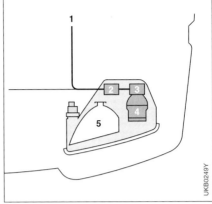

Figure 32

Figure 33

Litronic 2 System in projection headlamp (example)

1 Lens,
2 Gaseous-discharge lamp,
3 Plug,
4 Ignition unit,
5 ECU,
6 Vehicle electrical system.

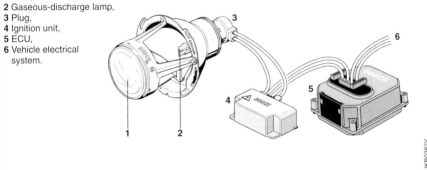

Figure 34

Litronic quad system in reflection headlamp with integral dynamic headlight leveling control (example)

1 Lens, with or w/o diffusion optics, **2** Gaseous-discharge lamp, **3** Ignition unit, **4** ECU, **5** Stepper motor,
6 Suspension travel sensor,
7 Vehicle electrical system.

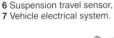
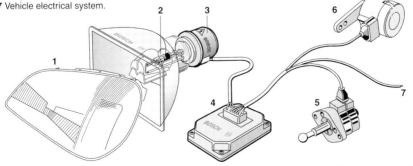

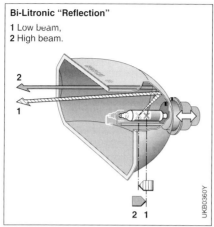

Bi-Litronic "Reflection"
1 Low beam,
2 High beam.

UKB0360Y

Figure 35

Bi-Litronic "Projection"
1 Low beam,
2 High beam.

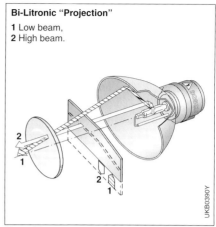

UKB0390Y

Figure 36

This layout endows Bi-Litronic with the following primary assets:
– Xenon light for high beam,
– Visual guidance provided by the continuous shift in light distribution from close to extended range,
– Substantial reduction in space requirements as compared to a conventional 4-chamber system, and
– Lower costs through the use of just one gaseous-discharge bub and one ballast unit.

Bi-Litronic "Projection"
The basis for the Bi-Litronic "Projection" is the PES Litronic headlamp. It shifts the position of the shutter for the light/dark cutoff to provide Xenon light for high-beam operation (Figure 36).
With lens diameters of 60 and 70 mm, the Bi-Litronic "Projection" is the most compact combined low and high-beam headlamp on the market, yet it still provides superb illumination.

Headlamp adjustment

Specifications for low and high beam

Aiming conditions
(Europe)
– Tires must be inflated to the prescribed pressures.
– Depending upon vehicle type, this may also have to be laden.

Passenger cars: one person or 75 kg in the driver's seat;
Trucks: unladen;
Single-track vehicles and single-axle tractors: one person or 75 kg in the driver's seat.
– The vehicle should be allowed to roll several meters in order for the loaded suspension to settle.
– The vehicle must be parked on a level horizontal surface.
Aiming without a visual/optical adjustment device proceeds using a test screen set up at a distance of 10 meters from the vehicle, with the center mark positioned forward of the headlamp being adjusted (Figs. 37 and 38).
– Each headlamp is adjusted individually, with all other headlights covered.
– If the headlamp unit features a manual range (vertical aim) adjuster, then the switch should be set to the position indicated by the vehicle's manufacturer.

Aiming conditions
(North America)
While Europeans have always used visual/optical aiming methods for headlamp alignment based on the position of the light beam, the mechanical adjustment procedure had been progressing to predominance in the US since the mid-50s. Here each headlamp lens is equipped with three protrusions, one for each of the three adjustment planes. The adjustment device

is pushed-up against these protrusions and the aiming process proceeds using the spirit-level principle.

The VHAD (Vehicle Headlamp Aiming Device) is employed to adjust the headlamps relative to an invariable reference axis defined by the vehicle itself. This procedure also relies on a bubble-balancer, this time firmly attached to the headlamp. The three lens pips are then redundant.

However, since mid-1997, or more specifically since official approval was issued on 1 May, 1997, visual/optical (vertical only) adjustment procedures have also been gaining rapid ground in the US. There is no requirement for horizontal adjustment.

Aiming information (Europe)

The reference marks and lines are arranged for main headlamps at standard installation height, with the center mark set to the height H, corresponding to the middle of the lamp. The aiming dimension e (usually 10 cm) represents the vertical distance between the center mark and the separation line (Figure 37).

On headlamps with $e > 10$ cm the separation line is lowered to the required level. Obviously, the center mark is then no longer in alignment with the center of the headlamp, but it can still be used to check the position of the high beam.

The separation line serves as the prescribed reference for aligning the left (horizontal) section of the light/dark cutoff. The regulations contain the adjustment specifications.

The headlamps must be readjusted following any modifications to the vehicle, or service work, with the potential to affect the aim of the headlamps (suspension work, etc.). Headlamp alignment checks and adjustments are also prescribed after bulb replacements.

If the high beam is incorporated in a single unit along with the asymmetrical low beam then it will be aimed automatically when the low beam is adjusted. Separate high-beam units are aligned horizontally and sym-metrically using the center of the headlamp and the center mark as references.

Figure 37

Headlamp-pattern test screen

1 Delineation line,
2 Center mark,
3 Test screen,
4 Break point,
H Height of headlamp center above support surface in cm.
h Height of test-screen delineation line above support surface in cm.
$e = H - h$ Aiming dimension.

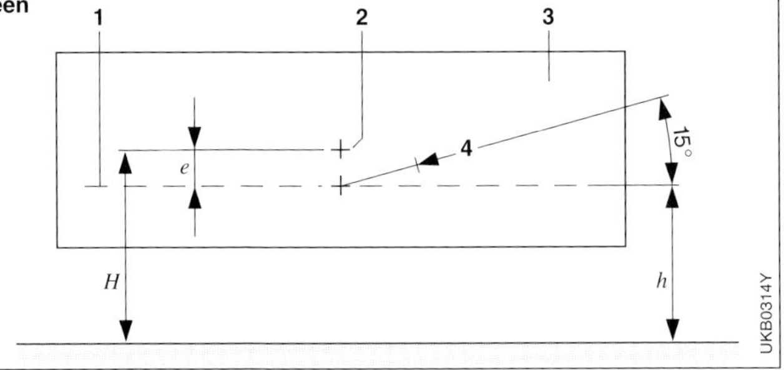

Figure 38

Relative positions of test surface and vehicle longitudinal axis

1 Center mark,
2 Test screen,
A Distance between headlamp centers.

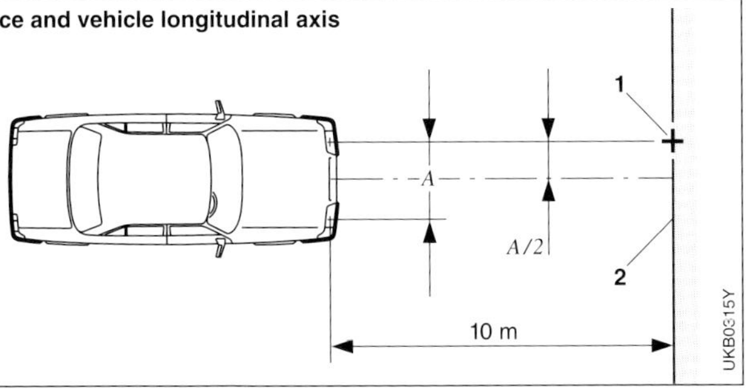

Headlamp alignment equipment

Purpose

It is essential that automotive headlamps be aimed to avoid blinding approaching vehicles. The vertical and lateral beam angles must therefore be adjusted in accordance with the official specifications enumerated in the "Directive governing motor-vehicle headlamp adjustment" as contained in §50 of the StVZO. Headlamp adjustment is usually conducted with the aid of visual/optical aiming devices.

Equipment design

Headlight aiming devices are in fact mobile imaging chambers containing a focal lens and a collector screen. This screen is located in the focal plane of the lens, to which it is solidly attached, and has markings to facilitate correct headlamp adjustment. The operator can view it using suitable appendages such as windows and adjustable refraction mirrors (Figure 39).

The prescribed headlamp adjustment, i.e., the declination relative to the center line of the headlamp in cm at an invariable distance of 10 m, is set by rotating a knob to move the collector screen.

The aiming device is aligned with the vehicle axis using a sighting device such as a mirror with an orientation line. The device is rotated to bring the orientation line into uniform contact with two external vehicle-reference marks. The imaging chamber can be adjusted vertically to the level of the vehicle's headlamps prior to being locked in position.

Checking the headlamps

Headlamp aim can be assessed once the equipment has been correctly positioned in front of the lens. An image of the light distribution pattern emitted by the headlamp appears on the collector screen. Some test devices are also equipped with photodiodes and a display to facilitate measurement of illuminance.

On headlamps with asymmetrical low beam patterns the light/dark cutoff should

Figure 39

Headlight aiming device

1 Alignment mirror, 4 Refraction mirror,
2 Handle, 5 Marks
3 Photometer, for lens center.

UWT0067Y

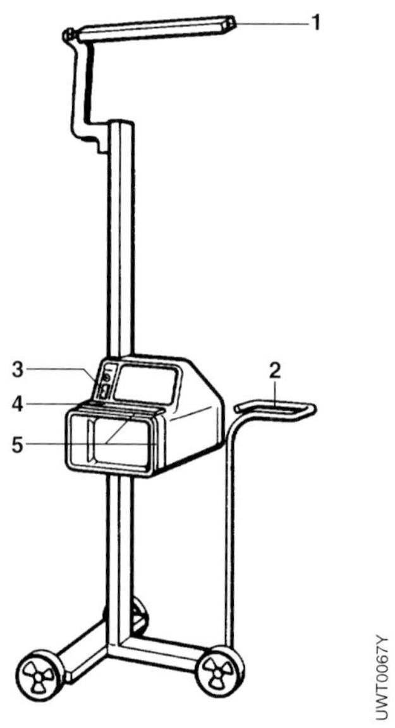

Figure 40

Sight window in headlight aiming device
Delineation line for light/dark cutoff
with asymmetrical low beam.

UWT0069Y

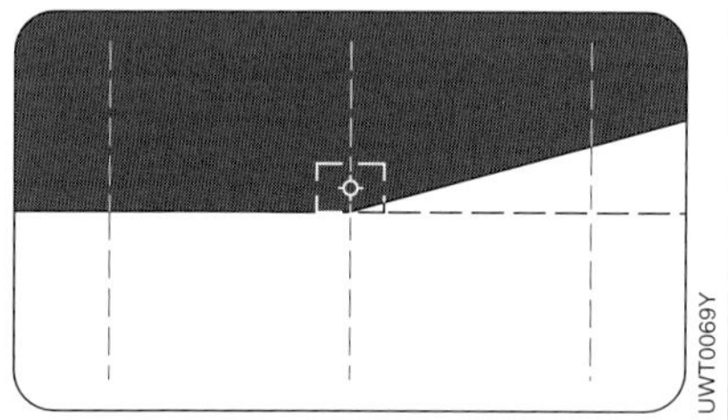

Figure 41

Sight window in headlight aiming device
Central mark for middle of high beam.

UWT0068Y

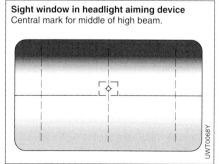

contact the horizontal separation line with the horizontal and vertical components intersecting on a vertical line extending through the center marking (Figure 40). Following prescribed adjustment of the low beam's light/dark cutoff, the center of the high beam (when both settings are being adjusted simultaneously) should be within the boundary marks that encompass the center marking (Figure 41).

Headlight leveling control

Purpose

The function of the headlight leveling control (also known as vertical aim control) is to compensate for changes in vehicle load, and maintain consistently satisfactory visual ranges while at the same time preventing oncoming traffic being blinded by excessive glare. The headlight leveling control executes this function by adjusting the inclination of the low-beam headlamp. Without headlight leveling control, visual range is subject to continuing variations due to changes in vehicle load (Figure 42).

Regulations

Table 2 indicates the geometrical ranges corresponding to various declination angles for headlamps installed at a height of 65 cm. The inspection tolerance extends to include angles of up to 2.5 % (1.5 % below standard setting). The EU specifications governing the base setting for headlight leveling controls stipulate: the basic setting relative to dimension e is 10...15 cm at a distance of 10 meters with one person in the driver's seat. The specifications for this setting are provided by the vehicle manufacturer. Within Germany, an automatic or manually-operated headlight leveling control device is mandatory on all new vehicles as a condition for initial registration. The sole exception to this requirement is when the vehicle is equipped with some other device capable of ensuring that the light beam remains within specified vertical tolerances (such

Table 2
Geometrical range for horizontal component of low-beam light/dark cutoff
Headlamp installed at height of 65 cm.

Inclination of light/dark cutoff ($1\% = 10\,cm/10\,m$)	1 %	1.5 %	2 %	2.5 %	3 %
Aiming dimension e (cm)	10	15	20	25	30
Geometrical range for horizontal component of low-beam light/dark cutoff	65 m	43.3 m	32.5 m	26 m	21.7 m

Figure 42
Illumination range on level surface without headlight leveling control
a Unladen steady-state operation, b Under acceleration or with rear load, c During braking.

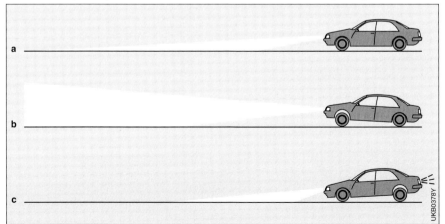

as automatic self-leveling suspension). Although this equipment is not mandatory in other countries, its use is permitted. A pan-European directive mandating installation of headlight leveling controls may be expected at some future date.

Designs

All headlight leveling controls feature actuators that move the headlamp reflector (housing-type design) or headlamp unit up and down. Automatic systems rely on sensors that monitor suspension travel as the basis for generating proportionate signals for transmission to the aiming actuators. Manual layouts employ a driver-operated switch to control height adjustment.

Automatic headlight leveling control

Automatic headlight leveling control systems fall into two categories: static and dynamic. While static systems compensate for load variations in the luggage and passenger compartments, dynamic systems also correct headlamp aim during acceleration – both from standing starts and when underway – and when braking.

The components within a typical headlight leveling control system include (Figure 43):
– Sensors on the vehicle axles to precisely measure the vehicle's pitch angle,
– An ECU that uses the sensor signals as the basis for calculating the vehicle's attitude. The controller compares these data with the specified values and responds to deviations by transmitting

appropriate control signals to the headlamps' servomotors,
– Servomotors to adjust the headlamps to the correct angle.

Static system

Along with the signals from the suspension sensors, the static system's controller also receives a velocity signal from the ABS control unit's electronic speedometer circuit. The controller relies on this signal to decide whether the vehicle is stationary, undergoing a dynamic change in velocity, or proceeding at a constant speed. Automatic systems based on the static concept always feature substantial response inertia, so the system corrects only those vehicle inclinations that are consistently registered over relatively extended periods.

Each time the vehicle moves off the system corrects the headlamp angle to compensate for any load variations, with a second correction cycle being initiated once the vehicle assumes steady-state operation. The static system employs the same servomotors found in manual systems to compensate for the deviation between the specified and monitored vertical angles of the headlamps.

Dynamic system

The dynamic automatic system relies on two distinct operating modes to ensure optimal headlamp orientation under all driving conditions. Supplementary capabilities in speed-signal analysis endow the system with the ability to differentiate

Figure 43

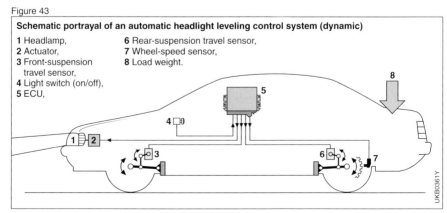

Schematic portrayal of an automatic headlight leveling control system (dynamic)

1 Headlamp,
2 Actuator,
3 Front-suspension travel sensor,
4 Light switch (on/off),
5 ECU,
6 Rear-suspension travel sensor,
7 Wheel-speed sensor,
8 Load weight.

between acceleration and braking; this represents a distinctive advance beyond the static aim control concept.

During steady-state operation the dynamic system operates in the same manner as its static counterpart with large rates of response inertia, but as soon as the controller registers acceleration or braking, the system immediately switches to its dynamic mode. This mode's distinctive feature is its faster signal processing and the higher servomotor adjustment speed, thus allowing beam range to be readjusted within fractions of a second. The ultimate result is that the driver consistently enjoys the visual range needed to efficiently monitor traffic conditions. Following acceleration or braking, the system automatically reverts to operation in its delayed-response mode.

Manual headlight leveling control

This type of control is activated by the driver. A detent is prescribed for the standard setting, which also serves as the reference point for setting the beam to its base position. Regardless of whether they feature infinitely-variable or graduated control, all manual units must incorporate hand switches in the close vicinity of which there are visible markings corresponding to the various vehicle-load conditions for which different correction settings are prescribed.

Adjustment mechanisms

Hydromechanical systems use a fluid medium to initiate motion in the headlight leveling control mechanism. Vacuum systems exploit the negative pressure in the intake manifold to achieve the same end. Both methods rely on an interior-mounted switch for control.

Electric stopper motors are used as the actuators, or servo elements in the electrical systems. These can be activated either by hand or by signals from the suspension travel sensors.

Front fog lamps

Purpose

Fog lamps are designed to improve road-surface illumination under conditions marked by limited visibility (fog, snow, heavy rain and dust).

Optical concept

Paraboloid

A parabolic reflector featuring a light source located at the focal point reflects light outward along a plane parallel to its axis (high-beam geometry). The lens extends the beam to form a horizontal band, while a screen prevents the emerging light rays from being projected upward (Figure 44).

CD technology

New calculation techniques (CAL, or Computer Aided Lighting) can be employed to create reflectors capable of directly shaping the light beam without the aid of specially profiled lenses. It is also possible to produce a well-defined

Figure 44

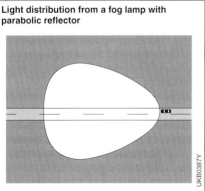

Light distribution from a fog lamp with parabolic reflector

UKB0387Y

Figure 45

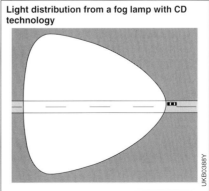

Light distribution from a fog lamp with CD technology

UKB0388Y

light/dark cutoff (delineation line which separates rays of light projected in different directions) without resorting to separate screens. This multiple orientation of light rays forms the basis of CD (converging-diverging) technology.

The fact that the lamp bulb is extensively enveloped leads to an extremely high volume of light as well as maximum scatter width (Figure 45).

PES fog lamp

This concept minimizes backglare for drivers in fog. The lens projects the image of a screen onto the road to furnish maximum contrast along the light/dark cutoff line.

Designs

External-fitted fog lamps incorporating their own optical elements can be mounted above the bumper (Figure 46) or suspended below it (Figure 47). Stylistic and aerodynamic considerations are leading to increased use of built-in fog lamps adapted to the shape of the bodywork, or fog lamps integrated within larger light assemblies (with adjustable

Figure 46
Compact 100 upright-mount aftermarket fog lamp

reflectors when the fog lamps are combined with the main headlamps).

Most fog lamps project a white beam; there is no substantive evidence that yellow lamps provide any special physiological benefits. A fog lamp's effectiveness depends upon the size of the illuminated area and the focal length of the reflector. Assuming identical illuminated areas and focal points, from

the phototechnical viewpoint any differences between round and rectangular fog lamps will be insignificant.

Regulations

Design is governed by ECE-R19, while mounting/installation specifications are set forth in 76/756/EEC, ECE-R48-01, StVZO § 52.

Two fog lamps projecting white or yellow light are permitted. Grouped design configurations together with other front lighting assemblies are allowed. Combined configurations including the fog lamp and other lighting equipment are prohibited. The primary fog-lamp control circuit must be separate from the circuits for low and high beams, i.e., it must be possible to switch the fog lamps on and off independently. The (German) StVZO allows fog-lamp installation in positions more than 400 mm from the widest point on the vehicle's periphery provided that the switching circuit ensures that operation is possible only in conjunction with the low beams. The adjustment procedures for front fog lamps mirror

Figure 47
Pilot suspended-mount aftermarket fog lamp

those used for the main headlamps. The setting dimensions e are contained in the specifications.

Auxiliary driving lamps

Purpose
Auxiliary driving lamps enhance long-range visibility with dual, quad and six-headlamp assemblies by generating a concentrated – and thus far-reaching – beam of light.

Optical concept
Operation relies on an essentially parabolic reflector with a light source situated at its focal point. Optical properties may be enhanced with supplementary lens patterns designed specifically for extended light projection.

Mounting/installation, regulations
Installation, technology and the adjustment procedures all correspond to those employed for high beams. Auxiliary driving lamps are also subject to the same regulations defining maximum levels of luminous intensity, according to which the sum of all reference numbers is not to

Figure 48

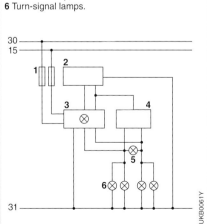

Passenger-car hazard-warning and turn-signal flasher system

1 Fuses,
2 Hazard-warning and turn-signal flasher,
3 Hazard-warning switch with indicator,
4 Turn-signal switch,
5 Indicator lamp,
6 Turn-signal lamps.

UKB0061Y

exceed 75. The number 10 is used for general assessment purposes on older lamps for which no approval number is available.

Turn signals and flashers

Purpose
These lamps generate signal images to alert other road users to changes in direction (turn signals) and potential danger (hazard warning flashers).

Regulations
The StVZO (FMVSS/CUR) and EU Directive 76/756/EEC stipulate that the basic illumination equipment will be supplemented by optical signaling devices to indicate changes in direction and to serve as hazard flashers on all vehicles with a maximum speed in excess of 25 km/h (Figure 48).

Flasher signals
The flasher signals are generated at a frequency of 60...120 pulses per minute and a relative illumination period of 30...80 %. Light must be emitted within 1.5 sec of initial activation. Should one bulb fail the remaining lamps must continue to produce visible light.

Turn signals
Turn signaling is in the form of synchronized signals which are generated by all flashers on one side of the vehicle. Lamp performance is monitored by an electronic circuit. Malfunctions are indicated by a warning lamp or a substantial change in the signal's flash rate.

Hazard-warning flashers
This mode is in fact synchronized flashing of all turn-signal lamps, and must remain available when the engine is switched off. An operation indicator is mandatory.

Turn-signal and flasher circuits for trailerless vehicles

The electronic hazard-warning and turn-signal flasher includes a pulse generator designed to switch on the lamps via relay, and a current-controlled monitoring circuit to modify the flash frequency in response to bulb failure. The turn-signal stalk controls the turn signals, while the hazard flashers are switched on with a separate switch.

Turn-signal and flasher circuits for vehicles with/without trailers

This type of hazard-warning and turn-signal flasher differs from that employed on vehicles without trailers in the way that turn-signal operation is monitored.

Single monitoring circuit
Tractor/towing vehicle and trailer share a single monitoring circuit that triggers two indicator lamps designed to flash at system operating frequency. If the first turn signal on the tractor or trailer fails, the first indicator lamp remains off. If the second turn signal also fails, the first and the second indicator lamp remain off. This configuration does not indicate the location (tractor/towing vehicle or trailer) of defective lamps. The flashing frequency remains unaltered.

Dual monitoring circuit
Tractor/towing vehicle and trailer are equipped with separate monitoring circuits. Faulty flasher lamps can be located depending upon which indicator lamp remains off. The flashing frequency remains unchanged.

Front and side flashers

Purpose
The flashers indicate intended changes in direction (turn-signal function) and potentially dangerous situations (hazard-warning-flasher function). Design and location must be selected to ensure reliable signal perception for other road users regardless of lighting and operating conditions.

Regulations
76/759/EEC, ECE-R6, StVZO § 54.
Group 1 turn-signal indicators (front-mounted), Group 2 (rear-mounted), and Group 5 (side-mounted) are specified for dual-track vehicles.
Group 5 side-mounted turn-signal indicators may be omitted if the vehicle is less than 6 m in length. The dashboard-mounted indicator lamp may be in any color desired. The flash frequency is defined as 90 ± 30 cycles per minute.

Front turn signals
Two amber-colored lamps are required. Approved are:

Figure 49

Front turn-signal lamp integrated in a front headlamp assembly
1 Turn signal

UKB0361Y

– Grouped design configurations with one or several other lamps (Figure 49),
– Combined design configurations only with turn-signal lamps from other groups, and
– Nested designs, allowed only with parking lamps,

An inside indicator to monitor operation is required.

Side turn signals

Two amber-colored lamps are required. Conditions for installation in nested and grouped designs are the same as those for forward-mounted turn signals. Combined designs are allowed only with turn signals from other groups.

Front side-marker and clearance lamps

Purpose

Side-marker and clearance lamps alert other traffic to the presence of large vehicles.

Regulations

76/758/EEC, ECE-R7, StVZO §§ 51 and 53. Vehicles and trailers more than 1600 mm in width must be equipped with (forward-facing) side-marker lamps. Vehicles wider than 2100 mm (such as trucks) must also be equipped with clearance lamps visible from the front.

Side-marker lamps

Two white-light side-marker lamps are specified. Yellow is allowed when the units are combined in a nested design with headlamps projecting yellow light (France). Positions are the same as those specified for front turn signals.
Grouped and nested designs combining the side-marker lamps with all other front-mounted lamps (including headlamps) are approved, and nested units integrating the position lamp in the headlamp structure are widespread. Combined designs with other lamps (or headlamps) are not approved.

Clearance lamps

Two white-colored clearance lamps are specified for installation at the front. Positions: as close as possible to the side extremities of the vehicle and at the greatest practicable height. Combined and nested designs with other lamps are not approved.

Front parking lamps

Purpose

Parking lamps are intended to ensure that stationary vehicles are visible for other road users. They must be suitable for operation without the need to switch on the headlamps or any other lighting equipment. The parking-lamp function is usually assumed by the side-marker lamps.

Regulations

77/540 EEC, StVZO § 51, ECE-R77. Vehicles may be equipped with two parking lamps at front and rear or one parking lamp on each side. White light is prescribed to the front.
Grouped designs are approved in combination with any other type of lamp. Combined designs together with other lights are not approved. Front-facing nested designs are approved with:
– Side-marker lamps,
– Headlamps and fog lamps.

Daytime running lamps

Daytime running lamps (using either the main headlamps on low beam or special units) have been approved for use throughout the EU since 1.1.98, and are mandatory in Norway, Sweden, Finland, Denmark, Poland (October through April) and Canada.
Low-beam headlamps or fog lamps can also be used to comply with the requirements for daytime driving provided that they supply light of the specified intensity. It is safe to assume that other countries will mandate daytime running lamps or constant low-beam operation in coming years.

Lighting devices Vehicle rear

Various lamps and projection devices are installed at the rear of the vehicle; their respective functions are described below. This section also describes the European regulatory framework applicable to both OEM units and aftermarket equipment intended for mounting or installation on the vehcile.

Backup (reversing) lamps

Purpose
These lamps are intended to illuminate the area to the rear of the vehicle when the vehicle is reversing.

Regulations
77/539/EEC, ECE-R23, StVZO § 52.
One or a maximum of two white-light lamps may be installed (Figure 50).

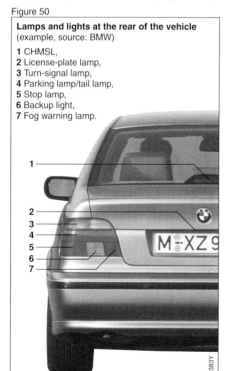

Figure 50

Lamps and lights at the rear of the vehicle (example, source: BMW)

1 CHMSL,
2 License-plate lamp,
3 Turn-signal lamp,
4 Parking lamp/tail lamp,
5 Stop lamp,
6 Backup light,
7 Fog warning lamp.

UKB0383Y

Bosch offers a range of aftermarket lamps for DIY and dealer installation.
The backup lamps may be installed together with any of the other rear lamps. Combined and nested designs including other lamps are prohibited.
The switching circuit must be designed to ensure that the reversing lamps operate only with reverse gear engaged and the ignition on.

Rear flashers

Purpose
The flashers indicate intended changes in direction (turn-signal function) as well as potentially dangerous situations (hazard-warning-signal function). Design and location must be selected to ensure reliable signal peception by other road users regardless of lighting and operating conditions.

Regulations
76/759/EEC, ECE-R6, StVZO § 54.
Group 2 (rear) flashers/turn signals are specified for dual-track vehicles (Figure 49). The dashboard-mounted monitoring lamp may be in any color desired.
The flash frequency is defined as 90 ± 30 cycles per minute.
Regulations call for two amber lamps.
The basic position mirrors that of the front turn signals. The only additional restriction is that the horizontal clearance to the tail lamp may not exceed 50 mm when the vertical clearance from the tail lamp is less than 300 mm.
The rules governing use of grouped, combined and nested designs are the same as those for the front flashers/turn-signal lamps.

Tail and clearance lamps

Purpose
Tail lamps and clearance lamps are intended to provide following road users with early warning of the vehicle's presence even when the brakes are not being applied.

Regulations

76/758/EEC, ECE-R7, StVZO §§ 51 and 53 (Figure 50).

Rear-facing tail lamps are mandatory equipment on vehicles of all widths. Supplementary front and rear clearance lamps are also specified for vehicles wider than 2,100 mm (e.g., trucks).

Tail lamps

Two red-light tail lamps are mandatory equipment (Figure 50).

The positions are the same as those prescribed for the rear turn-signal lamps. Approved are:

- Grouped designs along with any other rear lamps,
- Combined designs together with the rear license-plate lamps, and
- Nested designs together with the stop lamp, parking lamp and the fog warning lamp.

When the tail and stop lamps are joined in a nested assembly the luminous-intensity ratio for the individual functions must be at least 1:0.25 mm. Tail lamps are to operate together with the side-marker lamps.

Clearance lamps

Two red-light lamps visible from the rear are stipulated.

The clearance lamps must be positioned as far outward and as high as possible. Grouped designs with other lighting equipment are approved. Combined and nested designs with other lamps are not approved.

Rear parking lamps

Purpose

Parking lamps are intended to enhance the visibility of stationary vehicles. They must be capable of operating when all other lamps are off. The parking-lamp function is usually assumed by the tail and side-marker lamps (Figure 50).

Regulations

777/540/EEC, StVZO § 51, ECE-R77.

Two parking lamps at front and rear or one parking lamp on either side may be installed. The color prescribed for the rear unit is red. Yellow (amber) is approved for use at the rear when the parking lamps are grouped with the side-mounted turn signals. The positions are the same as those prescribed for flashers and turn-signal lamps.

Grouped designs may be used together with any other lamps. Combined designs including other lamps are not approved. Also approved are nested rear designs uniting the parking lamps with any of the following:

- Tail lamps,
- Stop lamps,
- Fog warning lamps, and
- Side-mounted flasher/turn-signal lamps.

Stop (brake) lamps

Purpose

The purpose of the stop lamps is to alert following drivers to the fact that a vehicle is being braked.

Regulations

76/758/EEC, ECE-R7, StVZO § 53.

(Main) stop lamps

Two red-light stop lamps are prescribed as mandatory equipment on all vehicles (Figure 50).

Approved are:

- Grouped designs together with the tail lamps,
- Combined designs with the license-plate lamp if the stop lamp forms part of a nested assembly within the tail lamp,
- Nested designs together with the tail and parking lamp.

When a nested design incorporating both the stop and tail lamps is used, the luminous-intensity ratio distinguishing the two functions must be a least 5:1.

Supplementary high-mount stop lamps

Installation of a supplementary raised stop lamp situated at the vehicle's center (CHMSL, or Center High-Mount Stop Lamp) is mandatory for new vehicle models in Europe.

These center high-mount units must operate together with the conventional main stop lamps. Bosch also has after-market versions of this lamp available (Figure 51).

Fog warning lamps

Purpose

The rear fog warning lamps are used to enable following drivers to detect un-braked vehicles when visibility is ex-tremely restricted.

Regulations

77/538/EEC, ECE-R38, StVZO § 53 d.
The countries of the European Union prescribe two red-light fog warning lamps on all newly registered vehicles (Figure 50). Bosch also has such fog lamps in its program for aftermarket installation.

Grouped designs with any other rear lamps are allowed. Combined designs with other lamps are prohibited. Nested designs also incorporating the tail or parking lamps are allowed.

The visible illuminated area along the reference axis is not to exceed 140 cm^2. The electrical circuitry must ensure that the fog warning lamp operates only in conjunction with the low beam, high beam or front fog lamp. A provision must also be available for switching off fog warning lamps while the front fog lamps remain in operation.

The prescribed color for the dashboard-mounted indicator lamp is yellow (green also approved for vehicles initially registered prior to January, 1981).

License-plate lamp

Purpose

The license-plate lamp is designed to make the vehicle's license plate visible to other road users.

Regulations

76/760/EEC, ECE-R4, StVZO § 60.
The lamp must ensure legibility of the rear license plate at a minimum distance of 25 m at night. Approved configurations are:

– Grouped designs together with any other rear lamps, and
– Combined designs with tail lamps.

Nested designs are not approved.

Figure 51
CHMSL – Supplementary Center High-Mount Stop Lamp. Aftermarket version.

Lighting devices Passenger compartment

Interior lamps

Because there are no legal mandates specifying the character of interior lighting, vehicle manufacturers are free to equip their vehicles as they want. The result is a wide range of different lighting layouts.

Ceiling lamp

The 3-position ceiling (dome) lamp with settings for "on," "off" and "on with front door open" has become the virtually universal standard. Also available are supplementary rear ceiling lamps switched on either by contact switches in the door pillars or by a dashboard-mounted switch.

Glove-compartment lighting

A contact activates the lamp(s) when the glove-compartment door is opened; there is no internal lighting when the door is closed.

Trunk lighting

Trunk lighting is now a standard feature on passenger cars. The lamp is switched on and off by a contact switch operated by the trunk lid.

Instrument-panel illumination

Controls and display instruments in the dashboard and instrument cluster are illuminated to ensure that they are legible in the dark. The illumination for display instruments features a provision for automatic or manual adjustment, making it possible to adapt lighting intensity to actual conditions and avoid dazzling the driver with backglare.

Indicator lamps in various colors are often used to monitor various operating conditions. Some of the colors are specified (such as blue for high beam,

yellow for fog warning lamps, etc.). The identification symbols conform to a uniform symbol code valid throughout the ECE (refer to Figure 52 for examples).

Controls and switches

In the interests of safety, official guidelines govern the design, installation and control arrangements of all vehicular lighting equipment except the interior lamps. The object is to arrange the controls and switches to faciliate efficient operation with minimal driver distraction.

Illumination for primary and secondary control elements

Direct or indirect illumination must be available for the primary controls and for

Figure 52

Switch and indicator symbols

1 Brake-system defect,	11 Headlamp wipe/wash,
2 Front fog lamp,	12 Spot lamp,
3 Hazard-warning flashers,	13 Windshield washer system,
4 Instrument-panel illumination,	14 Interior lighting,
5 High beam,	15 Windshield defroster,
6 Fog warning lamps,	16 Floodlamp,
7 Windshield wiper and washer,	17 Windshield wiper,
8 Rear-window wiper,	18 Ventilation/heater fan,
9 Main headlamp switch,	19 Heated mirror,
10 Rear-window defroster,	20 Rotating beacon.

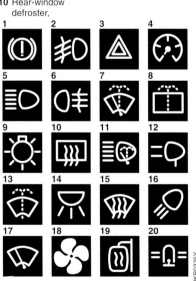

UKB0335Y

the comfort and convenience equipment (ventilation blower, heater and air conditioner, ashtray). This is necessary so that they can be operated or reached in the dark, without extended searches that would distract the driver.

Switch illumination

Switch illumination fulfills two functions in the dark:

- It allows the driver to identify specific switches immediately (e.g., hazard warning flasher),
- Together with the ECE symbols it supports driver orientation (Figure 52).

Frequent switching operationg

Switches frequently used during normal driving must be designed to be within easy reach of the driver without it being necessary for him/her to take a hand off of the steering wheel. This stipulation is especially important for items such as the turn signals, the horn, the headlamp dimmer switch and the wipe/wash systems for windshield and headlamps. This is why in all vehicles these functions are united in combination stalks or switches on the steering column or within the steering wheel. There are still no standards specifying precise placement.

Occasional switching operations

Logical disposition of those switches that are used only rarely but still need to be accessible for operation while the vehicle is being driven (headlamps, headlight leveling control, hazard-warning flashers, front fog lamps and rear fog warning lamps) makes a valuable contribution to active safety. The driver can locate these control elements "blindfolded" and can identify their function by touch alone, while remaining concentrated on the road and traffic.

Display elements

While illuminated switches can be employed to indicate operating conditions or equipment control status, another option is to provide this information through indicator lamps or as a direct readout in a display panel.

Illuminated colored lamps and LEDs can furnish status reports (parking brake, lights, diesel preglow, etc.), while displays (liquid crystal) provide not only status messages but also the accompanying quantified data (mileage, travel time, fuel consumption, remaining fuel, average speed, etc.).

Light-emitting diodes

The light-emitting diode, or LED display is an active (self-luminous) device based on a semiconductor element with a PN junction. During operation in the forward

Figure 53

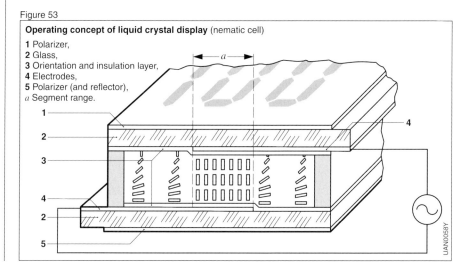

Operating concept of liquid crystal display (nematic cell)

1 Polarizer,
2 Glass,
3 Orientation and insulation layer,
4 Electrodes,
5 Polarizer (and reflector),
a Segment range.

UAN0058Y

direction the charge carriers (free electrons and holes) recombine, releasing energy. Specific semiconductor substances then convert this energy into electromagnetic radiation with infrared wavelengths and in the visible portion of the spectral scale. Frequently used conductor materials are: galllum arsenide (infrared), gallium arsenide phosphide (red to yellow) and gallium phosphide (green).

Liquid-crystal display
The liquid crystal display, or LCD is a passive display element. Visible contrast relies on supplementary illumination. The most widely used type of LCD is the twisted nematic, or TN cell (Fig. 53).

The liquid crystal medium is held between two glass plates. In the immediate vicinity of the display segment these glass plates are covered by a transparent conductive layer. Voltage can be applied to generate an electrical force field between the layers. An additional orientation layer causes the light's plane of polarization passing through the cell to rotate. The cell's initial reaction to addition of polarizers acting at mutual right angles on the outside surfaces is to become

translucent. Voltage application at the two opposed electrodes causes the liquid crystal molecules to line up with the electrical field. This suppresses rotation of the polarization plane and induces opacity to form the actual display.
Separately controlled segment ranges can be used to portray numbers, letters and defined symbols. The only disadvantage is that a supplementary light source is needed to produce a readable display.

Light sources
Incandescent bulbs
Conventional systems rely on bulbs for illumination of passive display elements. Color filters can be used to modify the hue of the bulbs to meet operational and design requirements.

LED
Continuing advances in miniaturization and modular design techniques are causing the durability and spatial advantages of the LED to appear in a new light. LEDs are available in red, green and yellow as well as blue.

Fluorescent lamps
Recent advances in the configuration of fluorescent lamps permit bright and extremely consistent backlighting for displays.

Figure 54
Instrument cluster with various monitors and displays (example)

Special-purpose lights

Special-purpose lighting equipment for motor vehicles makes a substantial contribution to highway security and occupational safety. Emergency identification lamps, floodlamps, and spot lamps alert other road users to potentially critical traffic situations and also make it possible to carry out vital and pressing jobs and operations in the dark.

Compliance with the regulations contained in the Highway Traffic Code (StVO) and the StVZO (FMVSS/CUR) is absolutely essential.

Emergency identification lamps

Emergency identification lamps must project what appears to be an intermittent, flashing beam of light around a 360° radius at a flashing frequency of between 2 and 5 Hz. Blue emergency flashers are approved for use on officially designated vehicles such as police cars, fire trucks and ambulances. Yellow flashers are intended to warn of potential hazards (e.g., construction sites) and dangerous loads on trucks (e.g., overwidth or overlength cargo).

A basic distinction is drawn between mechanically-actuated rotating emergency beacons and electronically-controlled high-intensity discharge (HID) flashers. While the mechanical emergency beacons rely on a reflector for their flashing effect, the HID flashers use a high-intensity discharge tube to generate signal images with periodic gaseous-discharges. Because HID devices remain fully operational under severe operating conditions, and resist extreme cold and contamination (no moving mechanical components), they are ideal for heavy-duty applications. The only critical element is the electronic control circuitry, which must also be capable of withstanding severe conditions (impervious to vibration, condensation and water spray; high level of corrosion resistance).

Minimum luminous intensities are specified for emergency identification lamps: 20 cd for blue and 40 cd for yellow lights in a plane parallel to the road surface. In their light beams, blue emergency flashers must achieve at least 10 cd at +4° and yellow lights a minimum of 20 cd at +8°.

Floodlamps

Floodlamps provide consistent illumination for stationary or mobile job sites by lighting extended surface areas. Robust designs make these units particularly suitable for installation on heavy commercial vehicles as well as for a wide variety of applications (construction, search and rescue, agriculture and forestry, shipping, etc.).

The use of floodlamps with the vehicle underway is approved only when vehicle travel is a part of the actual work process; examples are night-time road work and recovery of accident vehicles.

Spot lamps

Spot lamps project a concentrated, high-intensity luminous beam in applications where recognition of objects at extended ranges is of prime importance (rescue operations, police and fire department activity, technical support).

Spot lamps with transitionless focus are particularly effective in maintaining exact beam alignment at any range extending to approximately 225 m, allowing optimal exploitation of the unit's lighting power.

Cleaning systems

Windshield cleaning

Assignments and requirements

Dirty windshields impede the driver's vision. Windshield cleaning systems are therefore of major importance for road-traffic safety (Fig. 1).

Even when used under extreme conditions, the wiper system must still operate efficiently after 1.5 million wiping periods (wiper blade: 500,000). This corresponds to a wiped area of about 200 football fields.

As far as the driver is concerned, the following demands are made on the wiper system:

– The wiped area, and therefore the cleaned windshield surface must be as large as possible and in particular must permit the unimpeded view of the edge of the road, as well as of traffic signs and traffic lights.
– The wiping quality is also an important factor, and as far as possible must rule out scattered light and the associated effects of glare resulting from approaching traffic. The wiper system must run silently, and must operate without trouble for long periods. It should permit intermittent wiping or be coupled to a rain sensor.

It is the job of the wiper and wipe/wash systems to remove rain, snow, and dirt from the windshield (and in some cases also the rear window).

Fig. 1

Windshield cleaning system in an automobile

1 Wipe/wash system for headlamps and windscreen,
2 Pumps with water reservoir (front),
3 Nozzle (windshield),
4 Rear-window wiper,
5 Nozzle (rear window),
6 Wipe/wash system for the rear window,
7 Pump with water reservoir (rear),
8 High-pressure cleaning system (headlamps),
9 Water reservoir,
10 High-pressure pump.

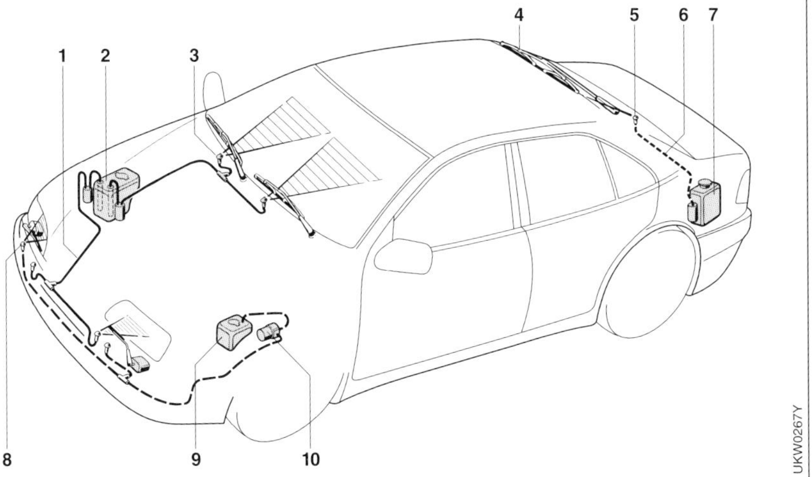

Design

A wiper system comprises an electric motor as the drive, a link mechanism, the wiper arms together with the wiper blades, and a switch which is usually combined with the steering column. The so-called "Twin" wiper blades from Bosch are characterized by their special two-component technology, in which the soft wiper-element spine guarantees quiet and even passage across the windscreen, and the hard wiper-element lip with its micro double-edge ensures thorough cleaning throughout the wiper blade's service life. The wiper system can be expanded by the addition of an intermittent-wipe relay, a time-delay relay, and a rain sensor.

The supplementary washer system is comprised of a pump with drive motor, a reservoir for the washer liquid, the nozzles through which the washer liquid is sprayed onto the windshield, the hose connections, and a switch which is usually integrated in the wiper switch.

Method of operation

Wiper systems

The driver switches the electric wiper motor on and off with a switch on the steering column. There are a number of versions depending upon the equipment fitted.

The wiper can be run at different speeds, whereby usually only two are provided. The switch lever has the positions "Off", "Speed 1", "Speed 2", and often "Intermittent" is also incorporated. These positions select the wiping speed as well as the pauses between intermittent wiping actions. According to legal regulations, in the "Speed 1" setting, at least 10 wiping movements must take place every 10 seconds in Europe (20 in the USA) and at the "Speed 2" setting at least 45.

Continuous wiping of the windshield when only very little rain or snow is falling poses a problem since wiper movement across a dry windshield surface causes excessive wear of the wiper-element lip.

Fig. 2

Schematic diagram of a wipe/wash system (extract from the complete circuit diagram)

M4 Windshield-wiper motor,	**M7** Headlamp-washer motor,	**S28** Headlamp wipe/wash switch,
M5 Windshield-washer motor,	**K8** Intermittent-wiper relay,	**S29** Windshield-washer switch,
M6 Headlamp-wiper motor,	**S27** Wiper switch,	**F** Fuse.

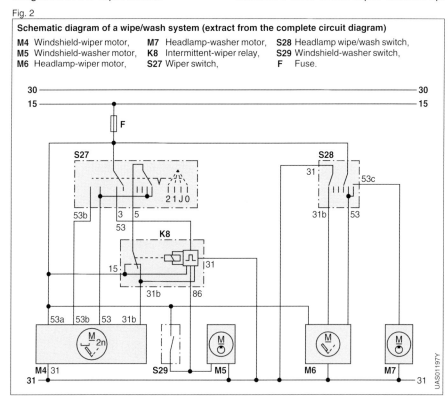

UAS01197Y

With an <u>intermittent-wipe relay</u>, wiping can be interrupted at specified intervals so that the wiper blade does not have to travel over a dry surface. The intermittent-wipe relay is a pulse generator which generates a pulse train which can be varied as a function of time. Every pulse which triggers the wiper motor via a relay causes a single back-and-forth movement of the wiper arms. On conventional intermittent-wiper switches, intermittent wiping can be finely adjusted with a knob. The range of adjustment provides for between 2 and 20 wiping cycles per minute. With the programmable intermittent-wipe relay, the selection lever is used to program the interval between wiping cycles. As soon as vision is impaired again and the windshield must be wiped, the selector switch is moved to the "intermittent wiping" position. The time between switching off and on again is the interval between the wiping cycles. It is adjustable between 2 and 45 seconds.

Washer systems

Washer systems are imperative for efficient cleaning of the windshield. Electric centrifugal pumps are used to spray the water, to which a special detergent has been added, onto the windshield in the form of a concentrated jet. The washer system pumps are also actuated from the steering-column selector switch.

Wipe/wash systems

The wiping and washing processes can be combined. Operating the washer-system switch causes the washer system to spray water onto the windshield, and after a brief interval the wiper system starts (Fig. 2). After the washer system has switched off, the wiper wipes across the windshield a few more times until it is dry. When the washer system switches on, the wiper comes into operation after maximum 1 second, and runs on for 3...5 seconds after washer switch-off.

Headlamp cleaning

Assignment

Headlamp cleaning systems remove the dirt from the headlamp lenses. With clean headlamp lenses there is no loss of light due to dirt, so that maximum illumination of the road is ensured. Furthermore, oncoming traffic is not dazzled at night.

Design and method of operation

Two different systems are used for headlamp cleaning:
The <u>wipe/wash system</u> is similar to that used for the windshield. But since plastic lenses, notwithstanding their scratch-proof coating, are too sensitive for this method, its application is limited to glass lenses.
The <u>high-pressure washer system</u> (Fig. 3) is coming more and more to the forefront since it can be used for both glass and plastic lenses. The cleaning effect is mainly due to the cleaning pulse of the water droplets. The following factors are decisive for cleaning quality:
– Distance between nozzle(s) and headlamp lens,
– Size, impact angle, and impact speed of the water droplets,
– Quantity of water sprayed onto the lens.

Fig. 3

Components of a high-pressure headlamp washer system

1 Water tank, **2** Pump, **3** Non-return valve, **4** T-fitting, **5** Nozzle holder, **6** Hose.

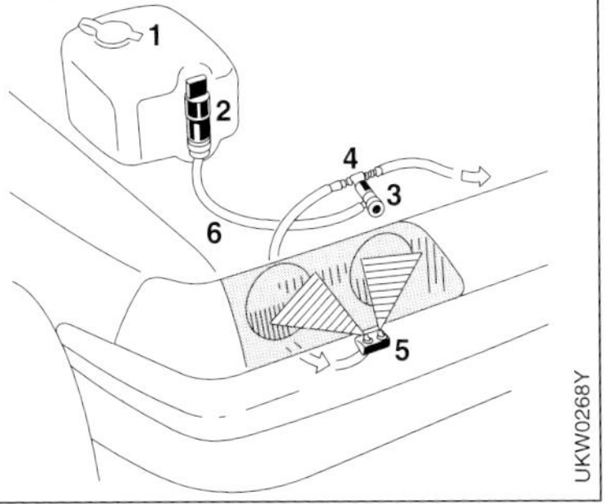

UKW0268Y

In addition to the spray-nozzle holders mounted in the bumper overriders, there are also versions on the market which move out on telescopic arms. Since the telescope arm can position the nozzle in the ideal location, this considerably improves the cleaning effect. A further advantage is that the nozzle can be retracted and concealed in the bumper when not in use.

The high-pressure washer system comprises the following components:
– Water tank, pump, hose and non-return valve, as well as
– Nozzle holder, which can also extend on a telescopic arm, each with one or more nozzles.

Legal requirements

The most important legal requirements for Europe are:
– Since 1996, stipulated by law for headlamps with gaseous-discharge lamps,
– Necessary for low beam and driving-lamp pairs,
– Water supply for 25 cleaning cycles (Class 25) or 50 cleaning cycles (Class 50),
– Cleaning efficiency ≥ 70 % on a headlamp whose luminous flux has been reduced to 20 % of the original value due to dirt on the lens,
– Fully serviceable from –35 °C to +80 °C, although impairment due to freezing is permitted.

Rain and dirt sensors

Basically, the <u>rain sensor</u> (Fig. 4) is composed of an optical transmission and receive path. The light from the light source is directed toward the windshield at an angle. If the windshield outer surface is dry, all of the light is reflected back (total reflection) from the outer surface to the receiver which is also mounted at an angle. If on the other hand there are rain drops or dirt on the windshield, a considerable portion of the light is reflected away so that only a weakened light signal is received and the wiper switches on automatically.

With the <u>dirt sensor</u> (Fig. 5), the reflected-light barrier comprises a light source (LED) and a light receiver, and is located on the inside surface of the windshield inside the cleaning area but not directly in the pattern of reflection from the driving light. With a clean windshield, or one covered by raindrops, the measurement beam from the light source, which is in the near infrared range, passes through the windshield practically without any reflection at all. If on the other hand, the measurement beam hits dirt particles on the windshield outer surface it is scattered back to the receiver to a degree which is proportional to the amount of dirt. Provided the headlamps are on, the headlamp cleaning system then switches on automatically.

Fig. 4

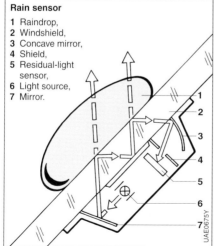

Rain sensor
1 Raindrop,
2 Windshield,
3 Concave mirror,
4 Shield,
5 Residual-light sensor,
6 Light source,
7 Mirror.

Fig. 5

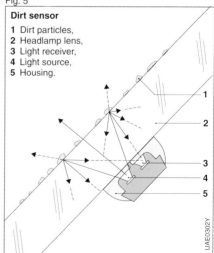

Dirt sensor
1 Dirt particles,
2 Headlamp lens,
3 Light receiver,
4 Light source,
5 Housing.

Theft-deterrence systems

The fact that theft and vandalism are increasing all over the world, necessitates the use of increasingly ingenious electronic systems to protect against intrusion and unauthorized use.

In addition to the protection against unauthorized use provided by ignition lock and the steering-wheel lock, the following devices are being applied to protect vehicles standing on private or public parking lots: Central door-locking systems, theft-deterrence alarm systems, and electronic immobilizers. For more details, please refer to the Tables on the right.

Central locking systems

Assignment

The central locking system prevents individual doors remaining unlocked inadvertently. With this system, all doors (and usually the trunk and the filler-neck cover as well) can be locked and unlocked by mechanically locking or unlocking the trunk or one of the doors by means of the key. Together with an integrated theft-deterrent system, the central door-locking system increases the protection against the vehicle being used without permission or even being stolen. The combination of infrared or radio remote control with a theft-deterrent system is a further step toward sophistication of operation.

Method of operation

Central locking systems are available with either pneumatic or electric-motor-driven actuators. In the pneumatic version, an electric-motor-driven dual-pressure pump generates the system pressure. It rotates in either direction to provide vacuum or overpressure as required. This system

Theft-deterrence systems	
Version	Protected areas
Basic system	Doors Trunk lid Central door-locking system Hood/trunk lid Car radio Car telephone Glove compartment Ignition & starting switch Windows
Passenger-compartment protection	Passenger compartment
Wheel and tow-away protection	Wheels Vehicle as a whole

Immobilizers	
Version	Protected areas
Electrical	Ignition system or diesel injection pump Fuel supply Starting system Central locking system
Electronic with coding device	Central locking system Engine management

can be switched on and off for instance using a central position switch inside the vehicle, or with the driver's door lock.

If required, locking and unlocking can take place from the driver's door, the front-seat passenger's door, or from the trunk-lid lock. The electric-motor version is more widespread than the pneumatic version. Signals are transmitted to the system's central control unit either from a central switch, which is triggered by remote control, or from the contacts of the various locks. The central control unit transmits the signals to operate the servomotors.

There are a number of different technical designs available depending upon functional scope and type of locking system. Principally though, all systems rely on the same basic concept: A small electric motor with step-down gearing shifts a lever which opens and closes the lock. In case of power failure, it must always be possible to open the door with the key or the inside door handle. In the case of central door-locking systems with integrated theft-deterrent system, manual operation from the "theft-deterrent system" position is possible either with the vehicle key or with remote control.

Alarm systems

Assignment

It is the job of a primed theft-deterrent system (car alarm) to trigger a warning signal when an unauthorized attempt is made to enter the vehicle or tamper with it. Best-possible vehicle protection is provided by combining and extending the already installed components (e.g. central door-locking system) with a car alarm (Fig. 1). The legal basis for safeguarding a vehicle against unlawful acts is laid down in §38b of the StVZO (FMVSS/CUR) and in European Directives ECE-R18.

For instance, a modern car alarm is provided with an electronic control which triggers the following (permissible) warning signals:

– Intermittent sound signal through the vehicle horn for a maximum of 30 s, and
– Visible flashing signals via the direction-indicator lamps for a maximum of 5 minutes (StVZO) or 30 s flashing of the dipped beam (ECE).

Fig. 1

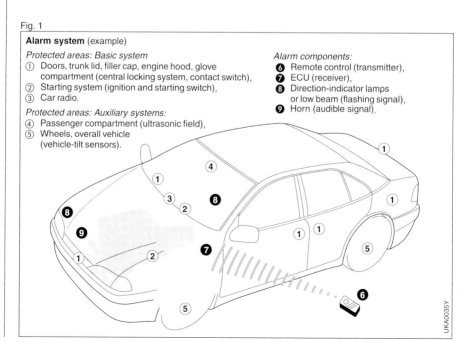

Alarm system (example)

Protected areas: Basic system
① Doors, trunk lid, filler cap, engine hood, glove compartment (central locking system, contact switch),
② Starting system (ignition and starting switch),
③ Car radio.

Protected areas: Auxiliary systems:
④ Passenger compartment (ultrasonic field),
⑤ Wheels, overall vehicle (vehicle-tilt sensors).

Alarm components:
❻ Remote control (transmitter),
❼ ECU (receiver),
❽ Direction-indicator lamps or low beam (flashing signal),
❾ Horn (audible signal).

UKA0035Y

Design and method of operation

Complex theft-deterrent systems (car alarms) are modular in design, and can be adapted or up-dated as required. Car alarms for instance, comprise the following modules (Table on Page 265):
– Basic system (the alarm is triggered as soon as a door, the trunk, or the hood is opened, an unauthorized attempt is made to remove the radio or start the engine, or window/windshield glass is broken),
– Passenger-compartment protection using ultrasonic detectors,
– Wheel and tow-away protection.

The wide variety of variants which can be implemented using this modular system permit adaptation to the specific equipment of a given vehicle and to the legal regulations of individual countries.

Basic system

The central control unit evaluates unauthorized intervention, or the signals at the inputs which result from such intervention. It activates the system, and the car alarm then triggers the alarm (acoustically with the horn, and visibly with the headlamps or turn-signal lamps).

Provided the alarm system is primed, the contact switches in the doors, trunk, or hood trigger the alarm immediately these are opened. The alarm is also triggered when the radio's monitoring loop is interrupted and when the ignition is switched on. Output Z activates the external auxiliary units (e.g. the tilt alarm) which trigger the alarm via input TZ. All inputs which are independent of each other are capable of triggering the alarm one after the other. The starter-disable function circumvents the ignition-and-starting switch and thus prevents the engine being started. An LED blinks to indicate that the system is primed. The switch-over "primed/unprimed" is by means of an infrared control or a remote radio control. These use individually coded signals. The above measures safeguard the vehicle against unauthorized use. The ignition must be switched on for the car alarm to be primed.

Ultrasonic field for passenger-compartment protection

1 Ultrasonic detector,   3 Side windows,
2 Windshield,   4 Rear window.

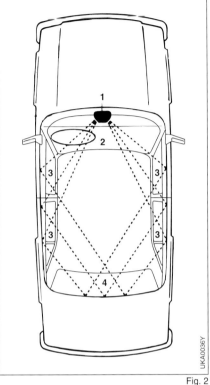

Fig. 2

Fig. 3

Design of the ultrasonic transmitter for passenger-compartment protection

1 Power supply,
2 Electrodes,
3 Crystal wafer,
4 Air,
5 Ultrasonic-sound radiating surfaces on the crystal wafer.

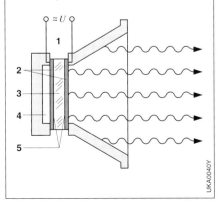

Ultrasonic passenger-compartment protection

An ultrasonic transmitter is used to generate and transmit oscillations (f = 20 kHz) in the vehicle's passenger compartment (Fig. 3) and set up an ultrasonic field. An ultrasonic detector detects movements due to fluctuations in this field (caused for instance by reaching into the vehicle or by breaking glass). As soon as the phase relationship, the frequency, or the amplitude of the oscillations change, the evaluation electronics trigger the alarm. In order to improve its effectiveness, the system's response threshold (sensitivity) is adjustable.

Prevention of wheel theft and tow-away

This module comprises position sensors and an evaluation unit (Fig. 4). When the vehicle is parked on the level or on a slope, its current position is programmed as the zero position when the car alarm is primed. If changes in position (longitudinal/transverse), and the speed of these changes, exceed specified limits the alarm is triggered. Normal changes in position (due to loss of air in the tires, rocking of the vehicle, softening of the standing surface for instance) do not trigger the alarm.

Fig. 4

Alarm system with electronic protection against wheel theft and tow-away

1 Transmitter (remote control), 2 Alarm-system ECU, 3 Receiver (remote control),
4 Microcomputer (**4a**), with battery connection (**4b**), and input stage for wheel-theft and tow-away protection device (**4c**), 5 Car radio, 6 Ultrasonic receiver, 7 Door-contact switch, 8 Contact switch for hood, trunk lid, and glove compartment, 9 Position sensor with evaluation unit, 10 Relays, 11 Starting system, 12 Horn, 13 Ultrasonic transmitter, 14 Direction-indicator lamps or low beam, 15 Satus display.

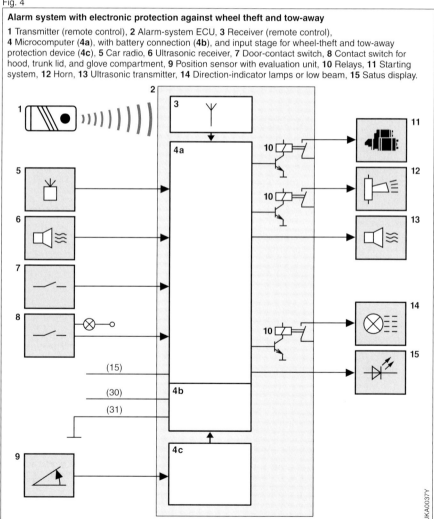

UKA0037Y

Immobilizers

Assignment

The vehicle immobilizer serves to prevent the vehicle being used by unauthorized persons. It must make it impossible to start or drive the vehicle without the coded access authorization, or some other legitimization (e.g. coded key). From the insurance standpoint, approved vehicle immobilizers which have been installed as original equipment, or retrofitted, are of major significance.

Design and method of operation

One differentiates between electrical and electronic vehicle immobilizers. Since 1.1.95, according to law, only electronic immobilizers may be fitted as automotive original equipment.

When the ignition is switched off, and at the latest when the vehicle is locked, both systems prime themselves automatically and render inoperative a number of the units which are imperative for vehicle operation.

In the majority of cases, the immobilizer is combined with a central door-locking system. With such a sophisticated installation, a single signal suffices to activate, or deactivate, both systems simultaneously.

Electrical vehicle immobilizer

The electrical vehicle immobilizer, as the name implies, uses conventional relays to immobilize a number of the electrical circuits which are essential to vehicle operation (Fig. 5).

Fig. 5

Immobilizer with interrupt circuits

1 Transmitter (remote control), **2** Immobilizer ECU, **3** Receiver (remote control), **4** Microcomputer (**4a**), with battery connection (**4b**), **5** Relay, **6** Central door-locking system, **7** Status display, **8** Starting system, **9** Engine-management ECU, **10** Electric fuel pump (gasoline engine) or fuel supply (diesel engine).

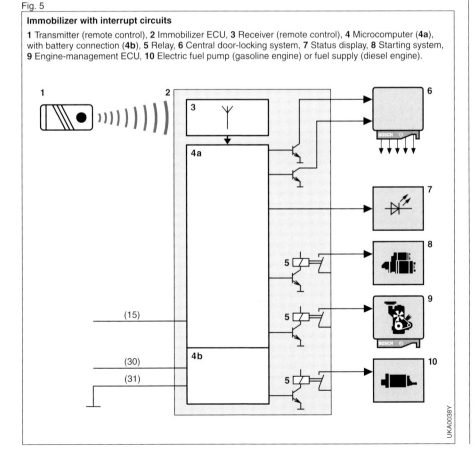

Usually, three main circuits are interrupted:
– Starter system,
– Fuel-supply system, and
– Ignition system, or the fuel supply to the diesel injection pump.

Electrical-type immobilizers in diesel-powered vehicles interrupt the supply of fuel using a conventional solenoid valve in the fuel-inlet line.

Electronic immobilizers

With the electronic-type immobilizer, a coded signal is used to block or release the operation of one or more functional units (starting system, fuel supply, ignition system). Blocking or releasing is implemented by the engine-management ECU (Fig. 6).

With the electronic immobilizers used in a "diesel environment", the diesel anti-theft device (DDS) which is attached directly to the injection pump activates (or deactivates) the electrical shutoff device so that the immobilizer becomes effective.

Activation and deactivation systems

A hand-held transmitter sends a radio or infrared signal in coded form to the immobilizer to deactivate it.

Transponder systems exchange coded signals with an electronic circuit which for instance is located at the ignition lock. The system deactivates the immobilizer as soon as the transponder, which can be integrated in the ignition key, enters the coil's receive area. The automatic exchange of data takes place without any action on the part of the driver and represents a further increase in operative comfort.

Using an electronic key (a system which is still available in only a few vehicles), the immobilizer is de-activated when conductive coupling takes place between the electronic key and the immobilizer system.

Using a code keypad, the driver enters the secret code number to deactivate the immobilizer system.

Fig. 6

Immobilizer with coded intervention

1 Transmitter (remote control), **2** Immobilizer ECU, **3** Receiver (remote control), **4a** Microcomputer, **4b** Microcomputer with battery connection, **5** Coding unit, **6** Central door-locking system, **7** Status display, **8** Engine-management ECU.

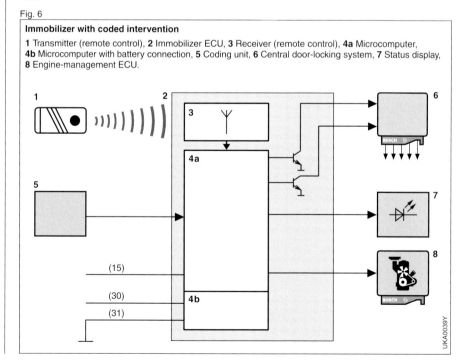

Comfort and convenience systems

Cruise Control (Tempomat)

Assignment

It is difficult and tiring for the driver to maintain a given speed for a long time. From the fuel-economy standpoint, it is impossible for the driver to optimally compensate for fluctuations in vehicle speed. The Cruise Control ("Tempomat") is intended to maintain a given speed as chosen by the driver for long distances and, as far as possible, completely independent of gradients.

Design

The Cruise Control system comprises the following components:
– Operator lever,
– Vehicle-speed sensors (for instance, using the ABS wheel-speed sensors),
– Electronic control unit (depending on system, sometimes integrated in the engine ECU),
– Electric motor as the actuator for operating the throttle valve (gasoline engine) or for controlling the injection pump on the diesel engine,
– Switches at the brake and clutch pedals.

Method of operation

In this example, operation is described for a gasoline engine on which the vehicle-speed control takes place via the throttle-valve setting.

Operator lever

Four functions can be selected using the operator lever:
1. Accelerate and desired-speed input. The vehicle accelerates as long as this button remains pressed. When it is released, the momentary speed is stored as the desired speed (the speed is set). It can be increased in steps by briefly pressing the key.
2. Decelerate and desired-speed input. With this key the vehicle decelerates as long as the key remains pressed. Upon releasing the key, the momentary speed is stored as the desired speed (it is "set" as above). Here too, briefly pressing the key leads to the desired speed being reduced in steps.
3. Switching off the Cruise Control. The Cruise Control is switched off by pressing the OFF key. The stored desired speed remains valid until the ignition is switched off.
4. Reactivate (WA). If this key is pressed after the Cruise Control has been switched off, the vehicle assumes the stored desired speed again.

Sensors
Sensors (for instance, the ABS wheel-speed sensors) provide the signals from which the actual vehicle speed is derived.

Electronic control unit (ECU)
On modern engine-management systems, the Cruise Control ECU is increasingly being integrated in the engine-management ECU, and can be sub-divided into the following function blocks (Fig. 1):
– Evaluating logic circuit (7),
– Acceleration controller (8),
– Speed controller (9),
– Control actuator (10),
– Output stage (11),
– Desired-speed memory (12),
– Switch-off logic circuit (15) with the thresholds for minimum speed (13) and speed difference (14),
– Monitoring unit (16).

The evaluating logic circuit (7) either converts the AC signal from the wheel-speed sensors (1) into a digital signal, or it inputs an alternative speed signal (e.g. from the ABS). This value is then compared with the value in the desired-speed memory.

When the desired-speed key is pressed, the vehicle's actual speed is stored as the desired speed in the desired-speed memory (12). The deviation of the actual speed from the desired speed forms the controlled variable. The output variable from the acceleration controller and the speed controller (8/9) is in the form of a signal for the throttle-valve angle and is the input signal for the control actuator. This controls the output stage (11) for the throttle-valve servomotor (17), and sets the throttle valve to the correct angle using the position feedback signal from the potentiometer (18). The speed controller (9) is active in the control range to maintain the selected desired speed at a constant level. The acceleration control-ler comes into effect if the actual and desired speeds differ excessively (for instance upon "reactivate"), and when the "accelerate and set" or "decelerate and set" functions are activated. Here, the throttle valve is adjusted in accordance with a defined characteristic. On auto-matic gearbox vehicles, in order to achieve more effective acceleration or deceleration figures, it is also possible to activate selected gear shifts. The driver can switch off the Cruise Control by pressing the OFF key (4), the brake (5), or the clutch (6). The input signal for the control actuator is outputted by the switch-off logic circuit (15). Here, the throttle-valve angle is returned to the setting defined by the driver with his/her accelerator pedal. If switch-off is by means of the brake or the clutch, throttle-valve adjustment is immediate, whereas if the OFF key is used, adjustment follows a defined characteristic.

The Cruise Control switches off auto-matically when:

Fig. 1

Block diagram of the Cruise Control ECU

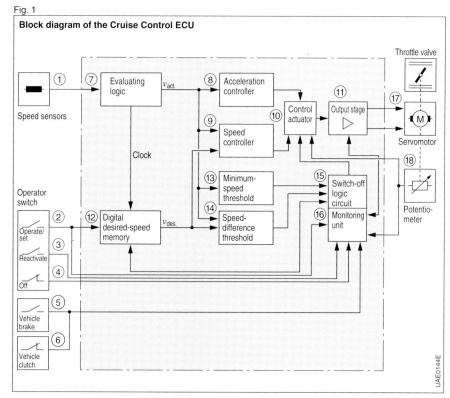

– The "minimum-speed" threshold (13) is dropped below. This is set to approx. 30 km/h (20 mph),

– The "speed difference" threshold (14) is dropped below when the actual speed has fallen far below the desired speed.

The monitoring unit (16) also switches off the Cruise Control under circumstances over which the driver has no control. For instance, when the traction control system (TCS) intervenes, or when "aquaplaning" occurs.

When the "reactivate" key (WA, 3) is pressed, the acceleration controller accelerates the vehicle up to the desired speed. The speed controller is then activated in order to maintain a constant speed. If the driver from time to time accelerates to above the speed dictated by the Cruise Control, this does not cause Cruise Control switchoff. The monitoring unit (16) is also responsible for monitoring the actuator's function, and the input and output signals. If an error is detected, the Cruise Control is switched off till the end of the journey.

Actuator

The servomotor (17) is triggered from the control actuator via the output stage, and shifts the throttle valve through a gear-set. Adjustment shift is continuous.

In case of servomotor malfunction (control-actuator error), this is switched off. If necessary, the fuel supply is reduced in order to provide reliable limp-home capabilities.

Power windows

Assignment

Power-window and sunroof drives enable the windows and sunroof to be opened and closed automatically simply by pressing a rocker switch.

Design

There are two systems in common use:

Geared linkage (less common)
Via a spur gear, the electric drive motor drives a toothed quadrant which is connected to a geared linkage (Fig. 2a).

Bowden-cable linkage (most common)
The drive motor drives a Bowden-cable linkage mechanism (Fig. 2b).

The cramped installation conditions in the vehicle doors dictate the use of flat designs (with flat motors). A self-locking worm-type reduction gear is used so that windows cannot open of their own accord, and unintentional and/or forced opening is prevented. A flexible claw coupling ensures effective damping during operation.

The electronic control can be combined in a central ECU or, in order to reduce cabling to a minimum, it can be decentralized and located in the power-window motors. These actuators permit multiple usage of the cables (use of multiplex systems).

Method of operation

Control is by means of a rocker switch. To increase operating convenience, power windows can be combined with a central door-locking system or with the individual door locks. There are also systems on the market in which the vehicle's windows close automatically upon leaving the vehicle, or move to a ventilation setting. A force-limitation device (excess-force limiter) is provided which comes into action when the windows close and prevents human appendages

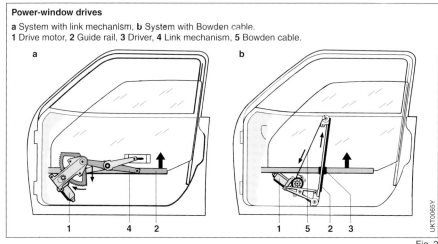

Power-window drives

a System with link mechanism, **b** System with Bowden cable.
1 Drive motor, **2** Guide rail, **3** Driver, **4** Link mechanism, **5** Bowden cable.

Fig. 2

being caught by the closing window. Paragraph 30 of the StVZO (FMVSS/CUR) stipulates that this protection mechanism remains effective while the window moves upwards through a 200...4 mm travel range (as measured from the upper edge of the window opening). Sensors are integrated in the power-drive mechanism to monitor the drive motor's speed during operation. The system responds to a reduction in engine speed by reversing the motor. In order to permit the window to always close completely as required though, the unit automatically overrides the anti-squeeze protection immediately before the window enters the door seal, thus allowing the motor to run to its end position and ensuring complete closure of the window. Feed-back always takes place of the window's respective position.

Power sunroofs

Design

Power sunroofs often combine the functions of a tilt-and-sliding sunroof. Here, special controls are required which can be either electronic or electro-mechanical. With the electromechanical control, mechanical interlocks on the

two limit switches ensure that the roof can be either tilted or opened from the closed position depending upon the polarity at the terminals. Once the sunroof has been opened or tilted, a polarity shift will initiate the closing or lowering process. If the sunroof is to be incorporated in a central-locking system, benefits are provided by an ECU featuring integral force limitation.

Method of operation

Drive for the roof is provided by Bowden cables or other torsion and pressure-resistant control cables. The drive motor is usually installed in the roof or at the rear of the vehicle (e.g. in the trunk). Permanently excited worm-gear motors with power ratings of approx. 30 W are used. As protection against overheating, software thermal protection is increasingly being used in place of thermal circuit breakers.

Electronic control is by means of a microcomputer which evaluates the signal outputs and the position of the sunroof. The extreme open and closed end positions are monitored by micro-switches or Hall sensors. The following supplementary functions can be incorporated at relatively modest expense:

- Preset position control,
- Automatic closing triggered by rain sensor,
- Motor-speed control, and
- Electronic motor protection.

Provision must also be made for ensuring that the roof can be closed with the on-board tools in the event of a failure in the vehicle electrical system.

Steering-column adjustment

Design

The steering-column adjustment device comprises a self-locking gearset and electric motor for each adjustment plane. It is located either in or on the steering column.

Method of operation

The position of the steering column is adjusted either by hand with a position switch, or by coupling to the programmable seat-adjustment facility. With the ignition switched off, getting into and out of the vehicle can be eased by swivelling the steering column upwards.

Seat adjustment

Assignment

The electric power-seat facility adjusts the seat to the required seating position by means of an electric motor. That is, to the appropriate setting of seat height and seat length, seat cushion, and backrest tilt. Programmable seat adjustment stores the data on individual person-specific settings so that these can be called up when needed. The steering-column adjustment facility rounds-off the system with a further sophisticated comfort feature.

Design

The seat-bottom frame of a conventional adjustable seat has four compact gearsets flanged to their respective drive motors. Motors and gearsets are connected by square shafts (Fig. 3). One of the gear-sets is for height positioning, and the second for combined and longitudinal positioning. The unit for adjusting the depth of the seat cushion is left out on simpler power-seat models. Another system comprises three identical motor-gear units with four height and two longitudinal-positioning gearsets. The gearsets are driven by the motor-gear units through flexible shafts. This is universally applicable and is not bound to a particular seat design.

In certain seat versions, not only is the lap belt attached to the seat frame, but also the shoulder belt, belt-height adjustment, inertia reel and seat-belt tightener are attached to the backrest. This type of seat construction permits best-possible belt positioning for all seat settings, and irrespective of the size of the person using the seat. As such, it makes a valuable contribution to occupant safety. The seat frame must be stiffened for such seat versions, stronger gearset components must be used and the connections to the seat frame must be reinforced.

Method of operation

As many as seven electric motors perform the following functions:
- Seat-cushion height adjustment, front/rear,
- Longitudinal positioning of the seat,
- Seat-cushion depth adjustment,
- Backrest tilt adjustment,
- Lumbar support (upper third of the backrest),
- Head-restraint height adjustment.

Electrical seat-adjustment system
The most favorable seating position for the particular driver is set by pressing the keys on a keypad.

Programmable seat adjustment (memory seat)

The programmable electrical seat-adjustment system is an optional extra and permits the data on previously selected seat positions to be stored and recalled repeatedly. Position feedback uses potentiometers or sensors, and reports back to the ECU on the momentary set-ting of backrest or head restraint. The ECU triggers the servomotors through transistor output stages and relays until the "fed-back" position is in line with the stored position.

On two-door vehicles, the front seat can be moved all the way forwards to provide better access to the rear seats.

Fig. 3

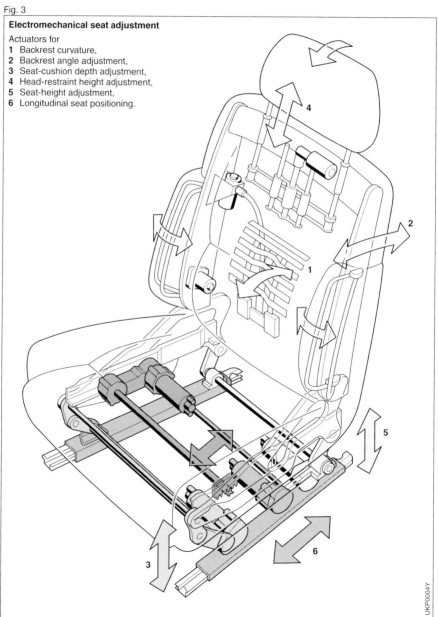

Electromechanical seat adjustment

Actuators for
1 Backrest curvature,
2 Backrest angle adjustment,
3 Seat-cushion depth adjustment,
4 Head-restraint height adjustment,
5 Seat-height adjustment,
6 Longitudinal seat positioning.

UKP0004Y

Electronic heating control

Assignment

The systems for vehicle heating and for climate control (air-conditioner) are faced with the following assignments:
- Maintaining a comfortable climate inside the vehicle for all occupants irrespective of outside temperatures (Fig. 4),
- Providing good visibility through all windows and windshield,
- Generating an environment for the driver which is calculated to minimise stress and fatigue,
- Cleaning the incoming air by filtering out particulate matter (pollen, dust) and even odors.

In many countries, the heater unit's performance is governed by legal regulations emphasizing the defroster's ability to maintain clear windows and windshield. In the EU for instance, these legal regulations include the EEC Directive 78/317, and in the USA the safety standard MVSS 103.

Method of operation

Inside the vehicle, temperature fluctuations are caused by changes in outside temperature and in vehicle speed. On systems without automatic control, these fluctuations can only be compensated for by continually readjusting the heater and ventilation settings by hand. The electronic heating control on the other hand maintains the vehicle's interior temperature at the set level. In the case of coolant-side heater controls, the temperature of the vehicle's interior and of the emerging air is measured by temperature sensors. The results are processed, and the controller compares them with the selected value. Meanwhile, a solenoid valve in the coolant circuit opens and closes at a defined frequency in response to the signals it receives from the ECU. The adjustments in the open/close ratio in the cycle periods regulate the flow rate from zero up to maximum. In air-side systems, an electric motor-gear unit (less often, pneumatic linear adjusters are used) is usually employed for the infinitely-variable adjustment of the temperature mixture flap. More sophisticated systems provide for separate adjustment on the left and right sides of the vehicle.

Electronic air-conditioner control

Task

The heater unit alone is not capable of providing a comfortable environment at all times. When the outside temperature rises above 20 °C, the air must be cooled by compressor-driven refrigerator units (Fig. 5) to achieve the required interior temperature. In addition, the moisture entrained with the cooled air is removed, thus reducing the air's humidity to the required level.

Since the constant monitoring and adjustment required to maintain a temperate climate present the occupants with a far too complicated task, automatic climate control is particularly useful for vehicles in which both air conditioner and heater are installed. This applies especially to bus drivers as they are exposed only to the temperatures at the

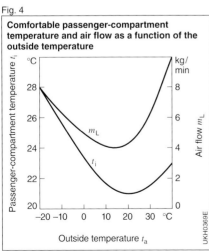

Fig. 4

Comfortable passenger-compartment temperature and air flow as a function of the outside temperature

Passenger-compartment temperature t_i °C

Air flow m_L kg/min

m_L

t_i

Outside temperature t_a

UKH0369E

Air-conditioner refrigerant circuit

1 Compressor, 2 Electromagnetic clutch (for compressor on/off), 3 Condenser, 4 Auxiliary fan,
5 High-pressure switch, 6 Fluid reservoir with dessicant cartridge, 7 Low-pressure switch,
8 Temperature switch or two-step control (for compressor on/off), 9 Temperature sensor,
10 Condensate drip pan, 11 Evaporator, 12 Evaporator fan, 13 Fan switch, 14 Expansion valve.

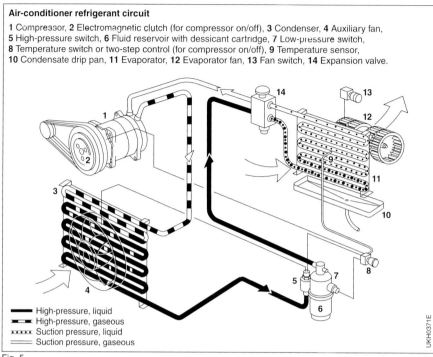

—— High-pressure, liquid
▬▬ High-pressure, gaseous
••••• Suction pressure, liquid
═══ Suction pressure, gaseous

UKH0371E

Fig. 5

front of the vehicle. Automatic climate-control systems incorporating program selection automatically maintain the correct inside temperature, air flow, and air distribution in the passenger compartment. These parameters are mutually interdependent, and changes to one will affect the others.

Method of operation

The compressor (Fig. 5, Pos. 1) compresses the vaporous refrigerant and heats it in the process. The refrigerant then cools in the condenser (3), whereby it returns to the liquid state and releases heat to the surroundings. An expansion valve (14) sprays the cooled refrigerant into the evaporator (11) where it evaporates and in the process extracts the heat required for evaporation from the incoming fresh air. Moisture is extracted from the cooled air in the form of condensation and reduces the air's humidity to the required level.

At the center of the system is the temperature-control circuit for interior tem-perature. The desired temperature is achieved by air-side (or coolant-side) control (Fig. 6), as already described under "Electronic heating control". Depending on temperature, the fresh air (a) drawn in by the blower (Pos. 1) is either cooled by the evaporator (2), or heated by the heating element (4), and then directed to where it is needed in the vehicle's interior by flap adjustment (b, c, f).

Using a variety of different temperature sensors (3, 5, 7), as well as the most important influencing variables and disturbances, and the temperature selected by the occupants with the setpoint control (6), the ECU (8) continuously generates the desired (setpoint) value. This is then compared with the actual temperature and the resulting difference used to generate the reference variables for the heater (4, 11), the cooling (2, 10) and the air-quantity control (1). Depending on the program that the occupants have selected, a further function activates the flap control for the air distribution (b, c, d, e, f). All servo loops can be influenced by manual input.

278

Infinitely variable, or graduated, blower control is used to adjust the air flow to the stipulated level. This adjustment is generally an open-loop control without setpoint processing. This type of arrangement is inadequate for dealing with the increases in flow rate caused by aerodynamic pressures at high speeds. Here, a special control function can compensate by responding to increasing vehicle speeds, initially by reducing blower speed to zero, and then, should the flow continue to rise, by using a restriction flap to throttle the stream of incoming air. Either manual control, programmed-control, or fully automatic control is used to distribute the air to the three levels – defroster, upper compartment and footwell. Especially popular are units featuring program control buttons in which each button selects a specific distribution pattern for the three levels.

The defroster ("DEF") represents a special case. In order to clear the windows of misting or ice as quickly as possible, the temperature control must switch to maximum heat and maximum blower speed, and the air must be directed through the upper defroster outlets. In the case of program switches and fully automatic units, this mode is selected with a single button. At temperatures above 0 °C, the air-conditioner is also switched on to extract humidity from the air. After cold starts in winter, in order to avoid the still unheated air causing drafts, the blower is switched off electronically until the coolant has reached mid-temperature. This blower switch-off does not apply to the "DEF" and cooling modes. Whereas, the above variations apply to passenger cars and trucks, climate control is particularly complicated in buses. Here, the inside of the bus must be divided into a number of control zones in which the temperature is controlled individually by regulating the speed of the particular zone's water pump.

Fig. 6

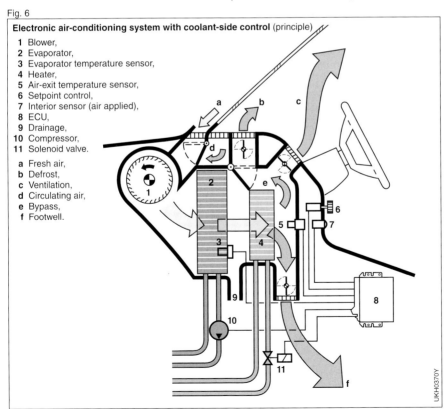

Electronic air-conditioning system with coolant-side control (principle)

 1 Blower,
 2 Evaporator,
 3 Evaporator temperature sensor,
 4 Heater,
 5 Air-exit temperature sensor,
 6 Setpoint control,
 7 Interior sensor (air applied),
 8 ECU,
 9 Drainage,
10 Compressor,
11 Solenoid valve.

 a Fresh air,
 b Defrost,
 c Ventilation,
 d Circulating air,
 e Bypass,
 f Footwell.

Information systems

Vehicle navigation systems

Assignment

It is often difficult for the driver to find his (her) way when in unfamiliar surroundings or when a diversion must be followed. In their search for the correct route to take, they are distracted from the traffic around them, or they are forced to stop and consult a road map. Vehicle navigation systems help the driver when he or she is in unfamiliar surroundings. Possibilities range from simple orientation aids to guidance systems with automatic route selection. Taking into account the traffic-jam situation, these systems recommend the optimum route, and if the driver leaves the suggested route or is forced to avoid a traffic jam that has not yet been reported to the system, they work out a new route. These systems are at present in the development stage under the general heading of "traffic telematics" and they require an infrastructure such as RDS/TMC[1]).

The driver is provided with the information in an easily understandable form and in good time, so that he (she) is hardly distracted at all from the traffic situation and has plenty of time to react. This rules out the risk of danger to other road users.

The most important task of these systems is to define the actual position of the vehicle (self-localization), since this is the prerequisite for determining the precise information for the remaining route.

[1] RDS Radio Data System
TMC Traffic Message Channel

Design

To determine the vehicle's current position, all systems are equipped with a localization device together with the appropriate sensors; a destination input device with priority and optimization criteria; and an output unit which informs the driver of the favorable routes to take to his (her) destination. The type and number of components depends upon the configuration of the particular system which can be extended by the addition of special units. These include, for instance, a data memory for digital road maps.

Method of operation

In order to define their own position, localization devices continually compare the stored road network with the vehicle's movements and automatically correct any deviations. Systems with sporadic position calibration were not successful on the market. Modern installations are equipped with the components as described below and, taking the Bosch TravelPilot as an example, have a localization accuracy of $\pm$ 5 meters.

Wheel-speed sensors

Wheel-speed sensors register the rotation of the wheels on an axle. This data is used cyclically to calculate distance travelled, and changes in direction as calculated from the differences in the rotation between the inside-curve and outside-curve wheels. In order that the installation doesn't "get lost" during stop-and-go or when maneuvering, these sensors must be capable of registering movements which take place at even the lowest speeds.

Geomagnetic sensors

Such sensors comprise an annular core in which an excitation winding generates a triangular alternating field upon which the earth's constant magnetic field is superimposed. The voltage pulses which are induced in the two sensor coils permit definition of the horizontal component of the earth's magnetic field (flux-gate principle), and thus enable the vehicle's own position to be calculated.

Before the navigation system is commissioned, it is necessary to determine the effects on the magnetic field due to iron masses and load currents in order that the ECU can compensate for these in its calculations.

Satellite positioning system

Alternatively, the receivers for the satellite positioning system GPS[2]) can use additional sensors, or sensors which operate in parallel.

The combination of wheel-speed sensors, geomagnetic sensor, and GPS, is increasingly being superseded by speedometer signals, yaw sensors, and GPS.

Destination selection

Destination input using coordinates is a complicated matter and is therefore uncommon. With systems featuring a large enough memory for name and street directories, the driver can directly input postal addresses. Conversion to the coordinates as required for localization is performed by the system itself. When selecting the route to be followed, the system automatically takes such auxiliary information as one-way and limited-access roads into account, as well as the information contained in previously received traffic-jam reports or in such reports which are currently coming in.

On sophisticated systems, the possibility of selecting specific destinations such as workshops, gas and filling stations, restaurants, hotels, or points of touristic interest, increases their usefulness considerably.

Road-map memory

The road-map memory must have sufficient capacity to provide all the data required within a vehicle's operational radius. A change of storage medium should seldom be necessary and when it is this should be an easy matter. The CD-ROM has a storage capacity of approx. 650 MB and can store the complete German road network together with all auxiliary roads. Due to the laser head's mechanical movement access can take seconds (Fig. 1).

Route calculations

Systems which have detailed road data at their disposal apply these to correct the localization and also to calculate a route. Here, the data must be extended by the addition of the time taken to drive through given sections of the road, under bridges, through one-way and limited-access roads. Route calculation must be fast enough for the driver to receive a route recommendation before the next crossing if he (she) failed to follow the last route instructions.

Fig. 1
Monitor of the "TravelPilot" guidance system
Display of the route using the road map, and other information.

2) GPS   Global Positioning System

Route and direction recommendations

The recommendation on the route to follow, or the direction to take, results from the computer comparing the actual position and the calculated route. Simple, less sophisticated systems merely indicate the linear distance and the linear direction. Without having an overview of the local road network at his/her disposal, the driver must then interpret this information when working out the route to take.

Such navigation components as the "TravelPilot RG05" which fits the Blaupunkt "Berlin" car radio, or the stand alone system either display a calculated route in color on the screen (Fig. 1), or output acoustical or visual recommendations on how the destination can be reached. The display on the screen is reduced to information which can be understood at a glance, that is the distance to the next junction and the direction to take upon arriving there. Acoustical outputs are only given when the driver is recommended to turn off. All in all, this reduces the distraction for the driver to a minimum.

Vehicle information systems

Assignment

In addition to the display and operator elements for the mainly vehicle-related applications, more and more information, communication, and comfort and convenience functions are being introduced in the vehicle. A car radio is practically standard nowadays, and telephone, and navigation systems etc. are following this trend. Each of these additional functions would need its own display and special operating elements, as well as having a different operating method. This multiplicity would burden the driver with even more work and, if only with respect to traffic safety, is no longer in line with modern-day requirements.

It is the job of the vehicle information system therefore to provide the driver with one standard "user environment" for a variety of different applications.

Fig. 2

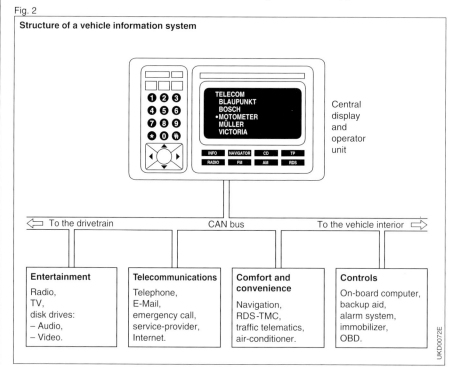

Structure of a vehicle information system

Entertainment	Telecommunications	Comfort and convenience	Controls
Radio, TV, disk drives: – Audio, – Video.	Telephone, E-Mail, emergency call, service-provider, Internet.	Navigation, RDS-TMC, traffic telematics, air-conditioner.	On-board computer, backup aid, alarm system, immobilizer, OBD.

UKD0072E

Design

The driver information system combines the display and operation of a number of functions in a single central display and operator unit. All in all, the total number of input and output elements has been considerably reduced, and more attention can be paid to ergonomic considerations when locating them in the vehicle.

This information system also permits the vehicle instruments (notwithstanding additional functions) to be more clearly arranged for the driver, a point which in the long run certainly contributes to traffic safety. For control and presentation of the required information on the display, the display and operator unit uses a bus system (e.g. CAN) for the bidirectional exchange of information with the connected components

Method of operation

Input
Preferably, the input of the most important operator functions takes place through an input element which is situated within easy reach of the driver and which the driver can find "blindfold". A practical method is to use operator elements on the steering wheel. Safety considerations dictate that extensive assignments such as the drawing up of a quick-dial telephone directory only take place with the vehicle at standstill.

Information outputs
The central display serves to present the most varied forms of visual information such as text, still pictures/figures, and video recordings.

Information which is important for the driver during the journey (e.g. the name of the traffic-information transmitter currently being received, or the direction arrow needed for navigation instructions) is easily implemented in the form of an instrument-cluster display. A voice-output facility can be used in addition to the optical display.

Outlook

In future, a voice-input facility for the various system functions will serve to further relieve the driver.

Parking systems

Assignment

Many modern-day automotive designs severely restrict rearward vision, and frequently make it difficult or even impossible to discern obstacles behind the vehicle. Particularly in the case of low boundary stones or vehicles with very low hoods, it is often impossible for the driver to estimate the actual distance between his/her vehicle and the obstacle. The result is that the available parking space is not utilized efficiently. Parking aids and backup aids assist the driver by indicating the distance to the obstacle. The driver is informed either by means of a signal or using a direct display.

Design

Parking aids are available with only backup monitoring or with all-round monitoring. The difference is in the number and configuration of the sensors, and in

Fig. 3

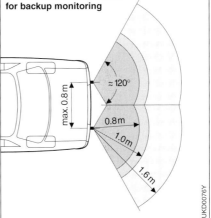

Sensing range of the ultrasonic sensors as used for backup monitoring

max. 0.8 m

≈ 120°

0.8 m

1.0 m

1.6 m

UKD0076Y

the display unit (LCD with digital display, LED or lamps, and in some cases also audible signalling). The ultrasonic sensors are connected to an ECU and incorporate an ultrasonic transmitter and receiver. The ECU provides the power supply for the sensors, as well as triggering them and evaluating the received signals in order to ascertain the distance to the obstacle. The ECU also informs the display unit of the results.

Method of operation

Measuring principle
The system operates using the echo depth-sounding principle. The sensors are periodically triggered one after the other and then transmit 30 kHz ultrasonic signals. The sensors then switch over to receive the signals reflected back from the obstacle. The echo signals' transit times permit calculation of the distance to the obstacle, and the definition of its location with respect to the vehicle.

Function
Following installation, the system is calibrated to the installation conditions using a calibration mode. The system switches on along with the ignition and starts a self-test which checks all sensor functions and all displays (by switching them on). The ready display indicates that the system is operating correctly. A warning is given in case of system fault (e.g. broken cable), or acoustic interference resulting for instance from a roadworks hammer.

Back-up monitoring system (Fig. 3)
In contrast to the all-round monitoring system, the back-up monitoring system is only switched on when reverse gear is engaged. The system is not in operation when forward gears are engaged.

All-round monitoring (Fig. 4)
With all-round monitoring, the front sensors are permanently switched on at speeds below 15 km/h (approx. 10 mph), and the rear sensors only when reverse gear is engaged. To avoid repeated warnings in stop-and-go traffic, all sensors can be switched off by hand.

Components

Ultrasonic sensors
Special sensors were developed for the parking-aid systems which are easily installed in the bodywork or bumper, and which above all are characterised by flush-fitting installation. The integration of the sensor's circuitry in an IC permits very compact design so that not only maximum precision and reliability are possible for this application, but also a reduction in the outlay required for wiring. The ultrasonic sensors are suitable for detection ranges up to max. 5 m (approx. 15 ft).

Design
The sensor is in the form of an aluminum housing with a piezo-electric wafer as wave generator and receiver, and all the electronic circuitry for the generation of

Fig. 4

Sensing range of the parking system with all-round monitoring facility

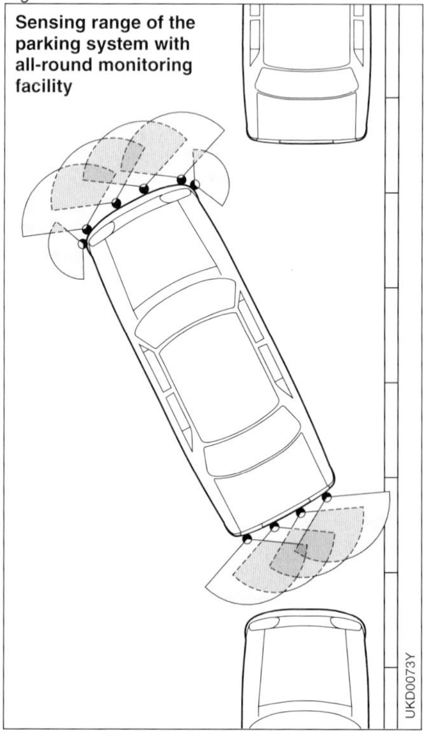

UKD0073Y

the ultrasonic waves and evaluation of the reflected waves. The transmitted levels correspond to the logic-circuit voltage and are therefore insensitive to disturbance so that it is unnecessary to screen the lines to the ECU.

Transmission and receive characteristics
Stipulations for parking-aid devices demand that in addition to a vertical area (approx. 50°), above all a wide horizontal area (approx. 120°) is scanned. This limits the monitoring distance to 2 m, and means that only four sensors are needed for almost complete scanning of the area to the side of the vehicle. "Complete" protection is provided by a system using 4 sensors for the rear area and 6 for the front (providing protection for the front corners, Fig. 4).

ECU

At the center of the ECU is a microprocessor whose amplified signals are used to control the sensors. Via a matching circuit, this microprocessor receives the sensors' echo signals and evaluates them. In addition to calculating the distance to obstacles and the management of the transmit and receive functions, the microprocessor also monitors all the system components. All errors, malfunctions, and disturbances which are detected are retained in a non-volatile data storage so that they can be read out and analyzed in the workshop. The ECU also provides the sensors and the display unit with power and, in addition to the signal amplifiers, also contains the circuitry needed for controlling the warning and display elements.

Warning and display elements

The driver must be informed in a suitable manner of the distance to the front or rear obstacle. Visual displays (lamps, LED, LCD) and audible warnings can be used to signal the distance to the obstacle. Since the parking aid does not absolve drivers from their obligation to take due care when reversing, he/she must concentrate their attention on what is going on behind the vehicle so that

they cannot continually observe the visual display in the instrument panel. Combinations of visual and audible warnings are therefore the ideal solution.

Visual displays
The visual display informs the driver of the operating state and directly indicates the distance to the obstacle. The most practical method uses a bar diagram or chart on an LCD display. For instance, a green segment lights up in the so-called distant sector (> 1 m), a yellow segment for every 5 cm in the warning sector ($\leq$ 1 m), and a red segment in the critical sector (< 30 cm).

Audible warning
In the warning sector, that is with distances of less than 1 m, an audible signal is given which warns unmistakably of the approaching obstacle. Common practice is to use an interrupted tone, whereby the nearer the vehicle gets to the object the shorter are the interruptions, so that the tone becomes continuous as soon as the vehicle enters the critical sector.
The distance sectors can be signalled adequately enough by suitable combination of the functions of a single light unit and a single tone generator.

Calculations of distance

Every 25...30 ms, the sensors are triggered one after the other for about 150 μs and – due to post-pulse oscillations – transmit an ultrasonic pulse of approx. 1 ms duration. All sensors then switch to receive so that they can "listen" to the reflected waves. The distance to obstacles can be calculated from the wave transit time.
In case of wide obstacles (e.g. a vehicle or a wall), the shortest measured distance corresponds to the actual distance. But even with individual obstacles such as a lamppost or a boundary stone, the fact that all sensors "listen" means that the system can reliably calculate the distance between the obstacle and the vehicle bumper.

Occupant safety systems

Vehicle safety

Active and passive safety

Active safety systems help to prevent accidents and therefore make a contribution to safety on the roads. The Bosch ABS antilock braking system is an example of active driving safety. In critical driving situations, ABS not only keeps the vehicle stable but also keeps it steerable. Passive safety systems, also known as passive restraint systems, serve to protect the vehicle's occupants against severe injury. In the first place, they reduce the danger of injury, as well as the severity of the accident injuries which sometimes occur nevertheless. The airbag is an example of passive safety. It protects the vehicle's occupants when an accident cannot be avoided.

Occupant restraint systems

Seat belts and seat-belt tighteners

Task

It is the job of the seat belts to hold the occupants in their seats when the vehicle hits an obstacle. With three-point seat belts, the tighteners improve the belts' restraining capabilities and protect the occupants against injury. In case of a head-on collision they pull the seat belt tighter around the upper part of the body and press it firmly against the seat's backrest. This prevents excessive forward displacement of the body caused by mass inertia (Fig. 1).

Fig. 1

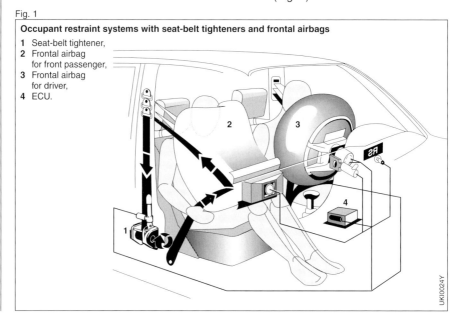

Occupant restraint systems with seat-belt tighteners and frontal airbags

1 Seat-belt tightener,
2 Frontal airbag
   for front passenger,
3 Frontal airbag
   for driver,
4 ECU.

UKI0024Y

Method of operation

On mechanical three-point seat belts, a fast-acting lock blocks the belt reel as soon as a given vehicle retardation is reached. In case of a head-on collision with a fixed obstacle at 50 km/h, the seat belts must absorb a level of energy which is comparable to the kinetic energy of a person in free fall from the 4th floor of a building.

Due to the slackening of the belt, the belt stretch, and the delay inherent in the seat-belt retractor (the so-called "film-reel effect"), in case of a head-on collision with a fixed obstacle at above 40 km/h, a three-point seat belt can only provide limited protection and cannot reliably prevent head and body impacting the steering wheel and/or instrument panel. As can be seen from Fig. 2, a body without restraint system moves forward a considerable distance when a collision takes place.

When a collision occurs, the shoulder-belt tightener compensates for "belt slack" and "film-reel effect" by retracting and tightening the belt strap. In case of a collision at 50 km/h, this system reaches full effectiveness within the first 20 ms following impact, and serves to supplement the airbag which will have inflated completely after about 40 ms. The body then moves forward slightly and is cushioned by the airbag which is in the process of emptying. In this manner, the occupants are protected against excessive forward movement.

To achieve maximum protection, it is necessary that the vehicle's occupants decelerate along with the vehicle after having moved forward out of their seats by a minimal amount. This is achieved by triggering the belt tightener immediately upon initial impact to ensure that effective restraint for the front-seat passengers starts as soon as possible. The maximum forward displacement with tightened seat belts should be approximately 1 cm, and the mechanical tensioning process should last 5...12 ms. Upon vehicle impact, a pyrotechnic propellant charge is ignited electrically. The rapid increase in pressure resulting from the explosive process is applied to a plunger which, via a steel cable, rotates the belt-tightener's reel so that the belt holds the upper part of the body tightly (Fig. 3).

Versions

In addition to the shoulder-belt tighteners which rotate the belt reel to tighten the belt strap, there are versions available which

Fig. 2

Retardation till standstill and forward displacement of a body at an impact speed of 50 km/h (approx. 30 mph)

1 Impact, **2** Ignition belt-tightener/airbag,
3 Belt tightened, **4** Airbag inflated.
– – – – without/ —— with restraint systems.

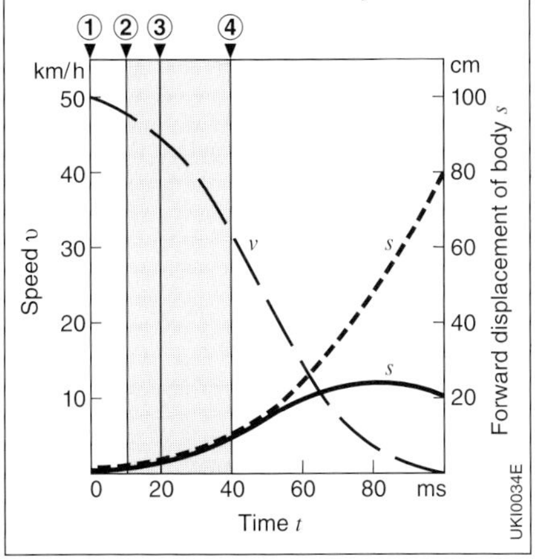

Fig. 3

Shoulder-belt tightener

1 Sensor connection,
2 Igniter squib,
3 Propellant charge,
4 Piston,
5 Cylinder,
6 Wire rope,
7 Belt reel,
8 Belt strap.

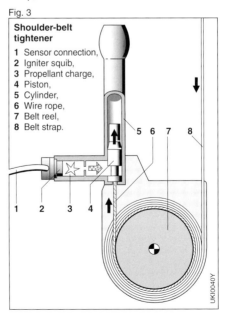

pull the belt buckle to the rear (buckle tightener) and thus tighten shoulder belt and lap belt simultaneously. This form of belt tightening using the buckle tightener not only improves the restraining effect but also provides enhanced protection against "submarining" (in which occupants slide forward beneath the lap belt). In both systems, the belt-tightening times are the same.

Frontal airbag

Assignment

The driver and front-seat passenger are each allocated their own airbag whose job it is to protect them against head and chest injuries when the vehicle hits a solid obstacle at speeds up to 60 km/h. Should two vehicles collide head-on, the airbags provide protection at relative speeds of up to 100 km/h. In case of a severe collision, a belt tightener alone will not prevent the driver's head smashing against the steering wheel. Depending upon the type of airbag concerned, its installation position, and vehicle type, airbags feature different fill quantities and pressure-build-up curves which are matched to the vehicle conditions.

Method of operation

Upon vehicle collision (as detected by a special sensor), the driver and front passenger are protected by their respective airbags being inflated in a sudden, explosive process by their individual pyrotechnic gas generator. To provide maximum protection, the airbag must be completely inflated before the vehicle occupant comes into contact with it (Fig. 4). The airbag responds to the upper body hitting it by deflating partially and, by means of non-critical surface pressures and decelerative forces, "gently" absorbing the energy with which the body makes contact with it. This procedure prevents head and chest injuries – or at least greatly reduces their probability and severity.

Fig. 4
Explosion-like deployment of a driver airbag
(Source: DaimlerChrysler)

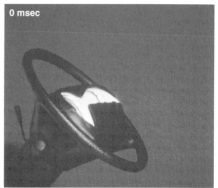

0 msec

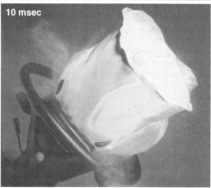

10 msec

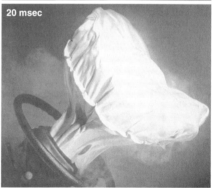

20 msec

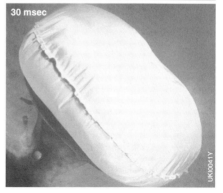

30 msec

Maximum forward displacement on the driver's side before the airbag has inflated fully is 12.5 cm, a figure which corresponds to approx. 40 ms after initial impact against a solid obstacle at 50 km/h. 80 ... 100 ms after the airbag has inflated fully, it starts to deflate again through the outlet openings and the porous textiles used in its manufacture. The whole procedure doesn't take more than about 1/10 of a second.

Detection of impact:
In case of frontal, offset, or oblique collision/impact, the optimum occupant protection is afforded by the coordinated interplay between the pyrotechnical, electrically fired frontal airbags and the seat-belt tighteners. In most cases, in order to maximise the effectiveness of the protective devices, a common ECU (triggering device) fitted in the passenger compartment activates them at the correct instant in time and in the correct order. Using either one or two acceleration sensors, the ECU measures the deceleration due to the collision and uses this to calculate the change in speed. Since the airbag must not be deployed due to a hammer blow in the workshop or a slight bump, or when the car is driven over a curbstone or pothole in the road, it is imperative that the collision itself is evaluated. To do so, the sensor signals are processed using evaluation algorithms whose sensitivity parameters have been optimised using crash-data simulation. Depending upon the type of collision, the triggering threshold is reached in 5...50 ms. The acceleration curve is different for every vehicle due to the differences in its equipment and in the way its bodywork reacts in case of an accident. This curve defines the adjustment parameters which are decisive for the sensitivity of the release algorithm (computational process) and in the long run for the deployment of the airbags. Depending upon the manufacturer's production concept, the ECU can be EoL (End of Line) programmed with data concerning the triggering parameters and the level of vehicle equipment.

Side airbag

Assignment
Side collisions account for about 20 % of total accidents. This means that the side collision is second only to the frontal, or head-on, collision as the most frequent form of collision. This is the reason for more and more vehicles being equipped with side airbags in addition to seat-belt tighteners and frontal airbags. Side airbags are intended to protect head and upper body. They are deployed from the roof cutout in the form of inflatable tubular systems, window bags, or inflatable curtains etc., and can also be deployed from the door or from the seat backrest (thorax bags). They are intended to "catch" the occupants gently and thus protect them against injury in case of a side collision.

Method of operation
Since there is no deformation zone available, and very little space between the occupant and the vehicle's side components, the correct timing of airbag deployment is extremely difficult. For instance, side-impact detection must take place within approx. 3 ms, and the inflation of the approx. 12 l thorax airbags must be completed within max. 10 ms. A standard for impact-detection systems and side-airbag triggering systems has not yet been approved.
Bosch though has two possibilities available for complying with the above-named requirements:
– Combined ECU which processes the input signals from peripheral acceleration sensors (mounted at a suitable point on the bodywork) which measure side impact, and which can then trigger the side airbags in addition to the seat-belt tighteners and the front airbags,
– Stand-alone ECU's which can trigger the side airbags independent of the ECU for the seat-belt tighteners and front airbags. These are termed "stand-alone sensing units" for side-impact detection.

Components

Acceleration sensors

The acceleration sensors for impact detection are integrated directly in the ECU (seat-belt tighteners, front airbag), or in addition at selected points on the right and left sides of the vehicle's bodywork (side airbags). Since it is vital that high-precision sensors are fitted, these acceleration sensors are of the surface micromechanical type constructed of fixed and movable finger structures and spring webs which form a "spring-mass system". Using a special process, this is applied to the surface of a silicon disk (wafer). Since these sensors feature a very low working capacitance (≈ 1 pF), their evaluation electronics must be installed in the sensor's housing in order to minimise the possibility of interference from the connecting lines.

ECU's for seat-belt tighteners and frontal airbags

According to the present state-of-the-art, the following functions are incorporated in the central ECU (airbag controller):

– Impact detection by means of acceleration sensor and safety switch, or by two acceleration sensors without safety switch (redundant fully electronic sensing).

– Correctly timed activation of the triggering circuits for airbags and seat-belt tighteners as a function of the type of impact (e.g. frontal, offset, or oblique, pole).

Here, the acceleration is detected at a central point in the passenger compartment and evaluated using a triggering algorithm.

– Voltage transformer and power stand-by in case the battery power supply should fail. Selective triggering of the seat-belt tighteners, depending upon the "belt-buckled?" enquiry: Triggering only takes place when the seat belt has been buckled.

– Setting of two airbag triggering thresholds, depending upon whether the occupant is wearing the seat belt or not.

– Adaptation to the vehicle's various characteristics (energy absorption of the vehicle forward structure, and the bodywork's vibration response).

– Diagnosis of functions and components (inside or outside the ECU).

– Storage in the crash recorder of types of fault and their durations. Parameter adjustment through the diagnosis interface.

– Triggering of a warning lamp.

Fig. 5

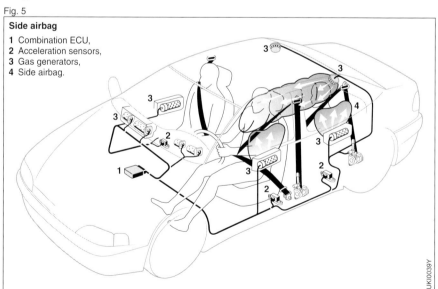

Side airbag

1 Combination ECU,
2 Acceleration sensors,
3 Gas generators,
4 Side airbag.

UKI0039Y

ECU's for seat-belt tighteners, frontal and side airbags

Combined ECU:

The combined ECU (Figs. 5 and 6) is located at a central point in the vehicle. It is based upon the previously described ECU for the seat-belt tighteners and the frontal airbag, although it is provided with additional final stages for triggering the side airbags. It uses an integrated sensor, comprising two sensor elements, for detecting the longitudinal and lateral acceleration. In addition (Fig. 5), there are sensors mounted on the ends of the seat cross-members or on the B-pillars, or four peripheral acceleration sensors on the bodywork's B and C pillars. These sensors detect the lateral-acceleration signals, evaluate them and send the triggering signal to the central ECU through a digital interface. In addition to the frontal airbags and the seat-belt tighteners, the central ECU also triggers the side airbags as soon as the internal lateral-acceleration sensor has passed the plausibility control and has confirmed a side impact.

Independent ECU's for side airbags:

The vehicle is equipped with a number of independent side-airbag ECU's which can trigger the side airbags independent of the central ECU for the frontal airbags and seat-belt tighteners. These are located at suitable points on the left and right side of the vehicle (preferably in the B-pillars) and with their own driver stages they control the gas generators of the respective airbags.

Gas generators

The pyrotechnical propellant charges for the gas generators needed for inflating the airbags and for seat-belt tightening are activated by an electrically activated firing element.

Following firing, the driver's airbag (volume 35...67 l), which is integrated in the steering-wheel hub, is inflated with nitrogen by its own gas generator within approx. 30 ms. The front-passenger airbag (volume 70...150 l) is installed in the glove compartment and is inflated within 40...50 ms. Its longer inflation time is acceptable since the front-seat passenger is further away from the glove compartment than the driver is from the steering wheel. That means that the front-seat passenger's permissible forward displacement exceeds that of the driver.

To avoid inadvertent triggering in case a triggering element is shorted to the vehicle electrical-system voltage (for instance, due to faulty wiring-harness insulation),

Fig. 6

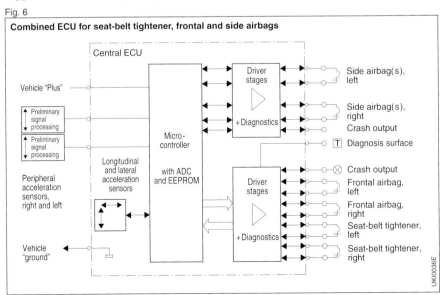

Combined ECU for seat-belt tightener, frontal and side airbags

Central ECU

Vehicle "Plus"

Preliminary signal processing

Preliminary signal processing

Peripheral acceleration sensors, right and left

Longitudinal and lateral acceleration sensors

Micro-controller with ADC and EEPROM

Driver stages

+Diagnostics

Driver stages

+Diagnostics

Side airbag(s), left

Side airbag(s), right

Crash output

Diagnosis surface

Crash output

Frontal airbag, left

Frontal airbag, right

Seat-belt tightener, left

Seat-belt tightener, right

Vehicle "ground"

UKI0036E

291

"AC firing" is applied using 100 kHz AC pulses. In the igniter circuit, there is a capacitor in the igniter plug which electrically isolates the igniter from DC. This isolation from the vehicle electrical supply prevents inadvertent airbag deployment. This also applies following a crash without airbag deployment, when the occupants must be cut out of the deformed passenger cell using rescue-service cutters.

Rollover protection systems for convertibles

Assignment
In case of a roll-over accident, open passenger cars (convertibles) and off-road vehicles are without the protection afforded by the roof construction of the fully closed sedan-type vehicle. Rollover protection systems have therefore been developed to protect the occupants of such open vehicles from injury in case of a roll-over accident.

Design
In some convertible-type cars, the windshield frame and a fixed rollover bar which is located behind the (rear) seats and spans the vehicle from one side to the other, form the rollover protection system. For classical convertibles without this fixed rollover bar, either a retractable rollover bar is located in the rear of the vehicle, or extendable rear head restraints are provided (Fig. 7). In the event of a rollover accident, these deploy to protect the occupants.

Method of operation
In case of vehicle rollover, the ECU triggers the high-speed deployment of the rollover protection devices. These passive safety systems are installed in a number of high-end convertible models and various off-road vehicles. They come into effect within approx. 150 ms and provide reliable rollover protection.

Fig. 7

The extendable head restraints "shoot out" during this vehicle rollover test
a Initiation of rollover,
b Head restraints are triggered,
c Rollover takes place,
d Vehicle lands on its wheels again.
(Source: DaimlerChrysler).

Triggering of the rollover protection

Since rollover is possible in each direction, the rollover detection must also be effective in all directions.

The various sensors detect the necessary data which the ECU then uses to calculate the probability of an impending rollover. Two criteria must be complied with in order for the danger of incipient rollover to be detected and for the triggering of the rollover protection:

– For the first triggering criterion, a sensor in the vehicle's longitudinal direction, and one its lateral direction detect the acceleration values. A microcomputer processes the sensor signals for all directions and compares the resulting acceleration with the programmed activation threshold of approx. $5\,g$. The system is triggered if this threshold is exceeded.

– For the second triggering criterion, the vehicle's tilt angle is evaluated by a tilt switch. As soon as this reaches or exceeds 27°, and at least one of the rear-wheel rebound sensors has outputted a signal to indicate that the rear axle has rebounded, the system is triggered. The microcomputer evaluates the second criterion independently of the first criterion in order to increase the functional reliability.

As soon as at least one of the two above criteria have been complied with and a pending rollover has been detected, the ECU gives the signal to trigger the rollover bar or the head restraints. A powerful solenoid then releases pretensioned springs which bring the protective devices into position. In other words, the rollover bar emerges from its recess, or the head restraints spring into the rollover position. At the same time, the central locking system unlocks all the doors.

Variants

Other forms of rollover detection use bubble (spirit-level principle) sensors located at an angle to the vehicle's longitudinal axis to measure the angle. In the transverse direction, the rollover protection system triggers as from an angle of 52° (roll angle) and in the longitudinal direction as from 72° (pitch angle).

A further sensor closes a spring-loaded reed contact when the vehicle looses contact with the ground.

Forecast

In future, on the so-called "intelligent airbag systems", in order to adapt the restraint effect (airbag pressure) to the accident situation, the frontal airbags will be deployed in two stages.
The inflation criteria are:
– Crash severity,
– Use of the seat belts,
– Occupant position and weight, seat position, and inclination of the backrest.

In such systems, a belt-force limiter is installed which reduces the danger of broken ribs or avoids it completely.

For instance, in case occupants are "out-of-position", the passenger-compartment sensors trigger only the first front-passenger stage, or even switch the airbag off completely if a rear-facing child seat is detected.

The CAN bus networking of the airbag control unit with other sensors and switches permits the evaluation of additional information so that all restraint devices can be triggered (time, stages) in line with actual requirements.

In order to reduce the size of the wiring-harness, notwithstanding the increase in the number of firing circuits, and to permit flexibility in varying the vehicle system, special bus systems were developed for occupant protection.

Improved concepts for the detection of pending rollover (side rollover) are being introduced to passsenger cars and commercial vehicles. These rely on the use of high-resolution sensors for measuring acceleration and angular velocity. To further improve the triggering function and the early detection of the type of crash (precrash detection), it is intended to use microwave radar sensors to register relative speed, spacing between vehicles, and impact angle in case of a head-on collision.

Driving-safety systems and drivetrain

Antilock braking system ABS

Assignment

In critical driving situations on slippery or wet roads, or when the driver suddenly hits the brakes (due to the appearance of an unexpected obstacle, or due to other road users making mistakes), without ABS the wheels can lockup during braking so that the vehicle is no longer steerable and goes into a skid. It can even leave the road. In such situations, ABS prevents wheel lockup so that the vehicle remains steerable and the risk of skidding is considerably reduced (Fig. 1).

Even in critical driving/braking situations such as panic or emergency braking, thanks to ABS it is still possible to take evasive action so that a collision can be avoided. For your personal safety though, it is nevertheless advisable to absolve an advanced safe-driving course.

Design

The antilock braking system (ABS) is comprised of the following components:

Wheel-speed sensors
The inductive wheel-speed sensors output their signals to the ECU which uses them to calculate the peripheral speeds of the wheels.

Electronic control unit (ECU)
Bosch only has 3 or 4-channel ABS systems in its program, since in comparison 2-channel systems have functional limitations.
The ECU shown in the block diagram (Fig. 2) belongs to a 4-channel ABS in-

stallation. It receives, filters, and amplifies the wheel-speed sensor signals from which it calculates the brake slip and the acceleration or deceleration of the individual wheels. For reliability reasons, it comprises 2 identical microcontrollers which are completely independent of each other. Each of the microcontrollers processes the information from 1 pair of wheels (channels 1+2 or 3+4), and executes the logical processes. Complex

Fig. 1
Braking effect with and without ABS
a Without ABS, the wheels lockup and the vehicle goes into a skid.
b With ABS, in case of panic braking the vehicle remains stable and steerable.

a
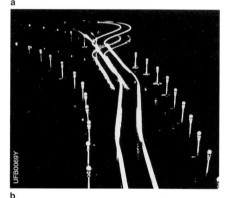

b
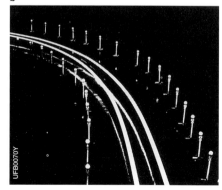

controller logic converts the control signals into control commands for the solenoid valves in the hydraulic modulator.

Hydraulic modulator

The hydraulic modulator transposes the ECU's control signals and, via the solenoid valves, controls the wheel-brake pressure for optimal braking, even if in an emergency the pressure selected by the driver is far higher. The hydraulic modulator is located between the ECU and the wheel-brake cylinders.

Wheel brakes

At the wheel brakes, the braking pressure transferred from the hydraulic modulator serves to force the brake linings against the brake drums, or in the case of disc brakes the brake pads against the brake discs.

Method of operation

When the driver hits the brakes hard in a panic situation, the ABS controls the braking pressure applied to the service-brake system. At the individual wheel-brake cylinders this takes place as a function of wheel-slip and wheel acceleration or deceleration. Developments in digital electronics were instrumental in permitting the complex procedures which are typical for the braking process to be monitored reliably so that it becomes possible to react in a fraction of a second to changing situations. This highly flexible system can be integrated without modifications being necessary in the existing basic braking system. It operates as follows:

Independent of the vehicle's actual driving status, with the ignition switched on, wheel-speed sensors at both front wheels and at the rear-axle differential (or on all 4 wheels) generate the signals which are passed on to the ECU for calculating the wheel's peripheral speed. If the ECU detects incipient lockup from the incoming signals, it triggers the recirculation pump inside the hydraulic module and the solenoid valves for the wheel(s) concerned. Each of the front wheel brakes is allocated a pair of so-

lenoid valves. These influence "their" brake so that, independent of the remaining brakes, it makes the best-possible contribution to the overall braking process (individual control). At the rear axle, the wheel with the lowest coefficient of slip between tire and road determines the joint pressure in each of the rear-wheel brakes (select-low principle). This leads to the wheel with the higher coefficient of slip being slightly under-braked during ABS. Although this leads to a slight increase in stopping distance, this is more than compensated for by the vehicle's increased stability. Each wheel's solenoid-valve pair can be switched to three different statuses by the ECU:

– In status 1 (both valves de-energized, inlet valve open, outlet valve closed), the master brake cylinder and the wheel-brake cylinder are connected and the wheel-brake pressure can increase.

Fig. 2

ECU (4-channel installation)

1 Wheel-speed sensors,
2 Diagnosis connection,
3 Battery,
4 Input circuit,
5 Digital electronic controller,
6 Microcontroller,
7 Non-volatile memory,
8 Voltage stabilizer/Error store,
9 Output circuits with driver stages,
10 Solenoid-valve pairs for compressed-air increase/reduction,
11 Relay,
12 Stabilized battery voltage,
13 Check lamp.

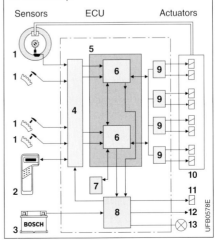

UFB0578E

– In status 2 (inlet valve energized and closed), the master cylinder is disconnected and the wheel-brake pressure remains constant.
– In status 3, the outlet valve is also energized so that it opens. This sets up a hydraulic connection between the wheel brake and the return pump so that the wheel-brake pressure drops.

This means that the increase or decrease of the braking pressure is not only continual, but can also be stepped by switched-mode operation so that moderate increase or decrease takes place.

Depending upon the road surface, 4...10 intervention cycles take place every second. The ABS achieves such a high reaction rate due to the electronic signal processing and the short response times. The ABS control loop (Fig. 3) comprises:

– The controlled system: The vehicle with wheel brake, wheel, and the coefficient of slip between tire and road,
– The disturbances: Road conditions, brake condition, vehicle loading, driving maneuver, and tires (e.g. insufficient pressure, state of the tread),
– The controller: Wheel-speed sensors and ABS-ECU,
– The controlled variables: Wheel peripheral acceleration or deceleration, and brake slip, all of which have been derived from the wheel speeds,
– The reference variables: The brake-pedal pressure (the driver's braking-pressure input),

Fig. 3

ABS control loop

1 Hydraulic modulator with solenoid valves,
2 Master cylinder, **3** Wheel-brake cylinder,
4 ECU, **5** Wheel-speed sensor.

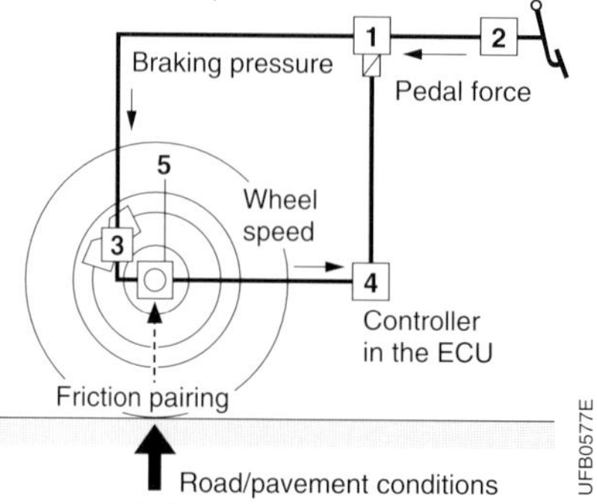

Braking pressure
Pedal force
5
Wheel speed
3
4
Controller in the ECU
Friction pairing
Road/pavement conditions

UFB0577E

– The manipulated variable: Braking pressure.

The processing of the individual controlled variables depends among other things on whether the wheels are coupled to the engine or not, and on the state of the road surface (high grip, or slippery). On certain ABS versions, the special conditions existing with all-wheel-drive vehicles are also taken into account, as are the effects of the yaw moment on small vehicles in particular (this occurs during braking on inhomogenous road surfaces).

Versions

The ABS 2S version of the ABS, in which the hydraulic modulator and the ECU are mounted separately, has proved itself in the field since 1978. The 3-channel hydraulic modulator for the front-axle/rear-axle configuration is comprised of a return pump (driven by an electric motor) and three 3/3 solenoid valves[1]. In the non-energized state these are switched to the "pressure increase" mode. The application of two defined voltage values switches them to either "pressure hold" or "pressure reduce". With front-axle/rear-axle configuration, a single solenoid valve controls both rear wheels, although two separate solenoid valves must be used with the diagonal configuration since each wheel is allocated to a different brake circuit. The ABS 5.0 is a further development of the ABS 2S version. Whereas the ABS 2S operates with 3/3 solenoid valves, the ABS 5.0 is equipped with 2/2 solenoid valves. With regard to the brake-circuit configuration, the ABS 5.0 and 2S are identical.

The ABS 5.3, in which hydraulic modulator and ECU are combined to form a single unit, was conceived for vehicles with "small-sized" braking systems. And although it has the same functional scope as the ABS 5.0 version, it is considerably smaller.

[1] The first number indicates the number of hydraulic connections, and the second the number of switch positions.

Traction control system TCS

Assignment

Critical situations (e.g. oversteering) not only occur when the brakes are applied, but also during drive-off and acceleration. This applies in particular on smooth/slippery roads on a gradient or when cornering. Such situations can make excessive demands on the driver and incorrect reactions can be the result.

Such situations can be mastered by the traction control system (TCS). This system brakes the driven wheel which shows a tendency to spin (or on all-wheel-drive vehicles, the driven wheels which show a tendency to spin), and/or adapts the engine torque to the drive torque which can be transferred to the road, so that the vehicle retains its stability.

The TCS is an extension of the antilock braking system (ABS). It relieves the driver of strain, and ensures that the vehicle remains steerable during acceleration (provided that physical limits have not been exceeded).

Design

Although TCS uses the same components as the ABS, some of these have had extra functions added (Fig. 4):

Wheel-speed sensors
The wheel-speed sensors provide the ECU with signals which enable it to calculate the peripheral wheel speeds.

ECU
The ABS electronic circuitry has been extended by the addition of a TCS stage. The same as with ABS, the signals from the wheel-speed sensors are applied to the ECU input circuits. Here, they are used to calculate the wheel slip of the individual wheels, and TCS intervention is initiated if the slip at one of the wheels is found to be excessive. The incoming signals are processed in two micro-controllers operating in parallel, and are converted in the output circuit into positioning commands for the solenoid valves and for the hydraulic-modulator supply pump which are responsible for implementing the brake-torque control.

Information for engine management (Motronic) is transferred to the ECU via an additional interface.

Fig. 4

Traction control system (TCS) with intervention at brakes and throttle valve

1 Wheel-speed sensors, **2** Wheel brakes, **3** ABS/TCS hydraulic modulator, **4** ABS/TCS electronic control unit, **5** Motronic ECU, **6** Throttle valve.

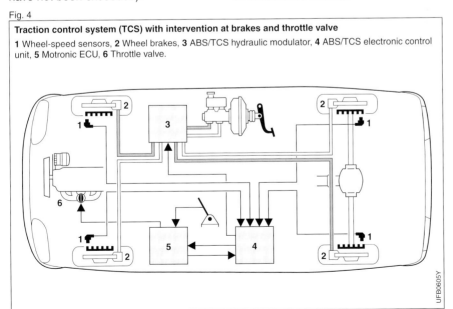

UFB0605Y

Hydraulic modulator

A TCS stage has been added to the ABS hydraulic modulator. It implements the orders from the ECU and, independent of the driver, controls the individual hydraulic pressures at the wheel brakes via the solenoid valves. During a single TCS control process, a supplementary pilot valve switches from the conventional braking mode to the TCS mode. The ABS return pump draws brake fluid from the master cylinder and generates the TCS system pressure. This means that braking pressure can be applied to the driven-wheel brake cylinders without any action on the part of the driver.

Wheel brakes

The braking pressure applied by the hydraulic modulator to the wheel-brake cylinders serves to force the brake linings against the brake drums or brake discs.

Method of operation

When the driven wheels show a tendency to slip, TCS controls the vehicle's propulsive force as a function of the wheel slip and the wheel acceleration or deceleration. The TCS is a highly flexible system and, without modifications being necessary, permits adaptation to the basic ABS-equipped braking system, and functions as follows:

When the vehicle is being driven, wheel-speed sensors at the 4 wheels generate signals which they pass to the ECU. When the driver presses the accelerator pedal the engine torque increases and along with it the drive torque, and provided that this can be transferred to the road completely, the vehicle can be accelerated without any problems. If though, the drive torque exceeds the torque which it is physically possible to transfer to the road surface, the speed of at least one of the driven wheels increases and it shows a tendency to spin. As a result, the tractive force which can be transferred decreases, and the vehicle can become instable due to the resulting loss of lateral guiding force. This

is where TCS intervenes and either controls the drive torque of the driven wheels, or brakes them, so that they no longer tend to spin and the vehicle remains stable.

In order that TCS can intervene independent of how far the driver has pressed the accelerator pedal, an electronic throttle control (otherwise known as "drive-by-wire" or ETC) must be fitted in place of the conventional mechanical linkage between the accelerator pedal and the IC-engine throttle valve (or diesel injection pump). Control commands from the TCS are allocated priority over the driver's wishes by the "drive-by-wire" system.

The accelerator-pedal sensor converts the pedal position into an electric signal. Taking into account the programmed parameters and the signals from other sensors (for instance, temperature, engine speed etc.), this is converted in the drive-by-wire ECU, or in a Motronic ECU, into a control voltage for an electric servomotor which actuates the IC engine's throttle valve, or the diesel injection pump's control lever, in order to influence the drive torque (Fig. 5).

Fig. 5

Electronic throttle control (ETC or drive-by-wire) for the TCS

1 Electronic control unit (ECU) for ABS/TCS,
2 Motronic ECU with ETC,
3 Accelerator-pedal sensor,
4 Servomotor,
5 Throttle valve (or diesel fuel-injection pump),
6 Wheel-speed sensor.
▬▬ Control circuit

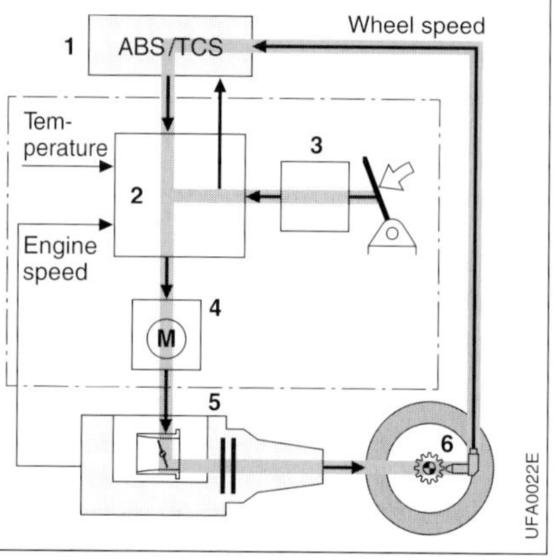

As soon as the ECU detects a serious deviation from the desired wheel speeds in the signals it has received from the sensors, the wheel that shows a tendency to spin is automatically braked without any action on the part of the driver. At the same time, the Motronic ECU, for example, intervenes through an electronic throttle-valve actuator in order to reduce the excess drive torque. The TCS controls the slip of the driven wheels to the optimum level. The braking action of the wheel with a tendency to spin is controlled by the braking-pressure modulation (increase pressure, hold pressure, reduce pressure) in the wheel-brake cylinders. Modulation takes place using the ABS valves and a number of other valves in the hydraulic modulator. On gasoline engines the drive torque is controlled by the ECU for ETC (drive-by-wire) or the ECU for Motronic with integrated ETC. The following components/systems are influenced in the process:
– Throttle-valve setting (adjusted by ETC),
– Ignition system (ignition timing adjusted by Motronic),
– Fuel-injection system (suppression of individual fuel-injection signals by Motronic).
On diesel-engine vehicles, the drive torque is adjusted through intervention at the injection pump's control lever (reduction of the injected fuel quantity). The TCS facility can be supplemented by an engine drag-torque control (MSR). When a change-down takes place, or the accelerator pedal is released suddenly on a slippery road surface, the engine's braking effect can lead to excessive brake slip. In such cases, the MSR briefly increases engine speed, and with it engine torque, so that the braking effect at the wheels is reduced to a level which still suffices for stability.

Versions

The version of the ABS/TCS unit depends upon the vehicle's engine (diesel or gasoline) and the type of

[1]) The first digit indicates the number of hydraulic connections, the second digit the number of switch positions.

intervention at brakes, throttle valve, ignition and/or fuel injection.
The TCS2 system comprises an ABS2 module with 3/3 solenoid valves[1]), and the TCS with an additional 3/3 pilot valve. The TCS5 family is a modular system based on the ABS5 system with 2/2 inlet and outlet valves, and it is supplemented by 2/2 intake and pilot valves.

Electronic stability program ESP

Assignment

In critical driving situations in the vehicle's longitudinal direction, ABS prevents wheel lock-up when the brakes are applied, and TCS prevents wheel-spin during drive-off. The electronic stability program ESP (otherwise known as vehicle dynamics control) also increases vehicle stability in the transverse direction so that the danger of skidding is drastically reduced. In limit situations it increases the directional stability and the vehicle remains on track when the driver slams on the brakes and during partial braking. This also applies when the vehicle is rolling freely, during engine-drag, and when the load is changed abruptly, and ESP is also able to cope with extreme steering situations (caused by shock and panic reactions on the part of the driver). Using additional sensors, the ESP registers the lateral acceleration, and the yaw behavior (yaw velocity) around a rotary axis perpendicular to the road surface (vertical axis). It processes the signals from these sensors in addition to the signals from ABS and TCS, and triggers the actuators in the hydraulic modulator accordingly. As a result, and this also applies to dynamic limit situations, the vehicle's stability is increased in straight-ahead driving as well as in bends, no matter whether retardation, braking or acceleration is taking place.
Vehicle steering stability depends on the extent to which the vehicle is able to precisely follow the particular course which is

defined by the steering-wheel angle, and whether the vehicle remains stable, that is neither "pushes out" when steering maneuvers take place nor becomes instable. To this end, the ESP controls the yaw velocity around the vehicle's vertical axis and the deviation of the driving direction from the vehicle's longitudinal axis. Figures 6 and 7 directly compare the courses taken by two vehicles when driving through a curve. Whereas the vehicle in Fig. 7 is equipped with ESP, the one in Figure 6 is not.

Design

In addition to the components which are common to the ABS and TCS, the ESP also requires a number of other sensors in order to register yaw velocity, steering-wheel angle, lateral acceleration, and brake pressure. The ESP (Fig. 8) components are:

- ECU (higher-level ESP, subordinate ABS and TCS),
- Hydraulic modulator (same as ABS/TCS),
- Wheel-speed sensors (same as ABS/TCS),
- Yaw sensor,
- Steering-wheel angle sensor,
- Lateral-acceleration sensor, and
- Brake-pressure sensor.

The two additional sensors for the lateral acceleration and for the yaw rate, together with the considerably expanded data processing in the vehicle, characterise the ESP system.

Fig. 6

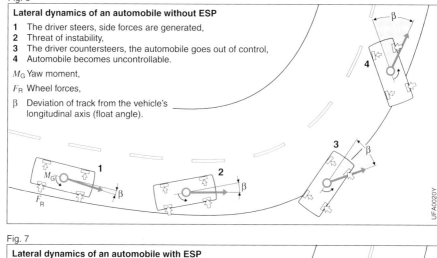

Lateral dynamics of an automobile without ESP

1　The driver steers, side forces are generated,
2　Threat of instability,
3　The driver countersteers, the automobile goes out of control,
4　Automobile becomes uncontrollable.

M_G Yaw moment,

F_R Wheel forces,

β　Deviation of track from the vehicle's longitudinal axis (float angle).

Fig. 7

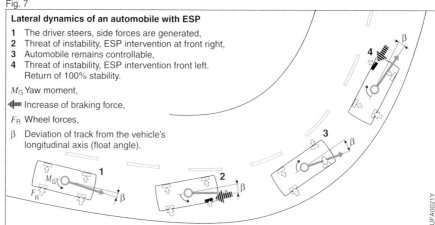

Lateral dynamics of an automobile with ESP

1　The driver steers, side forces are generated,
2　Threat of instability, ESP intervention at front right,
3　Automobile remains controllable,
4　Threat of instability, ESP intervention front left. Return of 100% stability.

M_G Yaw moment,

〰 Increase of braking force,

F_R Wheel forces,

β　Deviation of track from the vehicle's longitudinal axis (float angle).

Method of operation

Similar to ABS and TCS, the ESP also operates in dependent relationship to the tire forces (with ESP, the lateral tire force also plays a role), and to the tire slip.

In the limit situations defined by the dynamics of vehicle movement, the closed-loop control of the vehicle's behavior is intended to counteract the effects of "dangerous" accelerations. This must be implemented in a manner which complies with the driver's wishes and with the road surface. To do so, first of all the desired vehicle behavior in line with the driver's wishes must be ascertained (the desired status) and then the actual vehicle behavior (actual status). To reduce the difference between desired and actual status (= control deviation), the tire forces must be influenced in a suitable manner using actuators.

The vehicle becomes an integral part of the control loop. The braking and tractive forces at the individual wheels are set individually in line with the situation, so that the required handling is achieved to a great extent.

Sensors in the ESP control loop pick-up the quantities that need to be measured for ESP operation, and these are con-verted in the ECU into control com-mands which the hydraulic modulator and engine management (Motronic) imple-ment in the form of closed-loop control of the braking and drive torques. The run through the control loop takes place cyclically, and the higher-priority ESP stipulates the desired values for the subordinate slip controller in the form of a desired slip.

The desired behavior is defined using the signals which describe the driver's wishes, that is, those from the steering-wheel sensor (desired direction of vehicle travel), brake-pressure sensor (desired deceleration), and engine management (desired acceleration). The coefficient of slip between tires and road, as well as the vehicle road speed, are also taken into account when the desired behaviour is calculated. These are generated using the signals received from the wheel-speed sensors, and from the sensors for lateral acceleration, yaw velocity, and pressure. Depending upon the control deviation, the yaw moment is then calculated which is required to adjust the actual-state variable to the desired-state variable. To generate this nominal yaw moment, the ESP calculates the changes in the desired slip at the wheels concerned. The subordinate

Fig. 8

Overall ESP control system (installation points of the components)

1 Wheel brakes,
2 Wheel-speed sensors,
3 ECU,
4 Precharge pump,
5 Steering-wheel sensor,
6 Brake booster with master cylinder,
7 Hydraulic modulator with brake-pressure sensor,
8 Yaw sensor with lateral-acceleration sensor.

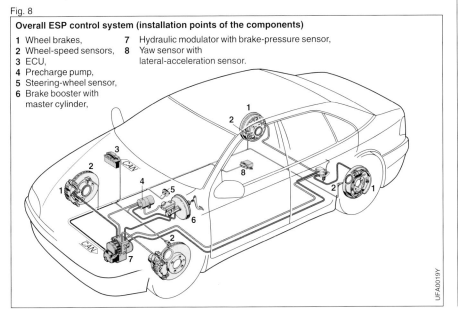

UFA0019Y

ABS and TCS, together with the "hydraulic modulator" and "engine management" actuators, adjust the required changes in slip by applying braking and tractive forces. In the process, the controller changes the longitudinal and lateral forces at each wheel.

Electronic transmission-shift control

Assignment

The automobile's drivetrain is responsible for providing the traction force and pushing power needed for propulsion in accordance with the equilibrium between drive and tractive resistance. This can be implemented in a great number of different ways by selected electronic assistance. Today, electronic controls for clutches and gearboxes make "drivetrain management" a practical proposition.

The electronic control of automatic gearboxes is primarily responsible for the selection of the most favorable gear and for the control of the actual gear shift. Gear selection depends upon a number of different parameters. One of the important parameters is engine torque, which in the case of Motronic engine-management systems can be made available directly (in Nm) for gear selection. In other systems, the torque is derived separately from other parameters (e.g. throttle-valve setting or air mass).

Further factors such as acceleration or deceleration, change up or change down, and vehicle road speed are important factors when the precise moment for applying pressure to the clutch via the electrohydraulic valves must be defined so that optimum gearshifts can be made.
The resulting advantages are:
– The possibility of choosing between several gearshift programs,

– More sophisticated gearshift,
– Flexible adaptation to various vehicle types, simplified hydraulic control, and
– Self-learning adaptation of the gear-ratio selection to the particular driving style and the traffic situation.

In addition, monitoring circuits are incorporated to protect the gearbox against damage resulting from faulty operation. In case of electrical-system malfunction, the system switches to a safe-running mode with limp-home function.

Design and method of operation

Sensors
A number of sensors are used to detect the following input parameters:
– Input and output speeds at the gearbox,
– Load status and engine speed,
– Selector-lever setting, and
– Position of the kick-down switch.

ECU
The ECU (Fig. 9), processes the sensor signals in accordance with a special program and calculates the output variables for the gearbox actuators:
1. Selection of the suitable gear.
2. Time available until pressure application.

Actuators
The interface between the electronics and the hydraulics is in the form of electrohydraulic valves (solenoid valves) and pressure regulators on the gearbox. With regard to their interaction, these can be regarded as an electrohydraulic converter.

The solenoid valves convert the positioning commands received as electrical pulses:
– Couplings in the gearbox are opened or closed,
– Pressure regulators are put into operation for the precise adjustment of the hydraulic pressure needed for the coupling/clutching operation.

Control ranges

Shifting-point control

This system shifts to the appropriate gear as a function of gearbox output speed and engine load. The shift to the "new" gear takes place when the solenoid valves are triggered.

So-called "intelligent" gearshift programs enable a whole range of other parameters to be processed in addition to those covered by the standard systems, and contribute to improving the vehicle's drive-ability. These parameters include:

– Longitudinal acceleration,
– Lateral acceleration,
– Rate with which the accelerator pedal is pressed, or
– Brake-pedal operation.

A complex control program is used to select the gear which is optimally suited to the particular driving situation, and to the operator's driving habits. Apart from suppressing upshifts when the accelerator pedal is abruptly released before a bend, and preventing gearshifts in the curve, this program can also se-lect a gear-shift program which shifts up at the lowest-possible engine speeds ("Economy Program"). It is also possible to have a manual input to the gearbox through the selector lever.

Torque-converter lockup

A mechanical lockup can eliminate the torque converter's slip and thus improve the gearbox efficiency. The converter clutch is triggered as a function of the engine load, the gearbox output speed, and gear change up or down.

Gearshift quality control

The quality of the gearshift is decisively influenced by the precise adaptation of the pressure in the gearbox clutches/couplings to the torque which is to be transferred (calculated from the load status and the engine speed). This pressure is adjusted by a pressure regulator. The brief reduction of engine output torque (for instance by retarding the ignition) at the instant of gearshift makes a further contribution to improving gearshift quality. This measure reduces the losses at the clutches/couplings and increases their useful life.

Fig. 9

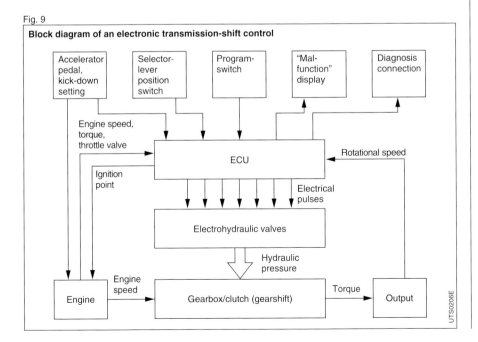

Block diagram of an electronic transmission-shift control

Gasoline-engine management

Technical requirements

The fuel supply and the ignition are what get the engine "running". Gasoline injection has in the meantime a 100-year history, but at that time due to the introduction of the carburetor it did not come to the forefront in automotive applications. It was not competitive enough considering the state-of-the-art in those days.

But things have changed drastically since then. In 1951 the first Bosch series-production direct-injection system was installed in a subcompact car, and in the following years progress pushed ahead with development of mechanical injection pumps. Although at the beginning, the sole emphasis was upon increasing the engine's output power, today other factors such as the necessity of complying with emissions-control legislation play an equally important role. The carburetor is incapable of complying with the legally· stipulated exhaust-gas limits.

Conventional ignition systems are also unable to meet the stipulations for engine power and exhaust-gas emissions. The mechanical (flyweight) ignition timing had to be superseded by electronic systems. The gasoline injection system injects the fuel into the intake manifold and onto the intake valves. The resulting air/fuel (A/F) mixture is entrained into the cylinder by the downward-going piston when the intake valve opens during the induction stroke. During the compression stroke it is compressed by the upwards movement of the piston and ignited at the ignition point by a spark at the spark plug. The resulting combustion energy forces the piston downwards and the conrod converts the piston's linear motion into rotary crankshaft motion.

Originally, the two individual subsystems "gasoline injection" and "ignition" controlled the individual parameters such as injected fuel quantity and ignition point completely independently of each other. Exchange of information between the two subsystems was either totally impossible or only possible to a very limited extent. This meant that the in part mutually contradictive requirements from each of these systems could only be implemented inside the respective system, but not in "system-overlapping" form. This ceased to be a problem when Bosch combined gasoline injection and ignition in a single system.

The combination of gasoline injection and ignition in the "Motronic" engine-management system enabled the optimisation of the control parameters for injection and ignition while taking into account the

Figure 1

Bosch gasoline-injection systems and ignition systems: History of developments.	
Gasoline-injection systems:	
D-Jetronic	1967 – 1979
K-Jetronic	1973 – 1995
L-Jetronic	1973 – 1996
LH-Jetronic	1981 – 1998
KE-Jetronic	1982 – 1996
Mono-Jetronic	1987 – 1997
Ignition systems:	
Coil ignition (CI)	1934 – 1986
Transistorized ignition (TI)	1965 – 1993
Semiconductor ignition	1983 – 1998
Combined ignition and gasoline-injection systems:	
M-Motronic	Since 1979
KE-Motronic	1987 – 1996
Mono-Motronic	Since 1989

various demands made on the combustion process. Fig. 1 shows the development history of the Bosch gasoline-injection, ignition, and Motronic systems.

Gasoline injection systems

The amount of air drawn in (inducted) by the piston moving downwards in the cylinder is a function of the throttle-valve setting. In conventional systems the accelerator pedal is connected to the throttle valve via a mechanical linkage. The induced air mass quantity determines the injected fuel quantity and is therefore decisive for the power developed by the engine.

The target is to achieve an A/F mixture with a ratio of 14.7 kg air to 1 kg fuel. With this so-called stoichiometric A/F ratio, the fuel combusts almost completely with the oxygen from the inducted air and forms water and carbon dioxide. Further components in the exhaust gas are generated by non-optimal combustion processes and fuel additives (such as sulphur).

Multipoint fuel injection systems using the
continuous-injection principle

The K-Jetronic mechanical-hydraulic gasoline injection system was installed in series-production vehicles from 1973 until 1995. K-Jetronic continually meters the fuel to the engine as a function of the intake air quantity. It was possible to extend the K-Jetronic by means of a Lambda closed-loop control in order to obtain low exhaust-gas values.

More extensive demands, which were not least of all concerned with better exhaust-gas quality, led to the K-Jetronic being expanded to form the KE-Jetronic by the addition of an ECU, a primary-pressure regulator, and a pressure actuator for control of the A/F mixture. The KE-Jetronic was installed between 1982 and 1996.

Injection systems with
intermittent fuel injection

The L-Jetronic, an electronic fuel-injection system with analog technology (1983 till 1986), intermittently injects the fuel as a function of the quantity of air drawn into the engine, the engine speed, and a number of other actuating variables.

The L3-Jetronic is a digital system. This means that it can take over additional control functions which otherwise would have been impossible with analog technology, the overall result being that the injected fuel quantity is better adapted to the engine's various operating requirements.

On the LH-Jetronic (1981–1998), instead of the air quantity drawn into the engine being measured, a hot-wire air-mass meter registers and measures the air mass. This enables even better A/F mixture formation independent of the environmental conditions.

Single-point injection system
with intermittent injection

In the electronic injection system Mono-Jetronic (1987–1997) fitted in small to medium-sized automobiles, a single injector is used which is located in the throttle body at a central point directly upstream of the throttle valve. This system is also referred to as throttle-body injection or TBI. Engine speed and throttle-valve setting are the controlled variables for fuel metering.

Ignition systems

The ignition is responsible for igniting the compressed A/F mixture at precisely the correct moment in time, and thus initiating the mixture's combustion. In the spark-ignition (SI) engine, ignition is by an electric spark in the form of a brief arc discharge across the spark-plug electrodes.

Correctly operating ignition is an absolute must if the catalytic converter is to operate efficiently. Misfiring results in damage to the catalyst (or its destruction) due to the afterburning of the mixture which was incompletely combusted as a result of misfiring.

In the course of time, electronic components have gradually replaced the ignition system's mechanical parts.

Coil ignition (CI)

The ignition distributor is provided with two spark-advance mechanisms. The centrifugal (flyweight) spark-advance mechanism adjusts the ignition point as a function of engine speed, and the vacuum advance mechanism as a function of engine load. The ignition spark is triggered by the distributor breaker points which interrupt the ignition-coil charging process at the ignition point. Wear at the breaker contacts and at the spark-advance mechanisms, due for instance to breaker-point burn and mechanical abrasion, lead to a shift of the ignition point, and can even cause misfiring.

Transistorized ignition (TI)

In the transistorized (coil) ignition, the contact breaker is superseded by a "magnetic pulse generator" which triggers the high-voltage ignition pulse. Breakerless interruption of the ignition-coil current takes place using a transistorized output stage. This means that the wear as described above for the conventional coil ignition no longer takes place.

Semiconductor ignition (breaker-triggered/distributorless)

The semiconductor ignition system calculates the ignition point from an optimized ignition map. The engine speed and the engine load as detected by sensors serve as the input quantities for the ignition-map calculations. The mechanical ignition advance has been dropped completely since its possibilities for adjustment are too restricted.

Breaker-triggered semiconductor ignition permits further intervention functions such as idle-speed ignition angle, timing adjustment via engine temperature, and knock control.

Distributorless semiconductor ignition (DLI) does away with the rotating mechanical high-voltage distribution. In this system, each cylinder is allocated its own ignition coil which is triggered by the ECU at the ignition point.

Motronic engine management

M-Motronic

The basic fuel-injection systems together with the basic semiconductor ignition systems (breaker-triggered and distributorless) provide the foundation for the process of integration to form an engine-management system.

The KE-Motronic is based on the continuous-injection system KE-Jetronic, the Mono-Motronic on the intermittent-injection Mono-Jetronic, and the M-Motronic on the intermittent-injection manifold injection system L-Jetronic.

ME-Motronic

The integration of the electronic throttle control (ETC) in the M-Motronic resulted in the ME-Motronic (Fig. 2). From the accelerator-pedal position, which reflects the driver's wishes for engine power, the ME-Motronic calculates the appropriate throttle-valve opening angle and generates the control signal.

With the M-Motronic, the control parameters are updated with the actual values as soon as the changes in operating conditions have been measured. Particularly the engine load can change rapidly. The ME-Motronic incorporates the possibility of making the control parameters available at the same instant in time as the load change takes place. It is even possible to make them available before the load change actually occurs, provided it is advantageous to do so. This means that the ME-Motronic stands for better handling and reduced emissions.

MED-Motronic

With the MED-Motronic, the gasoline direct injection has been added to the functional scope of the ME-Motronic. Gasoline is now injected directly into the combustion chamber instead of into the intake manifold onto the intake valves.

This means that due to the lean-burn operation, a considerable reduction in fuel consumption and exhaust emissions is possible during part load and idle.

Figure 2

Engine management Motronic ME7 (Example)

1   Carbon canister,
2   Shutoff valve,
3   Canister-purge valve,
4   Intake-manifold pressure sensor,
5   Fuel rail with injectors,
6   Ignition coils with spark plugs,
7   Phase sensor,
8   Electric secondary-air pump,
9   Secondary-air valve,
10 Air-mass meter,
11 Throttle device (ETC),
12 EGR valve,
13 Knock sensor,
14 Engine-speed sensor,
15 Temperature sensor,
16 Lambda oxygen sensor(s),
17 ECU,
18 Diagnosis interface,
19 Diagnosis lamp,
20 Immobilizer,
21 Tank-pressure sensor,
22 In-tank unit with electric fuel pump,
23 Accelerator-pedal module,
24 Battery.

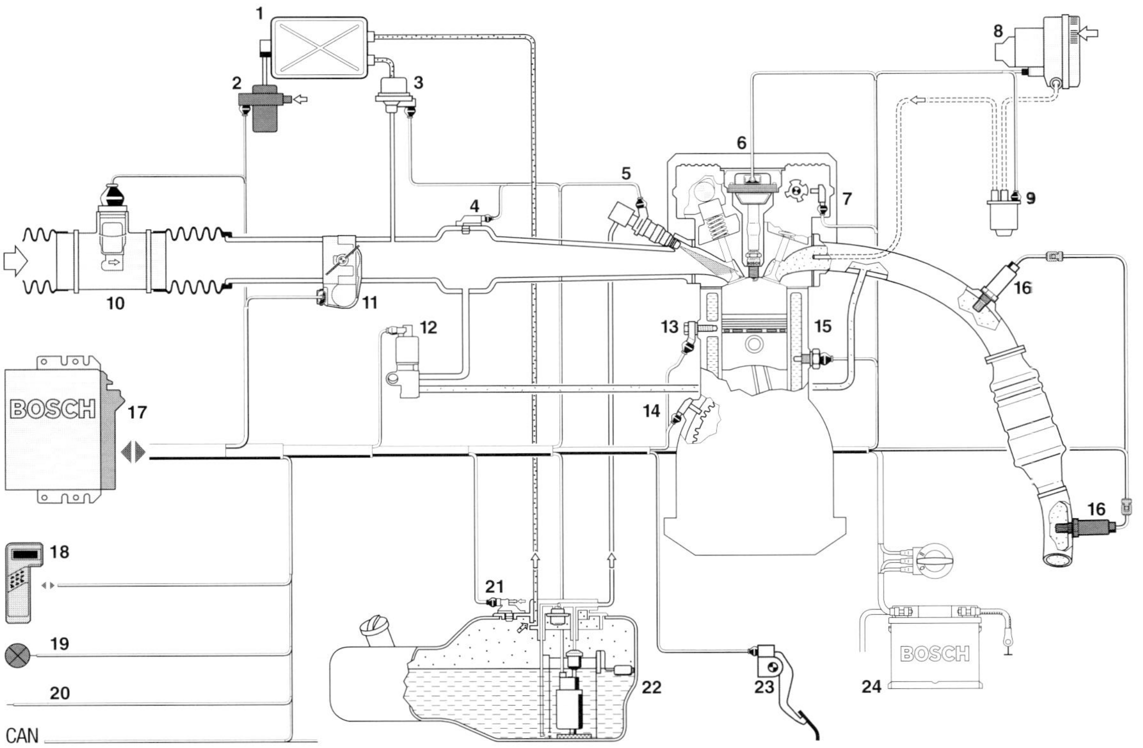

UKM1674Y

Diesel-engine management

Areas of application

Diesel engines are characterized by low fuel-consumption figures, low exhaust-gas emissions, little need for maintenance, and a very long service life. They are therefore used in a wide variety of different applications covering an extensive range of different power outputs.

Technical requirements

The need to comply with the increasingly strict regulations governing exhaust-gas and noise emissions, together with the wish for increased fuel economy, are placing more and more demands on the diesel engine's injection system. In contrast to the gasoline engine (spark-ignition [SI] engine), the diesel engine's intake air is not throttled. The manipulated variables for the diesel engine's speed and output power are the injected fuel quantity and the rate-of-discharge curve.

Depending upon the particular diesel injection process (direct or indirect injection/DI or IDI respectively), in order to ensure efficient A/F mixture formation not only must the system inject the fuel into the diesel engine's cylinders at very high pressures, but metering of the injected fuel quantity must also be as accurate as possible.

Electronic diesel control (EDC)

The diesel engine load and speed control takes place through the injected fuel quantity without throttling of the intake air. When unloaded, and if enough fuel is provided, the diesel engine can rev up to the self-destruction speed. It is thus imperative that some form of governor or controller is provided to limit the engine's speed. Stable diesel-engine idle speed also necessitates some form of govern-

ing or control. In the electronic diesel control (EDC), an electrical actuator supersedes the conventional mechanical (flyweight) governor mechanism as used for conventional diesel-engine governing. On the in-line injection pump, EDC shifts the control rack by means of a linear magnet, and on the distributor pump a rotary actuator intervenes at the pump's control collar. An ECU generates the signals for the electromagnets. Engine speed is registered by a sensor and transmitted as an electric signal to the ECU. In EDC, an analog angle sensor registers the accelerator-pedal position and sends a corresponding electrical signal to the ECU. This supersedes the accelerator-pedal linkage with which the driver formerly controlled the injected fuel quantity as a function of the load.

Progress in diesel technology led to the development of electronically controlled high-pressure solenoid valves which have now taken the place of the rotary actuator and permit more extensive variation of start of injection and injected fuel quantity. The high-pressure solenoid valves permit pilot injection so that the injection lag of the main injection phase can be shortened and the combustion pressure peaks reduced.

Both effects contribute to reducing combustion noise, and in many cases also to a reduction of exhaust emissions.

Compared to conventional mechanical (flyweight) governing, EDC can cope with additional requirements. By means of electrical measurement, flexible electronic data processing, and closed control loops incorporating electrical actuators, it is able to process influencing variables which the mechanical system was unable to handle. With EDC it is possible to exchange data with other electronic systems in the vehicle (for instance, TCS and electronic transmission-shift control), which means that EDC can be integrated completely in the overall vehicle system.

Injection-pump designs

In-line fuel-injection pumps

All in-line fuel-injection pumps have a plunger-and-barrel assembly for each cylinder. As the name implies, this comprises the pump barrel and the corresponding plunger. The pump camshaft integrated in the pump and driven by the engine, forces the pump plunger in the delivery direction. The plunger is returned by its spring.

The plunger-and-barrel assemblies are arranged in-line, and plunger lift cannot be varied. In order to permit changes in the delivery quantity, slots have been machined into the plunger, the diagonal edges of which are known as helixes. When the plunger is rotated by the movable control rack, the helixes permit the selection of the required effective stroke. Depending upon the fuel-injection conditions, delivery valves are installed between the pump's pressure chamber and the fuel-injection lines. These not only precisely terminate the injection process and prevent secondary injection (dribble) at the nozzle, but also ensure a family of uniform pump characteristic curves (pump map).

PE standard in-line fuel-injection pump
Start of fuel delivery is defined by an inlet port which is closed by the plunger's top edge. The delivery quantity is determined by the second inlet port being opened by the helix which is diagonally machined into the plunger.

The control rack's setting is determined by a mechanical (flyweight) governor or by an electric actuator (EDC).

Control-sleeve in-line fuel-injection pump
The control-sleeve in-line fuel-injection pump differs from a conventional in-line injection pump by having a "control sleeve" which slides up and down the pump plunger. By way of an actuator shaft, this can vary the plunger lift to port closing, and with it the start of delivery and the start of injection. The control sleeve's position is varied as a function of a variety of different influencing variables. Compared to the standard PE in-line injection pump therefore, the control-sleeve version features an additional degree of freedom.

Distributor fuel-injection pumps

Distributor pumps have a mechanical (flyweight) governor, or an electronic control with integrated timing device. The distributor pump has only <u>one</u> plunger-and-barrel asembly for all the engine's cylinders.

Axial-piston distributor pump
In the case of the axial-piston distributor pump, fuel is supplied by a vane-type pump. Pressure generation, and distribution to the individual engine cylinders, is the job of a central piston which runs on a cam plate. For one revolution of the driveshaft, the piston performs as many strokes as there are engine cylinders. The rotating-reciprocating movement is imparted to the plunger by the cams on the underside of the cam plate which ride on the rollers of the roller ring.

On the conventional VE axial-piston distributor pump with mechanical (flyweight) governor, or electronically controlled actuator, a control collar defines the effective stroke and with it the injected fuel quantity. The pump's start of delivery can be adjusted by the roller ring (timing device). On the conventional solenoid-valve-controlled axial-piston distributor pump, instead of a control collar an electronically controlled high-pressure solenoid valve controls the injected fuel quantity. The open and closed-loop control signals are processed in two ECU's. Speed is controlled by appropriate triggering of the actuator.

Radial-piston distributor pump
In the case of the radial-piston distributor pump, fuel is supplied by a vane-type pump. A radial-piston pump with cam ring and two to four radial pistons is responsible

for generation of the high pressure and for fuel delivery. The injected fuel quantity is metered by a high-pressure solenoid valve. The timing device rotates the cam ring in order to adjust the start of delivery. As is the case with the solenoid-valve-controlled axial-piston pump, all open and closed-loop control signals are processed in two ECU's. Speed is controlled by appropriate triggering of the actuator.

Single-plunger fuel-injection pumps

PF single-plunger pumps

PF single-plunger injection pumps are used for small engines, diesel locomotives, marine engines, and construction machinery. They have no camshaft of their own, although they correspond to the PE in-line injection pumps regarding their method of operation. In the case of large engines, the mechanical-hydraulic governor or electronic controller is attached directly to the engine block. The fuel-quantity adjustment as defined by the governor (or controller) is transferred by a rack integrated in the engine.

The actuating cams for the individual PF single-plunger pumps are located on the engine camshaft. This means that injection timing cannot be implemented by rotating the camshaft. Here, by adjusting an intermediate element (for instance, a rocker between camshaft and roller tappet) an advance angle of several angular degrees can be obtained.

Single-plunger injection pumps are also suitable for operation with viscous heavy oils.

Unit-injector system (UIS)

With the unit-injector system, injection pump and injection nozzle form a unit. One of these units is installed in the engine's cylinder head for each engine cylinder, and driven directly by a tappet or indirectly from the engine's camshaft through a valve lifter.

Compared with in-line and distributor injection pumps, considerably higher injection pressures (up to 2000 bar) have become possible due to the omission of the high-pressure lines. Such high injection pressures coupled with the electronic map-based control of duration of injection (or injected fuel quantity), mean that a considerable reduction of the diesel engine's toxic emissions has become possible together with good shaping of the rate-of-discharge curve.

Electronic control concepts permit a variety of additional functions.

Unit-pump system (UPS)

The principle of the UPS unit-pump system is the same as that of the UIS unit injector. It is a modular high-pressure injection system. Similar to the UIS, the UPS system features one UPS single-plunger injection pump for each engine cylinder. Each UP pump is driven by the engine's camshaft. Connection to the nozzle-and-holder assembly is through a short high-pressure delivery line precisely matched to the pump-system components.

Electronic map-based control of the start of injection and injection duration (in other words, of injected fuel quantity) leads to a pronounced reduction in the diesel engine's toxic emissions. The use of a high-speed electronically triggered solenoid valve enables the characteristic of the individual injection process, the so-called rate-of-discharge curve, to be precisely defined.

Accumulator injection system

Common-Rail system (CR)

Pressure generation and the actual injection process have been decoupled from each other in the Common Rail accumulator injection system. The injection pressure is generated independent of engine speed and injected fuel quantity, and is stored, ready for each injection process, in the rail (fuel accumulator). The start of injection and the injected fuel quantity are calculated in the ECU and, via the injection unit, implemented at each cylinder through a triggered solenoid valve.

Figure 1

System overview of a diesel injection system with VR radial-piston distributor pump and a variety of different system components

1 Fuel tank,
2 Fuel filter,
3 Injection pump,
4 Pump ECU,
5 High-pressure solenoid valve,
6 Timing-device solenoid valve,
7 Timing-device,
8 Engine ECU,
9 Nozzle-and-holder assembly, 3
  with needle-motion sensor,
10 Sheathed-element glow plug,
11 Glow control unit,
12 Coolant temperature sensor,
13 Crankshaft-speed sensor,
14 Intake-air temperature,
15 Air-mass meter,
16 Boost-pressure sensor,
17 Turbocharger,
18 EGR positioner,
19 Charge-pressure actuator,
20 Vacuum pump,
21 Battery,
22 Instrument panel with display of fuel
  consumption, engine speed etc.,
23 Pedal-travel sensor,
24 Clutch switch,
25 Brake contacts,
26 Vehicle-speed sensor,
27 Operator unit for the Cruise Control,
28 Air-conditioner compressor with
  switch,
29 Diagnosis display with connections
  for diagnostic tester.

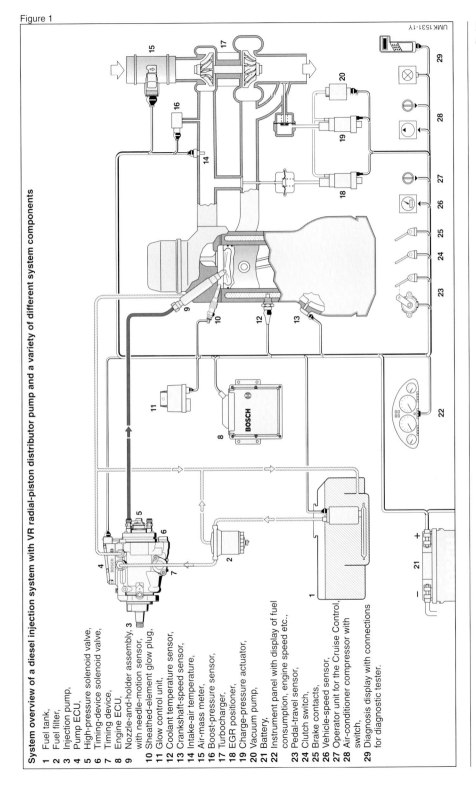

UMK1531-1Y

311

Index

Comprehensive information made easy

Bosch Technical Books

Automotive electrics and electronics

Vehicle electrical systems, Symbols and circuit diagrams, EMC/interference-suppression, Batteries, Alternators, Starting systems, Lighting technology, Comfort and convenience systems, Diesel-engine management systems, Gasoline-engine management systems.

Hard cover,
Format: 17 x 24 cm,
3rd updated edition,
314 pages, bound,
with numerous illustrations.
ISBN 0-7680-0508-6

Gasoline-engine management

Combustion in the gasoline (SI) engine, Exhaust-gas control, Gasoline-engine management, Gasoline fuel-injection system (Jetronic), Ignition, Spark plugs, Engine-management systems (Motronic).

Hard cover,
Format: 17 x 24 cm,
1st edition,
370 pages, bound,
with numerous illustrations.
ISBN 0-7680-0510-8

Diesel-engine management

Combustion in the diesel engine, Mixture formation, Exhaust-gas control, In-line fuel-injection pumps, Axial piston and radial-piston distributor pumps, Distributor injection pumps, Common Rail (CR) accumulator injection systems, Single-plunger fuel-injection pumps, Start-assist systems.

Hard cover,
Format: 17 x 24 cm,
2nd updated and expanded edition,
306 pages, bound,
with numerous illustrations.
ISBN 0-7680-0509-4

Driving-safety systems

Driving safety in the vehicle, Basics of driving physics, Braking-system basics, Braking systems for passenger cars, ABS and TCS for passenger cars, Commercial vehicles – basic concepts, systems and schematic diagrams, Compressed-air equipment for commercial vehicles, ABS, TCS, EBS for commercial vehicles, Brake testing, Electronic stability program (ESP).

Hard cover,
Format: 17 x 24 cm,
2nd updated and expanded edition,
248 pages, bound,
with numerous illustrations.
ISBN 0-7680-0511-6

Automotive terminology

4,700 technical terms from automotive technology, in German, English and French, assembled from the above Bosch Technical Books: "Automotive Electrics and Electronics", "Diesel-engine management"; "Gasoline-engine management" and "Driving-safety systems".

Hard cover,
Format: 17 x 24 cm,
1st edition
378 pages, bound.
ISBN 0-7680-0338-5

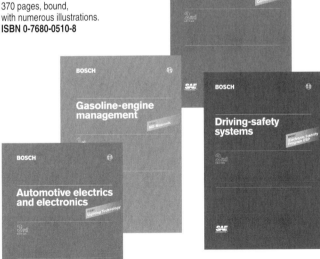